Genomic Designing of Climate-Smart Cereal Crops

Chittaranjan Kole
Editor

Genomic Designing of Climate-Smart Cereal Crops

Springer

Editor
Chittaranjan Kole
Raja Ramanna Fellow
Department of Atomic Energy, Government
of India, ICAR-National Institute for Plant
Biotechnology
New Delhi, Delhi, India

ISBN 978-3-319-93380-1 ISBN 978-3-319-93381-8 (eBook)
https://doi.org/10.1007/978-3-319-93381-8

This Springer imprint is published by the registered company Springer Nature Switzerland AG
The registered company address is: Gewerbestrasse 11, 6330 Cham, Switzerland

Dedicated to

Late Dr. Evgueni V. Ananiev

Formerly Head of the Laboratory of Plant Molecular Genetics and Genetic Engineering, N. I. Vavilov Institute of General Genetics, USSR Academy of Sciences, Moscow for mentoring me for my first research works on plant biotechnology through cloning, sequencing, and utilizing plant genes for phylogenetic analysis and evolution.
He remains as a model to us and the world as an outstanding scientist and a wonderful human being mingled in a single persona!!

Preface

The last 120 years have witnessed a remarkable evolution in the science and art of plant breeding culminating in quite a revolution in the second decade of the twenty-first century! A number of novel concepts, strategies, techniques, and tools have emerged from time to time over this period and some of them deserve to be termed as milestones. Traditional plant breeding, immediately following the rediscovery of laws of inheritance, has been playing a spectacular role in the development of innumerable varieties in almost all crops during this entire period. Mention must be made on the corn hybrids, rust-resistant wheat, and obviously the high-yielding varieties in wheat and rice that ushered the so-called green revolution. However, the methods of selection, hybridization, mutation, and polyploidy employed in traditional breeding during this period relied solely on the perceivable phenotypic characters. But most, if not all, of the economic characters in crops are governed by polygenes which are highly influenced by environment fluctuations and hence phenotype-based breeding for these traits has hardly been effective.

Historical discovery of DNA structure and replication in 1953 was followed by a series of discoveries in the 1960s and 1970s that paved the way for recombinant DNA technology in 1973 facilitating the detection of a number of DNA markers in 1980 onward and their utilization in construction of genetic linkage maps and mapping of genes governing the simply inherited traits and quantitative trait loci controlling the polygenic characters in a series of crop plants starting with tomato, maize, and rice. Thus new crop improvement technique called as molecular breeding started in the later part of the twentieth century. On the other hand, genetic engineering made modification of crops for target traits by transferring alien genes, for example, the *Bt* gene from the bacteria *Bacillus thuringiensis*. A large number of genetically modified crop varieties have thus been developed starting with the commercialization of "flavr Savr tomato" in 1994.

Meantime, the manual DNA sequencing methodology of 1977 was being improved with regard to speed, cost-effectiveness, and automation. The first-generation sequencing technology led to the whole-genome sequencing of Arabidopsis in 2000 and followed by rice in 2002. The next-generation sequencing technologies were available over time and used for sequencing of genomes of many other models and crop plants. Genomes, both nuclear and organellar, of more than 100 plants have already been sequenced by now and the information thus generated are available in public database for most of them. It must be mentioned here that bioinformatics played a remarkable role in handling the enormous data being produced in each and every minute. It can be safely told that the "genomics" era started in the beginning of the twenty-first century itself accompanying also proteomics, metabolomics, transcriptomics, and several other "omics" technologies.

Structural genomics have thus facilitated annotation of genes, enumeration of gene families, and repetitive elements, and comparative genomics studies across taxa. On the other hand, functional genomics paved the way for deciphering the precise biochemistry of gene function through transcription and translation pathways. Today, genotyping-by-sequencing of primary, secondary, and even tertiary gene pools; genome-wide association studies; and genomics-aided breeding are almost routine techniques for crop improvement. Genomic selection in crops is another reality today. Elucidation of the chemical nature of crop chromosomes has now opened up a new frontier for genome editing that is expected to lead the crop improvement approaches in the near future.

At the same time, we will look forward to the replacement of genetically modified crops by cisgenic crops through the transfer of useful plant genes and atomically modified crops by employing nanotechnology that will hopefully be universally accepted for commercialization owing to their human-friendly and environment-friendly nature.

I wish to emphatically mention here that none of the technologies and tools of plant breeding is too obsolete or too independent. They will always remain pertinent individually or as complementary to each other, and will be employed depending on the evolutionary status of the crop genomes, the genetic resources, and genomics resources available, and above all the cost–benefit ratios for adopting one or more technologies or tools. In brief, utilization of these crop improvement techniques would vary over time, space, and economy scales! However, as we stand today, we have all the concepts, strategies, techniques, and tools in our arsenal to practice genome designing, as I would prefer to term it, of crop plants not just genetic improvement to address simultaneously food, nutrition, energy, and environment security, briefly the FNEE security, I have been talking about for the last 5 years at different platforms.

Addressing FNEE security has become more relevant today in the changing scenario of climate change and global warming. Climate change will lead to greenhouse gas emissions and extreme temperatures leading to different abiotic stresses including drought or waterlogging on one hand and severe winter and freezing on the other. It will also severely affect uptake and bioavailability of water and plant nutrients and will adversely cause damage to physical, chemical, and biological properties of soil and water in cropping fields and around. It is also highly likely that there will be an emergence of new insects and their biotypes and of new plant pathogens and their pathotypes. The most serious concerns are, however, the unpredictable crop growth conditions and the unexpected complex interactions among all the above stress factors leading to a drastic reduction in crop yield and quality in an adverse ecosystem and environment. Climate change is predicted to significantly reduce productivity in almost all crops. For example, in cereal crops, the decline of yield is projected at 12–15%. On the other hand, crop production has to be increased at least by 70% to feed the alarmingly growing world population, projected at about 9.0 billion by 2050 by even a moderate estimate.

Hence, the unpredictability of crop growing conditions and thereby the complexity of biotic and abiotic stresses warrant completely different strategies of crop production from those practiced over a century aiming mostly at one or the few breeding objectives at a time such as yield, quality, resistance to biotic stresses due to disease-pests, tolerance to abiotic stresses due to drought, heat, cold, flood, salinity, acidity, or improved water and nutrient use efficiency. In the changing scenario of climate change, for sustainable crop production, precise prediction of the above limiting factors by long-term survey and timely sensing through biotic agents and engineering devices and regular soil and water remediation will play a big role in agriculture. We have been discussing on "mitigation" and "adaptation" strategies for the last few years to reduce the chances of reduction of crop productivity and improve the genome plasticity of crop plants that could thrive and perform considerably well in a wide range of growing conditions over time and space. This is the precise reason for adopting genomic designing of crop plants to improve their adaptability by developing climate-smart or climate-resilient genotypes.

Keeping all these in mind, I planned to present deliberations on the problems, priorities, potentials, and prospects of genome designing for development of climate-smart crops in about 50 chapters, each devoted to a major crop or a crop group, allocated under five volumes on cereal, oilseed, pulse, fruit, and vegetable crops. These chapters have been authored by more than 250 of eminent scientists from over 30 countries including Argentina, Australia, Bangladesh, Belgium, Brazil, Canada, China, Egypt, Ethiopia, France, Germany, Greece, India, Ireland, Japan, Malaysia, Mexico, New Zealand, Kenya, Pakistan, Philippines, Portugal, Puerto Rico, Serbia, Spain, Sri Lanka, Sweden, Taiwan, Tanzania, Tunisia, Uganda, UK, USA, and Zimbabwe.

There are a huge number of books and reviews on traditional breeding, molecular breeding, genetic engineering, nanotechnology, genomics-aided breeding, and gene editing with crop-wise and trait-wise deliberations on crop genetic improvement including over 100 books edited by me since 2006. However, I believe the present five book volumes will hopefully provide a comprehensive enumeration on the requirement, achievements, and future prospects of genome designing for climate-smart crops and will be useful to students, teaching faculties, and scientists in the academia and also to the related industries. Besides, public and private funding agencies, policy-making bodies and the social activists will also get a clear idea on the road traveled so far and the future roadmap of crop improvement.

I must confess that it has been quite a difficult task for me to study critically the different concepts, strategies, techniques, and tools of plant breeding practiced over the last 12 decades that also on a diverse crop plants to gain confidence to edit the chapters authored by the scientists with expertise on the particular crops or crop groups and present them in a lucid manner with more or less uniform outline of contents and formats. However, my experience gained over the last 7 years in the capacity of the Founding Principal Coordinator of the International Climate-Resilient Crop Genomics Consortium (ICRCGC) was highly useful while editing these books. I have the opportunity to interact with a number of leading scientists from all over the world almost on a regular basis. Organizing and chairing the annual workshops of ICRCGC since 2012 and representing ICRCGC in many other scientific meetings on climate change agriculture offered me a scope to learn from a large number of people from different backgrounds including academia, industries, policy-making, and funding agencies and social workers. I must acknowledge here the assistance I received from all of them to keep me as a sincere student of agriculture specifically plant breeding.

This volume entitled Genomic Designing of Climate-Smart Cereal Crops includes seven major crops including Rice, Wheat, Maize, Oat, Sorghum, Pearl Millet, and Finger Millet. These chapters have been authored by 52 scientists from 13 countries including China, Ethiopia, India, Ireland, Kenya, Mexico, Philippines, Spain, Sweden, UK, USA, Zambia, and Zimbabwe. I place on record my thanks for these scientists for their contributions and cooperation.

I feel myself proud that I could start my research works under the supervision of Late Dr. Evgueni V. Ananiev at his Laboratory of Plant Molecular Genetics and Genetic Engineering, N. I. Vavilov Institute of General Genetics, USSR Academy of Sciences, Moscow. As a member of his team, I could be a party for making some original contributions in the fields of cloning, sequencing, and utilizing some plant genes for phylogenetic analysis and evolution in cereals crops under the supervision of Late Dr. Ananiev, an outstanding scientist and wonderful human being. Hence, I have dedicated this book to (Late) Dr. E. V. Ananiev as a token of my respect, thanks, and gratitude. I wish to put on record my deep regards and acknowledgement to his wife, Dr. Olga N. Danilevskaya, for providing me the picture of Late Dr. Ananiev I have used in the dedication page of this book.

New Delhi, India — Chittaranjan Kole

Contents

Abbreviations

AATF	African Agricultural Technology Foundation
ABA	Abscisic Acid
ABCILs	Advanced backcross introgression lines
ABRE	ABA-responsive cis-elements
AFLP	Amplified fragment length polymorphism
AGCR	Agronomically important cloned rice genes
AGT	Advanced genomic technologies
AI	Artificial intelligence
AICPMIP	All India Coordinated Pearl Millet Improvement Program
AICRIP	All India Coordinated Rice Improvement Project
AM	Association mapping
*ANN*1	*Annexin*1 protein
ARCH	Agricultural Research Center-Hays
ARS	Agricultural Research Service
ASI	Anthesis-silking interval
B&MGF	Bill & Melinda Gates Foundation
BAC	Bacterial artificial chromosome
*Ban*II	Bacillus aneurinolyticus II
BC	Backcross
BGI	Beijing Genomics Institute
Bgt	*Blumeria graminis* f. sp. *tritici*
BLB	Bacterial leaf blight
BLUP	Best linear unbiased prediction
BMR	Brown midrib
BPH	Brown planthopper
BSST	Bioinformatics and statistical software tools
BYDV	Barley yellow dwarf virus
CA	Carbonic anhydrase
CAAS	Chinese Academy of Agricultural Sciences
CAH	Carbonic anhydrase

CBC	Customized breeding chip
CFT	Confined field trial
CGIAR	Consultative Group on International Agricultural Research
CIM2GTAILs	Second-generation tropically adapted inducer lines
CIMMYT	International Maize and Wheat Improvement Center
cM	CentiMorgan
CMIP	Climate Research Program's Coupled Model Intercomparison Project
CML	CIMMYT Maize Line
CMS	Cytoplasmic male sterile/ sterility
CNV	Copy number variations
COG	Cluster of orthologous genes
CRISPR	Clustered regularly interspaced short palindromic repeats
CRU	Climatic Research Unit
CS	Climate-smart
CSCWR	Single-copy genes between wheat and rice
CSH	Climate-smart hybrid
CSRH	Climate-smart rice hybrids
CSRVs	Climate-smart rice (inbred and hybrid) varieties
DArT	Diversity arrays technology
DH	Doubled haploid
DHN	Dehydrin
DM	Downy mildew
DP	Donor parent
DQP	Designed quantitative trait loci pyramiding
DREB	Dehydration responsive element binding proteins
DRESTs	Drought-regulated expressed sequence tags
DT	Drought-tolerant
DTMA	Drought-Tolerant Maize for Africa Project
Dwf	Dwarfing gene
*EcNAC*1	*EleusinecoracanaNo Apical Meristem*1
EDI	Ear digital imaging
EPA	Environmental Protection Agency
Eps	Earliness *per se*
ERF	Ethylene responsive factors
EST	Expressed sequence tag
F_2	Second filial generation
FAO	Food and Agriculture Organization
FAOSTAT	Food and Agriculture Organization Statistics
FAW	Fall armyworm
FHB	Fusarium head blight
GAB	Genomics-assisted breeding
GA	Gibberellic acid
GARSS	Genomics-assisted recurrent selection scheme
GB	Genetic background

GBLUP	Genomic best linear unbiased prediction
GBS	Genotyping-by-sequencing
GBTM	Genomics-based trait mapping
GBTS	Genotyping-by-target sequencing
GC	Guanine and cytosine content
GCA	General combining ability
GCA	Global climatic alterations
GCM	Global Climate Model
GEBV	Genome-estimated breeding value
GET	Genome editing tools
GIS	Geographic Information System
GISH	Genomic in situ hybridization
GLAD	Green leaf area duration
GLH	Green leafhopper
GM	Genetically modified
GO	Gene ontology
GP	Gene pool
GPT	Genomic prediction tools
GRIN	Germplasm Resources Information Network (USA)
GS	Genomic selection
GSR-BT	Green super rice-breeding strategy
GSR	Green super rice
GST	Glutathione-S-transferase
GT	Genotyping
GWAS	Genome-wide association study
GXE	Genotype-by-environment
GY	Grain yield
HapMap	Haplotype map
HG	Heterotic group
HHZ	Huanghuazhan
HIR	Haploid induction rate
His	Histidine-rich calcium-binding gene
HRDC	Hybrid Rice Development Consortium
HSP	Heat shock protein
HTL	*Hypersensitive to Light*
HTMA	Heat-Tolerant Maize for Asia Project
HTPG	High-throughput platform for genotyping
HTP	High-throughput phenotyping
IBPGR	International Bureau of Plant Genetic Resources
IBP	Interconnected breeding populations
ICMP	ICRISAT pearl millet
ICRISAT	International Crops Research Institute for the Semiarid Tropics
IITA	International Institute of Tropical Agriculture
ILs	Introgression lines
IMAS	Improved Maize for African Soils Project

InDels	Insertions–Deletions
INV	Invertase gene
IPCC	Intergovernmental Panel on Climate Change
IPM	Integrated pest management
IRD	Institute of Research for Development
IRGSP	International Rice Genome Sequencing Project
IRMA	Insect Resistant Maize for Africa Project
IRRI	International Rice Research Institute
ISSR	Inter-simple sequence repeat
IVDMD	In vitro dry matter digestibility
KAI2	KARRAKIN-INSENSITIVE2
KALRO	Kenya Agricultural and Livestock Research Organization
KASP	Kompetitive allele specific PCR
KBLUP	Kinship-enhanced best linear unbiased prediction
KSU	Kansas State University
LAMP	Leucine aminopeptidase
LD	Linkage disequilibrium
LEA	Late embryogenesis abundant
LG	Linkage group
LGS	Low germination stimulant
LIC	Low-input check
LiDAR	Light detection and ranging
LOX	Lipoxygenase
Lr	Leaf rust
LRR	Leucine-rich repeat
Ltn	Leaf tip necrosis
MABC	Marker-assisted backcrossing
MAGIC	Multi-parent advanced generation intercross
MARS	Marker-assisted recurrent selection
MAS	Marker-assisted selection
MCMV	Maize chlorotic mottle virus
MCR	Multi-copy rice genes
MDH	Malate dehydrogenase
ME	Malic enzyme
ME	Mega-environment
MET	Multi-environment trials
MLN	Maize lethal necrosis
mlo	Modulator of defense and cell death gene
MRI	Magnetic resonance imaging
*Msp*I	Moraxella species I
MSV	Maize streak virus
MTA	Material transfer agreement
NAC	No apical meristem
NAM	Nested association mapping
NARS	National Agricultural Research System (India)

NBS	Nucleotide-binding site
NCBI	National Center for Biotechnology Information
NDVI	Normalized difference vegetation index
NGS	Next-generation sequencing
NIL	Near-isogenic lines
NLR	Nucleotide-binding and leucine-rich repeat
NUE	Nitrogen use efficiency
OA	Osmotic adjustment
Opm	*Opaque2 modifiers*
OPVs	Open-pollinated varieties
PAGE	Parametric analysis of gene set enrichment
PAV	Presence–absence variants
PBCM	Process-based crop simulation model
PBMA	Prediction-based model approach
PCR	Polymerase chain reaction
PDLs	Pyramided lines
PEG	Polyethylene glycol
PEPC	Phosphoenolpyruvate carboxylase
PET	Positron emission tomography
PFT	Pore-forming toxin-like
PGD	Phosphogluconate dehydrogenase
PgDwarf8	*Pennisetum glaucum Dwarf8* gene
PgHd3a	*Pennisetum glaucum* heading date *3a* gene
PGM	Phosphoglucomutase
PgPHYC	*Pennisetum glaucum* phytochrome C gene
PGR	Plant genetic resources
Pgt	*Puccinia graminis* f. sp. *triticin*
PHG	Practical haplotype graph
PL	Pyramiding line
PMiGAP	Pearl millet inbred germplasm association panel
Pm	Powdery mildew
Ppd	Photoperiod
*Pst*I	Providencia stuartii I
Pst	*Pucciniastriiformis* f. sp. *triticin*
Pt	*Pucinia triticina*
PYT	Preliminary yield trial
QA/QC	Quality assurance/Quality control
QPM	Quality protein maize
QTL	Quantitative trait locus
QTLs	Quantitative trait loci
R4D	Research for development
RAD	Restriction site-associated DNAsequencing
RAPD	Random amplified polymorphic DNA

RCGS	Rapid cycle genomic selection
rDNA	Recombinant DNA
Rf	Fertility restoration gene
RFLP	Restriction fragment length polymorphism
RGB	Red, green, and blue
RGT	Rapid generation turn over
Rht	Reduced height
RILs	Recombinant inbred lines
ROS	Reactive oxygen species
RP	Recipient Parent
RP	Recurrent parent
Rr1	Resistance gene 1
R	Resistance
RRL	Reduced representation libraries
RRS	Reduced representation sequencing
RS	Recurrent selection
RTK	Real-time kinematic
RUBISCO	Ribulose 1, 5-bisphosphate carboxylase
SAM	Sequence alignment map
SAT	Semiarid tropics
SCA	Specific combining ability
SCAR	Sequence-characterized amplified region
SCA	Sugarcane aphid
SCMR	SPAD chlorophyll meter reading
SCMV	Sugarcane mosaic virus
SCR	Single-copy rice genes
SC	Salinity check
sgRNA	Single-guide RNA
SHW	Synthetic hexaploid wheat
SIM	Simple interval mapping
SKDH	Shikimate dehydrogenase
SMA	Single marker analysis
SME	Small-and medium-enterprise
SNP	Single nucleotide polymorphism
SOC	Soil organic carbon
SRAP	Sequence-related amplified polymorphism
Sr	Stem rust
SSA	Sub-Saharan Africa
SSCP	Single-strand conformational polymorphism
SSD	Single seed descent
SSR	Simple sequence repeat
STARP	Semi-thermal asymmetric reverse PCR
STMA	Stress-Tolerant Maize for Africa Project
STS	Sequence tagged site

TAILs	Tropically adapted inducer lines
TALENs	Transcription activator-like effector nucleases
TB	Transgenic breeding
tGBS	Tunable genotyping-by-sequencing
TKW	Thousand-kernel weight
TLB	Turcicum leaf blight
TOI	Target trait of interest
TPEs	Target populations of environments
TRAP	Target region amplification polymorphism
TSILs	Trait-specific introgression lines
TSW	Thousand seed weight
TTE	Target trait environment
UAV	Unmanned aerial vehicle
USAID	US Agency for International Development
USDA	United States Department of Agriculture
VIP	Vegetative insecticidal proteins
Vrn	Vernalization
WA	West Africa
WB	Wheat blast
WCA	West and Central Africa
WEMA	Water Efficient Maize for Africa Project
WGRS	Whole-genome re-sequencing
WGS	Whole-genome sequencing
WTR1	Weed-Tolerant Rice 1
WUE	Water use efficiency
XRF	X-ray fluorescent
Yr	Yellow rust/leaf rust
ZFNs	Zinc-finger nucleases
ZFP	Zinc-finger protein

Chapter 1
Genomics-Assisted Breeding of Climate-Smart Inbred and Hybrid Rice Varieties

Jauhar Ali, Anumalla Mahender, G. D. Prahalada, Ma. Anna Lynn Sevilla, Angelito Galang, Erik Jon De Asis, Madonna Dela Paz, Corinne Mira Marfori-Nazarea, Katrina Leslie Nicolas and Ricky Vinarao

Abstract Global climatic alterations (GCA) such as drought, flood, salinity, submergence, and high/low temperature pose serious threats to crop productivity in several Asian and African countries. As a mitigation option to GCA, we need to develop and deploy climate-smart rice (inbred and hybrid) varieties (CSRVs) resilient to biotic and abiotic stresses. Genomics-assisted breeding (GAB) by integrating advanced genomics tools and an improved green super rice (GSR) breeding strategy (GSR-BT) is one of the leading reliable strategies to develop CSRVs. This provides a high-quality genome sequence and SNPs as allelic variants from the advanced genomics tools that help in understanding the molecular and physiological mechanisms underlying trait expression. It also plays a key role in the quick introgression of the desired genetic variants with the highest precision and less or no genetic drag through an innovative GSR-BT. Under the GSR program at the International Rice Research Institute (IRRI), first, BC_1F_2 populations derived from Huanghuazhan (HHZ), Weed Tolerant Rice 1 (WTR-1), and TME80518 (recipient parents) and 16 donors were developed. Later, these introgression lines (ILs) underwent simultaneous screenings over three rounds for different abiotic and biotic stresses as well as evaluation under normal irrigated conditions, resulting in the identification of trait-specific ILs. A total of 1333 (HHZ-ILs) + 2232 (WTR-1-ILs) + 1408 (TME80518-ILs) ILs were developed, which were further used either for varietal improvement or as parental lines for the development of pyramided lines (PDLs). Furthermore, a designed QTL pyramiding (DQP) approach began for stacking traits/genes derived from the generated trait-specific ILs. These more productive efforts resulted in the development of 1280 (HHZ-PDLs) + 850 (WTR-1-PDLs) PDLs that showed a significantly superior performance over the tolerant checks. Notably, we developed and distributed

J. Ali (✉) · A. Mahender · G. D. Prahalada · Ma. Anna L. Sevilla · A. Galang · E. J. De Asis · M. D. Paz · C. M. Marfori-Nazarea · K. L. Nicolas · R. Vinarao
Rice Breeding Platform, International Rice Research Institute, 4031, Los Baños, Laguna, Philippines
e-mail: J.Ali@irri.org

C. Kole (ed.), *Genomic Designing of Climate-Smart Cereal Crops*,
https://doi.org/10.1007/978-3-319-93381-8_1

more than 240 CSRVs with multiple abiotic and biotic stress tolerance to Asia and Africa without compromising on grain yield and quality. A total of 26 IRRI-bred CSRVs were released, and 91 such cultivars were nominated into national cooperative yield trials from three recipient parents within a short span of 7 years. The released CSRVs are being cultivated on more than 2.7 million ha on a cumulative basis in Asia and Africa.

Keywords Breeding strategy · Genomics · Climate change · QTL and GWAS · MAS · GSR varieties

1.1 Introduction

Providing sufficient nutritious food to feed the increasing world population is a significant challenge for agriculture (Abberton et al. 2016). The global population is predicted to reach 9.7 billion by 2050, thus requiring agricultural food production to increase by 60.0–110.0% to meet the additional food demand (Tilman et al. 2011). Moreover, global agricultural productivity is severely threatened by biotic and abiotic stresses, reduced arable land, and, most alarmingly, global climate change (Kole et al. 2015; Hampton et al. 2016). The fluctuation in rainfall regimes, day/night temperature on land and sea, and weather patterns; elevated carbon dioxide concentrations; and greenhouse gas emissions in particular regions globally are the impacts of global climatic alterations (GCA) (Batley and Edwards 2016; Beckage et al. 2018). The Intergovernmental Panel on Climate Change (IPCC) estimated that the global surface temperature of Earth is expected to increase by 1.4–5.8 °C by 2100 (IPCC 2007), thereby resulting in a decrease in precipitation in the subtropics and an increase in the possibility of frequent occurrence of extreme climatic events such as drought, high/low temperature, floods, and cyclones. Furthermore, Nelson et al. (2014), Rosegrant et al. (2014), and Iizumi et al. (2017) conducted a combining analysis of biophysical and economic models by using multiple climatic conditions. The analysis concluded that about 11.0% of the grain yield decline in the four major crops (maize, soybeans, rice, and wheat) by 2050 would eventually affect global food and nutritional security, especially in developing countries.

Rice (*Oryza sativa* L.) is one of the major staple food crops of the world that provide the daily calorie intake for more than 50.0% of the world population (Muthayya et al. 2014). This crop is grown on 159 million hectares annually, and more than 90.0% of this rice is produced by 200 million smallholding farmers in Asia and Africa (Mottaleb et al. 2012). It is a semiaquatic annual plant that includes 22 *Oryza* species, of which two species, Asian rice (*O. sativa*) and African rice (*O. glaberrima*), are well domesticated for human consumption, and the other 20 are wild species (Londo et al. 2006). Both the cultivated species have unique domestication histories and include subspecies/subpopulations with distinctive morphological and physiological traits (Choi et al. 2017; Singh et al. 2018). The majority of the rice from *O. sativa* is well described and further classified into five subpopulations: *aus*, *indica*, temperate *japonica*, tropical *japonica*, and aromatic rice (Garris et al. 2005). The origin of *O. sativa* was first observed in Southeast Asia (India, Myanmar, Thailand, North Vietnam, and China) between 8,000 and 15,000 years ago (Normile 1997; Wei et al. 2012; Civáň and Brown 2017), whereas *O. glaberrima* was domesticated from its wild ancestor *O. barthii* in the floodplains of the Niger River in Africa about 3,000 years ago (Portères 1976).

Global rice production is severely affected by global climate change, which in turn challenges the food security of at least 70.0% of the Asian population (Chang et al. 2013; Ali et al. 2017). It has resulted in drought, salinity, flood, submergence, elevated CO_2, and high/low temperature. Drought is one of the severe constraints that affect rice production. Nearly 23 million hectares of rainfed rice is facing drought stress by showing negative effects on plant growth, physiology, and grain development (Serraj et al. 2011; Fahad et al. 2017). Salinity is the second most abiotic stress after a drought that causes severe hindrance to rice production and productivity depending upon the level of lethality (Gregorio 1997). Similarly, about 20 million hectares of rice land is affected by flooding and submergence leading to complete crop loss (https://irri.org/our-impact/increase-food-security/flood-tolerant-rice-saves-farmers-livelihoods). Matthews et al. (1997) revealed that a 1 °C increase in mean daily temperature could lead to a decrease in rice yield by 5–7%, whereas an increase of 1 °C in night temperature could lead to a 10.0% yield reduction (Peng et al. 2004; Yang et al. 2017). Importantly, an increase in temperature during flowering and pollination can drastically reduce pollen viability and ultimately reduce seed set and grain yield (Hatfield and Prueger 2015; Kumar et al. 2015a). In addition to this, elevated CO_2 concentration in the atmosphere resulting from GCA leads to increased damaged rice grains and decreased grain quality (Madan et al. 2012; Liu et al. 2017). Moreover, it has been predicted that worldwide rice consumption would increase from 465 million tons to 487 million tons from 2012 to 2020 (Mohanty 2009) and 25.0% more rice might be required for ensuring food security in 2030 vis-à-vis rice consumption in 2010 (Seck et al. 2012). Hence, there is an urgent need for the introgression of climate-smart (CS) traits such as tolerance to drought, salinity, flood and submergence, elevated CO_2 and temperature, and other stresses like biotic stresses. It is essential to develop climate-smart rice varieties (CSRVs) with CS traits, which is crucial to make them withstand increase or decrease in day and night temperature, elevated CO_2, and resistant to multiple broad-spectrum stresses including biotic stresses with low input use efficiency to ensure high quality and quantity of rice production. Securing food under changing climatic conditions requires addressing four key elements. First is the development of CSRVs (inbreds and hybrids) that provide stable yields under both favorable and marginal environments. Second is all crop management technologies associated with this. The third is by tackling nutritional quality, including biofortification, vis-à-vis human health. Last is ensuring food availability, accessibility, and use with stability (Fig. 1.1). However, all these require conducive policies and an action plan to keep food accessible to all at all times. Research efforts to cope with climate change are currently integrated into several national agricultural research programs globally.

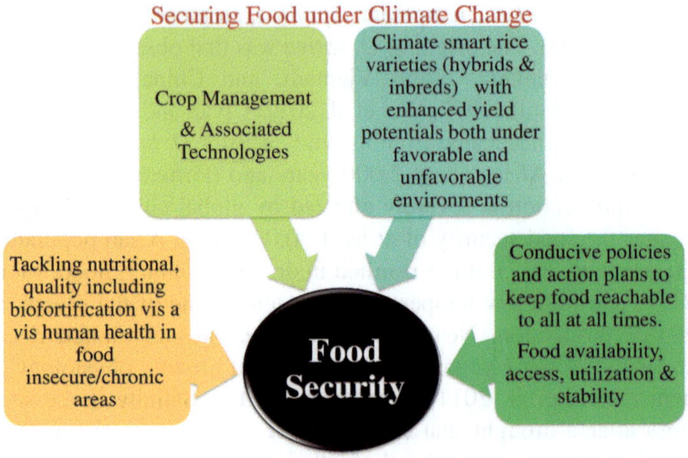

Fig. 1.1 Key strategies to secure food under climate change

The increase in rice yield along with resistance to biotic stress and tolerance abiotic stresses has been driven by conventional rice breeding approaches. Till date, most of the ruling mega-varieties were developed through conventional breeding approaches. These varieties were high yielding and possessed key traits to withstand major insect pests and diseases like IR62 and IR64 (Khush 1987). These varieties showed their resistance to major biotic stresses such as blast, bacterial leaf blight, tungro, brown planthopper (BPH), green leafhopper (GLH), and stem borer. Notably, these varieties are still being used as the biotic stress-resistant varieties and checks for BPH and GLH bioassay experiments. On the other hand, for the abiotic stress tolerance, several conventionally bred rice varieties were also released. One among them is Sahbhagi Dhan (IR74371-70-1-1) drought-tolerant rice variety released for cultivation in India, in 2010. It was introduced under direct-seeded upland as well as for lowland conditions. The same variety was released in Bangladesh (as “BRRI Dhan 56”) and in Nepal (as “Sookha Dhan 3”) (Mandal et al. 2009; Dar et al. 2012; Anantha et al. 2016).

Recent advances in genomics technologies, whole-genome sequencing (WGS) using next-generation sequencing (NGS) platforms, genotyping-by-sequencing (GBS), fixed Affymetrix genotyping platforms, and genomic selection (GS) have led to rapid success in discovery and deployment of novel genomic regions that confer useful traits. Understanding the molecular and physiological basis, especially the traits that influence climate resilience, is essential for developing an appropriate breeding strategy (Kole et al. 2015; Rasheed et al. 2017; Ali et al. 2018).

The development of CSRVs needs to explore the genetic variability of the target traits and introduce variation into genotypes of interest through rapid breeding cycles. Hence, there is a need to develop a novel breeding strategy that helps in generating breeding products within a short period by precise introgression of the expected genetic segments. In this review, we showcase the potential of the integration of novel efficient breeding strategies developed by green super rice (GSR) at IRRI and advanced genomics tools such as genomics-assisted breeding (GAB) to develop and deploy CSRVs in Asian and African countries.

1.2 Rice Genomics

Rapid advances in structural and functional genomics helped in cloning over 2000 genes controlling key agronomic traits and partially characterizing their molecular biological mechanisms (Li et al. 2018). Rice has an immense treasure of genomics and bioinformatics resources that can accelerate the development of useful products. It is the first monocot species whose whole-genome sequence (WSG) was known. Among the major world food crops, rice is a model plant for genomics studies (Tyagi and Khurana 2003). It has the smallest genome size (330–430 Mb) compared to other cereals such as sorghum (750 Mb), maize (3000 Mb), barley (5000 Mb), and wheat (16,000 Mb) (Tyagi et al. 2004). Improved de novo sequencing technologies helped in generating a draft rice genome sequence of two rice subspecies, *O. sativa* subsp. *japonica* and *indica*, using Nipponbare and 93–11 varieties, respectively (Goff et al. 2002; Yu et al. 2002). The WGS of Nipponbare published by the International Rice Genome Sequencing Project (IRGSP) with a 6X coverage is being used as a reference genome sequence for the sequencing of several varieties because of its high sequence quality (Yu et al. 2005). The reference genome sequence served as a powerful tool in the mining of useful traits by capturing exact allelic variants, presence–absence variants (PAV), and copy number variations (CNV), along with the dissection of complex traits (Huang and Han 2014). The availability of enormous information on the genomics of rice helped not only in understanding the predicted and annotated rice genes (30,000–50,000) but also in unraveling the genetic interaction controlling the essential agro-morphological traits under varied environmental conditions.

Advanced and cost-effective NGS methods facilitated the rapid dissection of the complex genetics of agronomically essential traits through quantitative trait locus (QTL) mapping and genome-wide association study (GWAS) by developing a large number of genetic markers, including insertion—deletion (InDel) and single nucleotide polymorphism (SNP) markers (Fig. 1.2). NGS platforms increase the resolution of the mapping, which in turn increases the success of precise localization of genetic loci/QTLs (Guo et al. 2014; Yang et al. 2017; Descalsota et al. 2018; Varshney et al. 2018). One of the most important applications of rice genomics is carrying out precise marker-assisted selection (MAS), and GAB approaches. This resulted in the successful introgression of several beneficial traits with less or no genetic drag. These remarkable capabilities of rice genomics studies helped in the rapid understanding of molecular genetics, enhancement of breeding selection strategies, functional genomics through characterizing the specific genes, and also in the use as a reference approach for different high-throughput genome

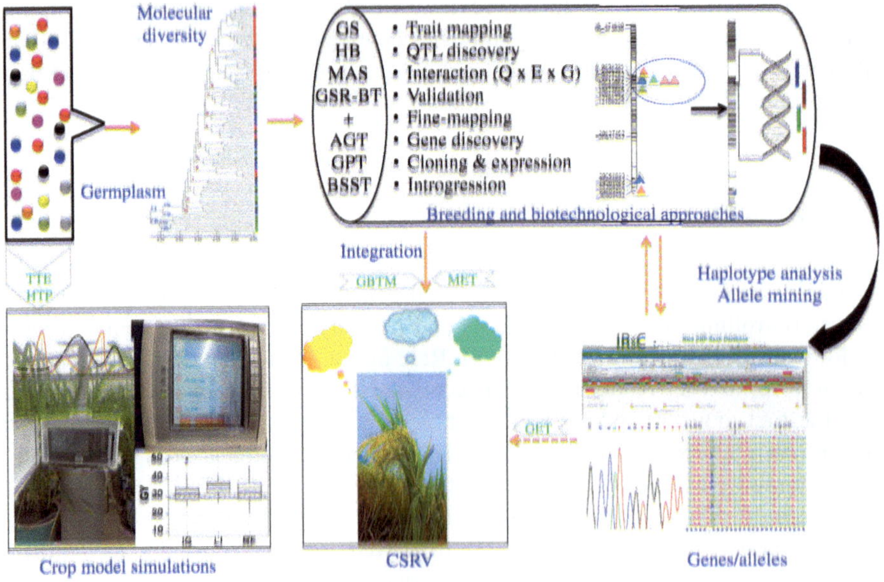

Fig. 1.2 Generalized schematic diagram showing different GAB strategies for developing CSRVs

sequencing technologies in other staple food crops (He et al. 2014; Das et al. 2017). In summary, rice genomics studies paved the way for increased genomics studies, particularly in exploring the hidden genetic diversity existing in a large number of cultivars, landraces, and wild accessions that are the richest sources of hidden genetic diversity and shaping it into farmers' chosen varieties. The available bioinformatics tools and databases which are being used to explore structural and functional genomics technologies are presented in Table 1.1.

1.3 Different Genomics Approaches Used for Rice Crop Improvement

Developing CSRVs is a major concern for ensuring global food security under erratic changes in climatic conditions (Tong et al. 2016; Varshney et al. 2018). To overcome these challenges, efficient and advanced genomics approaches have been deployed in recent decades to understand the molecular genetics and physiological responses of complex traits and in designing novel breeding strategies for improving genetic gains in rice (Takagi et al. 2013; Bhat et al. 2016; Sun et al. 2018). Although conventional breeding strategies put forward popular rice varieties, development of these varieties through traditional breeding strategies is time-consuming, and it is difficult to dissect the hidden genetic variation existing for essential agronomic traits, which mostly have low heritability and are highly

Table 1.1 Distribution of SNPs in different customized breeding chip (CBCs) used for molecular genetic studies and breeding programs in rice

Sl. no	SNP array	No. of SNPs	Function	Avg. SNP call rate (%)	No. of genotypes	References
1	50 K	51,478	Molecular genetic diversity and breeding	99.9	801[b]	Chen et al. (2014a)
2	50 K	50,051	Molecular genetics and phylogenetics	99.9	192[b]	Singh et al. (2015)
3	44 K	44,100	Molecular genetic diversity and breeding	80.0	500[b]	Tung et al. (2010)
4	44 K	44,100	Pyramiding for useful genes	92.8	31[b]	Kurokawa et al. (2016)
5	6 K	5,274	Molecular genetic diversity and breeding	80.0	258[b]	Thomson et al. (2017)
6	5 K	5,246	Heat stress tolerance	89.0	272[a]	Ps et al. (2017)
7	6 K	5,102	Genetics for yield-attributed traits	80.0	197[a]	Tan et al. (2013)
8	6 K	5,102	Grain shape	80.0	197[a]	Hu et al. (2013)
9	6 K	6,000	Salinity stress tolerance	89.0	220[b]	Kumar et al. (2015b)
10	6 K	5,291	Molecular genetic diversity	90.0	471[b]	Xu et al. (2016)
11	6 K	5,274	Heat tolerance	90.0	167[a]	Ye et al. (2015)
12	6 K	6,000	Salinity stress tolerance	89.0	94[a]	Gimhani et al. (2016)
13	6 K	5,000	Low-temperature stress tolerance	80.0	230[a]	Najeeb et al. (2019)
14	6 K	5,000	Weed-competitive ability	80.0	167[a]	Dimaano et al. (unpublished)

[a]Biparental mapping populations, [b]Diverse rice accessions

influenced by the environment. Hence, the application of advanced genomics strategies is necessary to achieve the colossal task of feeding the alarming growing global population more quickly. The following section describes different genomics approaches used for rice crop improvement.

1.3.1 Whole-Genome Sequencing (WGS) of Diverse Rice Genotypes

The rapid development of omics technologies provides novel opportunities to understand the complexity of molecular, physiological, and biochemical network pathways in crops. In recent decades, the WGS of rice accessions has held great potential to identify the genetic background of rice accessions, map the genetic locus associated with the target trait of interest (TOI), and understand the molecular function of the mapped locus (Li et al. 2014; Duitama et al. 2015; Rahman et al. 2017). The omics technologies including transcriptomics, metabolomics, and proteomics surely benefited from the advancement in NGS technologies as they give high-quality sequence information that helps in better understanding gene design and its mechanism (Rasheed et al. 2017). Increasing the genetic gain of yield and yield-associated traits and tolerance of the major stresses under GCA is crucial for the development of CSRVs. This can be achieved through the identification and incorporation of novel sources of genetic variation as genes/alleles from various genetic resources of rice germplasm (McCouch et al. 2013; Xie et al. 2015; Ali et al. 2017). The current trend in NGS platforms of WGS revolutionized the discovery of novel genomic regions derived from even wild *Oryza* species, which possess quite unique genome characteristics compared with *O. sativa* cultivated varieties, through linkage (QTL mapping) and association mapping (GWAS) approaches, which help in unlocking the hidden novel alleles in them (Varshney et al. 2014; Kole et al. 2015). Under the GSR breeding program, we used a set of 500 varieties (mini-core), which were whole-genome sequenced as a part of the 3K Rice Genomes project, to explore the genetic essence of the breeding lines developed. Among them, only 50 varieties were used for the identification of genomic regions conferring multiple stress tolerance and, eventually, for the development of CSRVs (Ali et al. unpublished). Simultaneous screening of BC_1F_2 bulk populations in Huanghuazhan (HHZ) recipient parent and eight donor parents over three rounds of multiple abiotic and biotic stresses in comparison with checks and parental lines led to the development of 495 backcross introgression lines (BC_1F_6 BILs) and these were sequenced using an NGS platform of WGS by the BGI (Ali et al. unpublished). Similarly, WGS platforms were also used for understanding the sequential variations among the genotypes, the diversity pattern of rice accessions, population structure studies, and trait association analysis for QTLs/genes for TOI (Guo et al. 2014; Tang et al. 2016). Genome-wide SNP discovery was initially undertaken from *japonica* (Nipponbare), and then *indica* (93–11) varieties and found 408,898 SNPs and PAV between the two subspecies (Feltus et al. 2004; Shen et al. 2004). Further, a total of 20 diverse varieties and landraces were sequenced in the OryzaSNP project, and 160,000 high-quality polymorphic SNPs were discovered (McNally et al. 2009). In addition, 23 million SNPs, from 94 *O. sativa* varieties and 10 wild relatives (Duitama et al. 2015); 22,682 polymorphic SNPs from 12 parental varieties (Tang et al. 2016); 3.6 million SNPs from 517 rice landraces (Huang et al. 2010); 8 million SNPs from 446 diverse *O. rufipogon* accessions; 794,297 SNPs from 576 introgression lines (ILs) from 11 donors (Ali et al. 2018); and 29 million SNPs and 2.4 million PAV from 3,010 diverse Asian cultivated rice varieties were discovered from the WGS (Wang et al. 2018). Finally, these large numbers of SNPs were used in analyzing the diversity pattern and further validation of these SNPs. The same information can be helpful to understand the functional genomics and transcriptomics for identifying superior alleles and their mechanism in future breeding programs. In summary, WGS provides the highest density of widely distributed SNPs, which enhances the resolution to achieve precise localization of the putative QTLs/genes conferring tolerance of several biotic and abiotic stresses, which resulted from climate change, and the introgression of these genomic regions would facilitate the development of climate-resilient rice varieties.

1.3.2 3K Genomes and Their Use in Developing Climate-Smart Breeding Products

Improving genetic gain by developing CSRVs is essential for increasing global rice production. Harnessing the hidden genetic diversity existing in cultivars through efficient NGS platforms is the prime objective of GAB to accelerate breeding cycles and enhance genetic potential (Seck et al. 2012; Fahad et al. 2014; Ali et al. 2018). Globally, about 780,000 accessions of *Oryza* are maintained in gene banks. Germplasm of Asian origin occupies a larger portion of the total accessions maintained in gene banks across the world (Jockson 2010; Jacob et al. 2015). To explore genetic variation and use more efficiently, the Chinese Academy of Agricultural Sciences (CAAS), Beijing Genomics Institute (BGI) Shenzhen, and IRRI launched a 3,000 (3K) rice genomes project for sequencing 3024 gene bank rice accessions (The 3000 Rice Genomes Project 2014). Among the total 3K rice accessions, 92.0–94.0% of the genome sequence aligned to the reference genome of Nipponbare, which indicated the conserved genomic regions among the total 3K rice accessions. A total of 17 terabytes of raw genomic sequence data generated through various bioinformatics pipeline databases, 40 million SNP variants, and 2.4 million short InDels (≤ 50 bases long) were retrieved from the whole 3K genomes analysis (Alexandrov et al. 2014). The 3K genome sequences provide high-quality, reliable information on the large set of valuable genes for the development of novel CSRVs for higher grain yield, higher tolerance of biotic and abiotic stresses, and higher nutritional grain quality.

Hence, to develop CSRVs by exploring genetic variation, a total of 500 elite varieties belonging to the mini-core collection and part of the 3K set were systematically used under the GSR-BT. The donor and recurrent parents of the GSR-BT underwent WGS using efficient NGS platforms to isolate a large set of reliable SNPs. Interestingly, among the several donor and recipient parents sequenced, only three recipient parents (HHZ, WTR-1, and TME80518) and 16 donor parents were fully used for the early backcross introgression-breeding program of GSR-BT. In this program, the BC_1F_2 populations derived from HHZ, WTR-1, and TME80518 (recipient parents) and 16 donors at IRRI were screened simultaneously over three rounds for different abiotic and biotic stresses and normal irrigated conditions. This resulted in the identification of 1333 (HHZ-ILs) + 2232 (WTR-1-ILs) + 1408 (TME80518-ILs) trait-specific introgression lines (ILs) and, further, by designed QTL pyramiding efforts, 2023 HHZ-PDLs + 661 (WTR-1-PDLs) pyramiding lines (PDLs) were developed and found superior to the checks for all of the traits studied (Sasaki 2017). The enormous genome sequence information on these lines was also used for the discovery of genomic regions conferring resistance to several biotic stresses, tolerance of multiple abiotic stresses, nutrient-use efficiency (NUE), and high CO_2 concentrations. Notably, several QTLs governing complex traits were identified by genome sequencing of 495 ILs in an HHZ background and tGBS for 575 ILs in a WTR-1 background (Garcia-Oliveira et al. 2009; Yorobe et al. 2016; Ali et al. 2017, 2018; Pang et al. 2017a; Taghavi et al. 2017; Feng et al. 2018). The same WGS information from the GSR breeding lines will be used again for characterizing the detected genetic loci for molecular and physiological characterization (Fig. 1.3). Interestingly, out of 50 breeding lines sequenced, only three breeding lines were used for gene discovery and varietal improvement. This information opens up the way forward for tapping the insights of the genetic value of the remaining breeding lines for identifying novel traits/ genomic regions and developing more efficient CSRVs.

In addition to the several improved breeding lines, we developed and distributed more than 240 rice materials with multiple abiotic and biotic stress tolerance to Asia and Africa without compromising on grain yield and quality. Most convincingly, 26 IRRI-bred climate-resilient materials were released, and 91 such cultivars were nominated into national cooperative yield trials within a short span of 7 years. These varieties are now being cultivated on more than 2.7 million ha on a seed distribution basis alone to farmers in Asia and Africa (Ali et al. 2017; Feng et al. 2018). Newly developed materials breaking three yield barriers and having multiple stress tolerance (GSR IR2-8-Y14-SU3-R2, GSR IR2-5-L10-Y1-Y2, and GSR IR2-1-R5-N1-Y3) are now being shared, which give an average of 5.9 and 18.3%

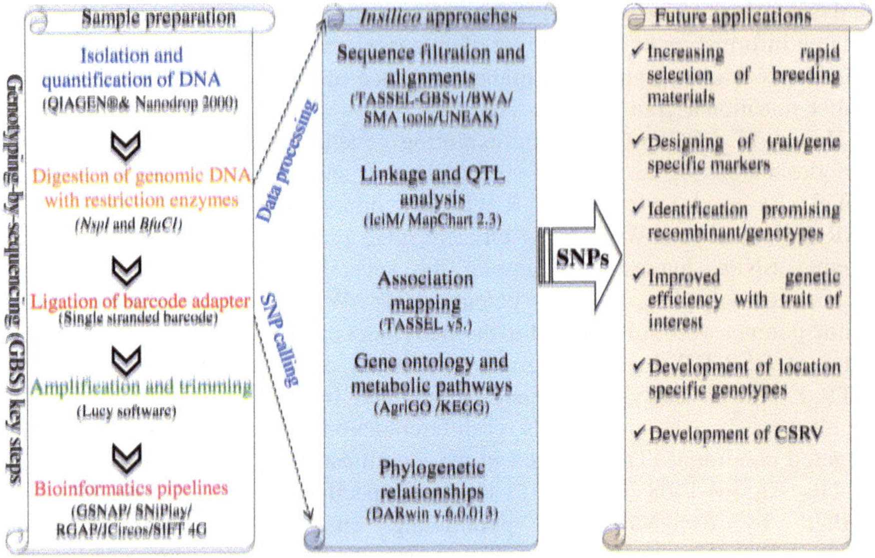

Fig. 1.3 Schematic representation of GBS technology, approaches, and applications in developing CSRVs

higher grain yield than the best hybrid check (Mestiso 6) and NSIC Rc38 under MET, respectively (Sasaki 2017). Such high-yielding materials would further augment the ongoing hybrid rice breeding program at IRRI by using them as parental lines. Success might be attributed to innovative introgression breeding and designed QTL pyramiding efforts combined with appropriate cross-tolerance screening and selection techniques that tapped the hidden genetic diversity from a large number of donors into an adaptable varietal background such as HHZ, WTR-1, and TME80518.

1.3.3 GBS and Its Use for QTL Discovery of Climate-Smart Traits

High-throughput genome sequencing technologies have paved the way to novel approaches to reveal genetic information on an unparalleled scale compared to Sanger sequencing (Pareek et al. 2011; Reuter et al. 2015). Numerous genotyping technologies are widely used by several researchers based on the cost of sequencing, time, genome coverage, reduction in sequence depth, and missing rate (Sonah et al. 2013). GBS, restriction site-associated DNA (RAD) sequencing, and reduced representation libraries (RRL) are a few of the high-throughput genotyping methods that reduce cost and retrieve high-quality SNPs (Davey et al. 2011; Sonah et al. 2013). As compared with RAD and RRL, GBS or next-generation genotyping (NGG) has revolutionized crop genotyping methodology to discover thousands to millions of SNPs across the wide range of species for the identification of genetic variations underlying promising agronomic traits (Huang et al. 2009; Torkamaneh et al. 2016; Scheben et al. 2017; Wickland et al. 2017). It was developed as a rapid, robust, and cost-effective genotyping approach that is capable of extracting numerous SNPs, which are markers of choice and abundantly distributed across the genome (Scheben et al. 2017). Tunable Genotyping-By-Sequencing (tGBS®) is a novel technique that involves the ligation of single-strand oligonucleotides and two restriction enzymes to produce overhangs in opposite orientations, which can be amplified and sequenced. As compared with numerous GBS methods, tGBS libraries produce higher average read depth, a low percentage of missing rate, and a higher rate of SNP calling accuracy (>97.0–99.0%) across populations (He et al. 2014; Ott et al. 2017; Pang et al. 2017a; Ali et al. 2018). Hence, the tGBS approach allows the generation of high-quality SNP markers, which are excellent sources in GAB and genetic applications for dissecting many complex traits (Torkamaneh et al. 2016; Bhatia et al. 2018; Feng et al. 2018). Figure 1.3 represents the methodology, different approaches, and applications in GBS.

GBS has the greatest application in gene discovery and its characterization. It has revolutionized association mapping, which relies on historical recombination events, as it increases the statistical power of detection of the rarest allele (Kole et al. 2015). Because of the enhanced resolution of the genetic map, GBS enables the identification of rare alleles conferring TOI through GWAS. GBS helps in identifying the rarest recombination, which happened during the several recombination events in history. The following are some of the case studies that proved the successful application of GBS in GWAS analysis. Huang et al. (2010) and Huang et al. (2012) discovered several QTLs using GBS as a genotyping platform in GWAS analysis (49 QTLs for 14 agronomic traits, 32 QTLs for flowering time, and 10-grain yield-related QTLs). Through combined gene-based association and haplotype analysis, Wang et al. (2017b) identified 72 QTLs using GWAS. The functional annotation of the detected loci indicated 19 candidate genes associated with seven important QTL regions that were significantly affecting grain quality traits.

On the other hand, GBS has also been used in discovering the genomic regions influencing TOI in biparental mapping populations. As compared to traditional QTL mapping using a low resolving marker system, GBS-based QTL mapping ensures the high-resolution mapping of the genomic regions conferring TOI (Guo et al. 2014), thereby enhancing the dependability of the detected putative QTLs. Considering these advantages, we have also employed GBS as a genotyping platform for the identification of genomic regions conferring resistance to multiple stresses, including biotic and abiotic stresses, NUE, Fe toxicity, and deficiency traits in GSR materials. The developed biparental population that was generated through unique GSR-BT underwent GBS genotyping. Numerous high-quality SNPs extracted from GBS were used for the marker–trait association study using QTL analysis. Ali et al. (unpublished) identified several promising novel loci for the biotic stresses BPH, GLH, blast, BLB, and RGSV. Furthermore, Pang et al. (2017a), Feng et al. (2018), and Ali et al. (unpublished) also proved the effectiveness of the detected QTLs through QTL pyramiding for broadening the genetic base by using the designed QTL pyramiding approach. Research is ongoing to dissect the detected QTLs to identify candidate genes through functional validation by a map-based cloning approach.

In summary, recent trends in molecular breeding and sequencing-based association studies (QTL and GWAS) derived from WGS and/or tGBS have become more popular for dissecting complex agronomic traits and identifying haplotypes in various crops because of the higher success rate and statistical power. Hence, the high-density GBS-SNP marker assay provides crucial insights into the novel genetic architecture that further helps us to identify potential candidate genes/alleles for the development of CSRVs in future breeding programs.

1.3.4 Allele Mining

Advanced genomics tools, WGS, and GBS platforms provide information about genome sequence variations as favorable alleles, which allows isolation of these alleles that contribute to trait expression. The favorable alleles within the detected novel genomic regions conferring multiple stress resistance could also be isolated through allele mining for molecular characterization and function analysis (Fig. 1.3).

The genetically diverse mini-core collection of GSR breeding materials available in the 3K genomes underwent allele mining combined with haplotype analysis. The allele mining results showed that the GSR breeding materials were found to carry several previously identified genes/loci along with novel genomic regions for biotic and abiotic stresses, nutritional and low input use efficiency traits. Hence, the CSRVs developed and released by GSR harboring these desirable genomic regions (alleles) exhibited a superior performance for several traits and were thereby well accepted by farmers throughout Asian and African countries.

In another case study Leung et al. (2015), used 3K rice genomes for allele mining of allelic variation and distribution of blast and bacterial leaf blight (BLB) resistance genes. Three blast resistance genes (*Pi9*, *Pi5*, and *Pikm*) were selected for studying their distribution as these genes were well used in building blast resistance in several blast resistance breeding programs. The results indicated that *Pi9*, *Pi5*, and *Pikm* were distributed in the rice genome 23.6%, 31.3%, and 33.4%, respectively. Interestingly, 75.0% of the 3K genomes had a *Pi9* allele that belongs to *indica* types, which was almost four times higher than for *japonica* types. On the other hand, for BB resistance, *SWEET* gene allelic variants were compared among the set of accessions. The comparison of the *SWEET* gene sequence variations revealed 0.2–11.0% of the mutation rate across accessions. However, the lowest mutation rate (0.1%) was observed in *aus* subpopulations compared with other subpopulations (Leung et al. 2015). Thus, allele mining helped in understanding and assessing the distribution and diversity pattern of R-genes in selected accessions and subsequent validation of these *R*-genes in 3K genomes. Further, this provides a valuable resource for the identification of novel alleles among 3K genotypes specific to geographic location and characterizing the novel alleles. Hence, allele mining, being an advanced genomics tool, helps in exploring the widely distributed SNPs across different genomes in the 3K set. This has opened the way to discover novel alleles (PAV/CNV) and functional candidate markers associated with TOI for the enhancement of yield productivity, nutritional grain quality, and tolerance of multiple biotic and abiotic stresses in molecular breeding programs.

1.3.5 SNP Breeders' Marker Chip Development and Use (Affymetrix Chip)

SNPs are the most abundant form of molecular genetic variations that reveal the genetic background of a mapping population, historical recombination, evolution, and associated genomic sequential variations with the phenotypic trait (McNally et al. 2009). Analysis of these variations through DNA sequencing of the identified gene of interest and detecting the causal alleles for the desirable traits are the best application of NGS-SNP-based molecular breeding (Huq et al. 2016). In recent decades, a tremendous application of advanced genomic sequence technologies, NGS, GBS, and SNP marker chip (SNPMC)/customized breeding chip (CBC) technologies has generated enormous volumes of nucleotide sequence data in a single run, thereby providing genome-wide sequence raw reads that have led to the discovery of thousands of SNPs with a density of approximately three to four SNPs per kilo-base pair (kbp) in the rice genome (Kumar et al. 2012; Xu et al. 2017b; Yang et al. 2017). These technologies increased the number of gene/trait discovery studies as these are fast-tracking and cost-effective genotyping platforms for a larger number of populations/accessions (McCouch et al. 2010; Varshney et al. 2014; Rasheed et al. 2017). The main advantage of CBS technologies is the use of SNPs for genotyping; they are abundant, biallelic, sequence tagged, and codominant. A customized SNP chip provides the foundation for highly informative genotyping assays as these SNPs are associated with the most important traits, and can be used in a breeding program directly (Davey et al. 2011; Rasheed et al. 2017).

A customized trait-associated breeding chip plays an important role in molecular diversity analysis, gene discovery (GWAS and QTL mapping), and MAS, which can be useful in the gene pyramiding and gene tagging during a crop breeding program (Table 1.2). The flexibility of a CBC gives a choice to breeders for selecting any number of markers and samples associated with desirable TOI. Additionally, the different resolution of SNP arrays has also been developed and successfully used in several crop species and domesticated animals (Rasheed et al. 2017). The high-resolution 50K Infinium array (RiceSNP50) and 44K Affymetrix array have been developed for rice SNP genotyping (Chen et al. 2014a; Singh et al. 2015; Xu et al. 2016) and also in other crops such as maize (Ganal et al. 2011; Xu et al. 2017a), sunflower (Bachlava et al. 2012), soybean (Song et al. 2013), wheat (Rimbert et al. 2018), barley (Bayer et al. 2017), and chickpea (Deokar et al. 2014). However, the major challenges of these arrays are to understand the allelic variations at desirable SNP loci and ensuring that the loci represent single-copy (SC) genes. CBC includes a large number of SNPs from both intronic and exonic regions. For designing the rice SNP genotyping chip, Singh et al. (2015) classified rice genes into four categories as unique single-copy rice genes (SCR), multi-copy rice genes (MCR), agronomically important cloned rice genes (AGCR), and single-copy genes between wheat and rice (CSCWR). Among these, the highest proportion of genes (79.0%) and SNPs (96.0%) can be retrieved from SCR. Notably, compared to a previous report of McCouch et al. (2010) and Chen et al. (2014b), Singh et al. (2015) identified the highest SNP call rate of 99.9% from genic regions with extremely low missing data (<0.1) and high assay reproducibility of 99.9%. The SNPs derived from the genic region will provide high-quality information for rice breeders for the characterization of rice germplasm for biotic and abiotic stresses, genetic background selection, phylogenetic analysis, genetic diversity, haplotype analysis, QTLs, and GWAS for promising key agro-morphological, physiological, and quality traits. On the other hand, the SNP array from the conserved intronic region will be useful in rice domestication and evolutionary studies from cultivated and wild rice relative accessions. These several

Table 1.2 List of major climate-smart traits associated QTLs/genes identified in different genetic backgrounds of GSR-BT strategies in rice

Trait	Mapping population	Genotyping platform	Genomic regions	Chromosome	Genetic background	Reference
Drought tolerance	TSILs and DQP	WGS-400K SNPs	9 QTLs	2, 3, 5, 6, 8, and 12	HHZ	Feng et al. (2018)
Low nitrogen	TSILs and DQP	WGS-400K SNPs	5 QTLs	1, 2, 3, and 8	HHZ	Feng et al. (2018)
Cold tolerance	IBP-GWAS	WGS-41,754 SNPs	6 QTLs	3, 4, and 12	HHZ	Zhu et al. (2015)
Salinity tolerance	DQP	tGBS-2,188 SNPs	*qSES2, qSES4, qChlo1* and *qChlo4*	1, 2, and 4	WTR-1	Pang et al. (2017a, b)
Drought tolerance	ABCILs	169 SSRs	14 QTLs	1, 3, 5, 6, 7, 8, 9, and 11	Tarom Molaei,	Wang et al. (2013)
Low-temperature tolerance	TSILs	6 K-704SNPs	82 QTLs	All 12	WTR-1	Najeeb et al. (2019)
Weed-competitive ability	TSILs	6 K-704SNPs	44 QTLs	Except 4 and 8	WTR-1	Dimaano et al. (2019) (unpublished)
Nutrient-use efficiency	TSILs	6 K-704SNPs	261 QTLs	All 12	HNG	Jewel et al. (2019)
Nutrient-use efficiency	TSILs	tGBS-2782 SNPs	13 QTLs	1, 2, 3, 4, 5, 9, 10, and 12	HNG	Mahender et al. (2019)
Nutrient-use efficiency	TSILs	tGBS-2005 SNPs	4 QTLs	4, 5, 6, and 8	Cheng-Hui 448	
Nutrient-use efficiency	TSILs	tGBS-1361 SNPs	2 QTLs	1 and 11	Zhong 413	
Brown planthopper and Green leafhopper resistance	TSILs	6 K-702 SNPs	One for each QTL	6 and 1	HHZ	Balachiranjeevi et al. (2019)

BS-Breeding strategies; **GT**-Genotyping, **GB**-Genetic background, **ST**-Salinity tolerance; **DQP**-Designed quantitative trait loci pyramiding; **ABCILs**-Advanced backcross introgression lines; **TSILs**-Trait-specific introgression lines (ILs); **PL**-Pyramiding line; **RS**-Recurrent selection; **IBP**-Interconnected breeding populations applications of the CSC Affymetrix chip were also witnessed in many GSR projects, especially for key traits such as drought, low nitrogen tolerance weed-competitive ability, and NUE.

1.3.6 Genome Editing Through Targeted Mutagenesis (TALENS and CRISPR/Cas-9)

Advanced genetic manipulation/editing tools such as T-DNA insertion/transposons, zinc finger nucleases (ZFNs), transcription activator-like effector nucleases (TALENs), clustered regularly interspaced short palindromic repeat (CRISPR)/ CRISPR-associated protein (Cas), and transgenics have played a significant role in studying the various molecular and physiological functions for crop improvement in the past few decades (Mishra et al. 2018). However, in transgenic approaches, the incorporation of transgenes into the specific host genome is mostly nontargeted and unstable. Additionally, significant concern arises when the transgene is inserted into edible transgenic crop species (Barakate and Stephens 2016). In contrast to the transgenic approach, genome editing tools (GET) provide novel opportunities to breed for any TOI by applying functional genomics tools (Shan et al. 2013; Kamthan et al. 2016; Mishra et al. 2018). The genome-edited crops have additional benefits over transgenic plants since they are target oriented, which cleaves at the intended region of the desired TOI, and are relatively easy for consumption acceptability issues as compared with genetically modified crops (Malzahn et al. 2017; Abdelrahman et al. 2018). As compared to ZFNs and TALENs, CRISPR/ Cas9 is a potential genome editing tool as it is less expensive and more flexible for targeted genome editing. This tool includes different guide RNAs targeting multiple sites in the genome and hence is successfully used in several cereal crops in a short period (Xie and Yang 2013; Liu and Fan 2014; Endo et al. 2016; Jaganathan et al. 2018; Mishra et al. 2018).

The introgression of desirable TOI (QTLs or genes) from natural diverse rice varieties or mutation rice panels into elite rice cultivars is a time-consuming process in a backcross breeding program. CRISPR technology can offer direct induction of novel mutations in elite rice cultivars for TOI, which can accelerate breeding programs for improving tolerance/crop productivity under adverse climatic factors (Bortesi and Fischer 2015). So far, more than 20 crop species have used a targeted genome editing approach (Song et al. 2016). In addition, several researchers reported using the CRISPR/Cas9 method for drought, salinity, and cold tolerance (Shan et al. 2013; Zhang et al. 2014; Shen et al. 2017); potassium deficiency tolerance (Mao et al. 2018); biotic stresses (Zhou et al. 2015; Wang et al. 2016); and agronomic trait-related genes that are involved in controlling responses to various molecular and physiological stress signaling pathways for the enhancement of yield productivity in rice (Shen et al. 2017; Jaganathan et al. 2018). Similarly, MAS breeding programs have always been experiencing a serious flaw with undesirable linkages that are difficult to break and are time-consuming. These genetic drags can be broken by using the CRISPR/Cas-9 approach. As this strategy possesses the highest success rate of targeted genome editing, it can knock out unwanted genetic segments that otherwise cause genetic drag. Hence, CRISPR technology can be used as a potential tool for developing CSRVs by improving traits that increase climate resilience and knock out undesirable linkages that otherwise lower the quality of MAS breeding programs.

Transgenic breeding (TB) is an attractive strategy for developing multiple stress tolerance crops as compared to conventional breeding as it can introduce one or multiple genes into the desired genome without affecting their genetic background (Jain and Jain 2000). During the last two decades, a set of biotic and abiotic stress tolerance genes have been identified and transferred into several rice cultivars for building tolerance to drought, cold, high temperature, salinity, iron efficiency, and nitrogen deficiency (Garg et al. 2002; Sun et al. 2005; Su et al. 2010; Sanghera et al. 2011; Kamburova et al. 2017). Recently there are two major approaches to achieve clonal seed propagation of hybrids in rice appeared (Khanday et al. 2018; Wang et al. 2019). The first approach involved the discovery and possibility of fixing heterosis through a male-expressed embryogenic trigger redirected for asexual popagation through seeds (Khanday et al. 2018). This discovery substitutes mitosis of meiosis (*MiMe*) by the genome editing of *REC8*, *PAIR1*, and *OSD1* genes and activation of BABY BOOM (BBM) like transcription factor of APETELLA 2/ETHYLENE RESPONSE FACTOR (*AP2/ERF*) resulted in normal embryogenesis even without fertilization thereafter, which they confirmed through genome editing of *bbm1*, *bbm2*, and *bbm3* allelotypes of dominant alleles of *BBM* gene. The second approach also involves the genome editing of *REC8*, *PAIR1*, and *OSD1* genes for engineering meiosis in addition to *MATRILINEAL* (*MTL*) gene (involved in fertilization) which could induce the formation of haploid seeds in hybrid rice

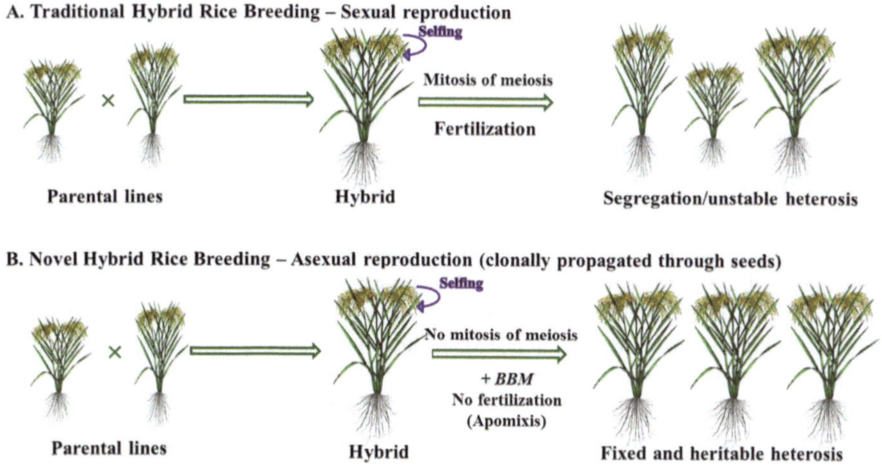

Fig. 1.4 Schematic diagram showing the comparison of traditional and novel hybrid rice breeding strategies

(Wang et al. 2019). These novel approaches open up more opportunities for placing heterotic combinations with high grain yield and quality. Thus, it addresses the market needs in asexually propagated seeds bringing down the cost of hybrid rice seeds and making it available at affordable costs to millions of resource-poor rice farmers across the globe (Fig. 1.4) Although proven technologies and products derived from the TB are readily available, the ethical and socioeconomic issues and policies around it makes it difficult for its further utilization and adoption.

1.4 A Strategy for Designing Climate-Smart Rice Inbreds and Hybrids in Rice

GCA is an emerging threat that affects global agricultural productivity (Team et al. 2014). The impact of GCA is mostly triggered by increasing temperature, changes in rainfall patterns, and differences in the occurrences and intensity of drought, floods, and salinity (Lobell et al. 2012; Prasanna 2014; Altieri and Nicholls 2017). Porter et al. (2014) estimated that yield loss can reach 60.0% for maize, 50.0% for sorghum, 35.0% for rice, 20.0% for wheat, and 13.0% for barley. Several strategies have been followed in agricultural practices and technologies, tillage, crop establishment, agronomic management, alternate wetting and drying irrigation, water, and NUE (Jat et al. 2014; Sapkota et al. 2015; Altieri and Nicholls 2017). However, there have been no significant achievements in developing rice varieties that can adapt to GCA variations. Hence, finding a genetic resource that can combat the erratic changes in GCA is imperative. This can be achieved by developing systematic breeding strategies by combining conventional and innovative molecular breeding (Fig. 1.5). These strategies are the most feasible options that can increase rice productivity and income security under changing GCA (Vermeulen et al. 2012; Lipper et al. 2014; Khatri-Chhetri et al. 2017).

Adaptation to GCA is mainly achieved by adjusting allele frequencies at different genetic loci through rapid breeding cycles that deliver stable and improved cultivars. This accomplishment requires exploring broader ranges of rice germplasm from different regions across the world, a high-throughput screening facility, shortened breeding cycles, multi-environment testing (MET) systems, and advanced genomics tools that are adequate to the target population and environments (Atlin et al. 2017). Several breeding methods and strategies such as pedigree, modified bulk, single seed descent (SSD), doubled haploid (DH), and MAB have been used in plant breeding programs to develop rice varieties (Mackill 1996; Khush 2005; Collard and Ismail 2013). In addition to these strategies, is the GSR breeding strategy involved in the simultaneous screening of early backcross BC_1F_2 bulk populations over three successive rounds with stringent selection sieves to develop climate-resilient rice varieties. GSR-BT ensures the fixation of homozygous inbred cultivars within a short duration of 4–5 years as compared with a conventional breeding program with 9–10 years. The GSR breeding strategy mainly involves three activities. First is the development of segregating BC_1F_2 bulk

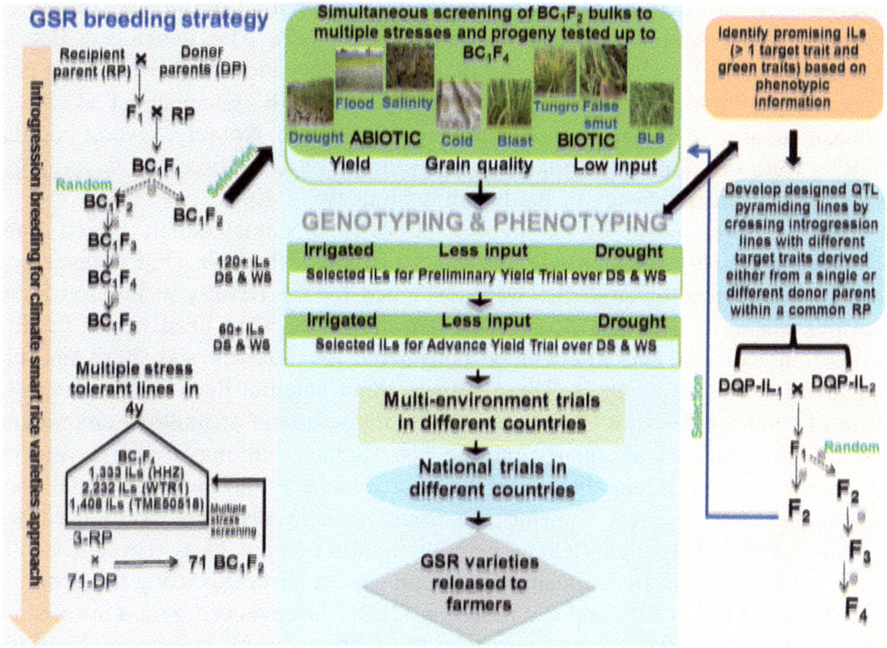

Fig. 1.5 Schematic representation of the GSR-BT for improving multiple stress tolerance in rice for climate resilience

populations by crossing an elite widely adaptable variety as a recipient parent and a diverse set of donors. Second is highly stringent phenotypic screening for selected BC_1F_2 bulk populations and testing in desired target environments and for specific traits. Last is the identification of genomic regions for the traits that are influenced by climate fluctuations and their characterization for unraveling the molecular and physiological basis of the detected genomic regions (Ali et al. 2017; Dimaano et al. 2017; Taghavi et al. 2017; Wang et al. 2017a). Our strategy is proven and has helped us to develop more than 240 nutrient- and water-use-efficient lines and also targets multiple biotic and abiotic stress tolerances in different ecosystems, mostly in Asian and African countries. The efficiency of this strategy mainly depends on choosing the adaptable recipient parents, especially the background parent, which must possess the least undesirable traits that should not later hinder the quality of the developed breeding products in the form of BILs. It helps to restrict the accumulation of more undesirable genetic segments from the donor parent and only allows positive favorable genetic segments to improve the target trait while keeping the background parent genome constant (Ali et al. 2017). The ILs developed will have a high recovery of recipient parental genome and the target locus from the donor parent but with less donor segment noise. Similarly, other ILs were developed in different cross combinations aiming a different set of traits/genes. Finally, these ILs specific to a single target trait/locus will be used for pyramiding multiple traits/genes to broaden the genetic basis of withstanding multiple stresses through a designed QTL pyramiding (DQP) approach (Fig. 1.5). The developed ILs or inbreds possessing a single trait/gene or several traits/genes showed a better performance for biotic stresses, abiotic stresses, and NUE (Chauhan et al. 2015; Dimaano et al. 2017; Pang et al. 2017a; Feng et al. 2018). Notably, several popular CSRVs were developed using these ILs developed by GSR breeding technology, which proved the efficiency of the breeding strategy.

For the development of CS hybrids (CSH), the climate-smart inbred lines (possessing single gene/multiple genes and stacked through DQP) developed through GSR breeding strategies were evaluated for the fertility restoration locus and possession of desirable floral traits. The lines that were positive for fertility restoration loci *Rf3* and *Rf4* will be considered as restorers and those entirely negative for it as maintainers. These climate-smart parental lines for the development of CSH are essential to complement the introgression of multiple abiotic and biotic stress-tolerant QTLs/genes into the F_1s. The heterotic pools were developed by genetic distance among the hybrid rice source nursery, and the CS parental lines (maintainer and restorer) showing maximum genetic distances were used for identifying promising high-yielding heterotic hybrid combinations. These heterotic pools are being carefully maintained based on genetic distances using the genomic information. MAS, GBS, and GS based on GEBV of advanced genomic research will be used for improving and purifying the heterotic pools. Similarly, heterotic maintainer pools will be generated for CMS line diversification for several beneficial traits that bring about climate resilience. Finally, the CSH was generated from the heterotic pools of the restorers and CMS lines, which will be highly heterotic and withstand the challenges posed by GCA. Validation of the CSH under different abiotic stress conditions helped us identify several promising CSH.

Additionally, depending upon the selection of a variety with desirable traits in the target environment and reducing the breeding cycles, some vital points need to be considered for the various strategies for developing CSRVs in future breeding programs.

- Advanced molecular marker technologies are the most promising tools for the structural characterization of rice germplasm and hence help in identifying superior genetic resources of rice germplasm/ILs for adapting to GCA. The association between DNA markers and useful agronomic and morphophysiological traits through MAB approaches will allow plant breeders and biotechnologists to accelerate the improvement of pure lines from crosses as compared with the conventional pedigree method and this will help to design crop ideotypes for future GCA (Collard et al. 2005; Semenov and Stratonovitch 2013).
- The process-based crop simulation model (PBCM) and prediction-based model approach (PBMA) play a promising role in the study of interactions among genotype, environment, and management practices. These are extensively useful to evaluate the impacts of GCA on crop yield potential and other factors such as water and NUE, and the structure of phenology (Fumoto et al. 2008; Gu et al. 2014; Rötter et al. 2015; Das et al. 2016; Li et al. 2016; Challinor et al. 2018).

- Developing integrated breeding and genomics strategies for developing site-specific varieties with high grain yield and nutritional quality is a major concern (Kole et al. 2015; Li et al. 2016; Rasheed et al. 2017). To predict better performances of individual rice accessions/mapping populations, GS is a key for the estimation of breeding value for selection and forwarding to the next generation by use with genome-wide molecular markers, and it also provides traits associated with/controlled by QTL or gene effects (Fig. 1.2). Hence, GS is most preferable for breeding and tackling major polygenic traits (Grenier et al. 2015).
- Recently, GSR-BT was extended to develop high nutrient- and water-use efficiency and tolerance of biotic and abiotic stresses in different ecosystems. GSR-BT improves multiple traits simultaneously and helps speed up breeding cycles to fix breeding lines and identify promising high-yielding rice varieties (Ali et al. 2006, 2012, 2017; Marcaida et al. 2014; Dimaano et al. 2017; Sasaki 2017; Feng et al. 2018). GSR varieties responded well under varying environments and had a relative yield advantage of 31.0–36.0% (Marcaida et al. 2014).
- GSR-BT-produced cultivars could be further exploited through a genomics-assisted recurrent selection scheme facilitated by a dominant male sterility line, "Jiabuyu" (Pang et al. 2017b). A set of 31 founder lines from the GSR-BT in an HHZ background with multiple abiotic stress tolerance and another 25 multiple-stress-tolerant restorer lines were used. This approach allowed us to breed several promising lines, especially combining for complex traits such as drought, salinity, and high grain yield under irrigated conditions.
- Extensive rice genomics information facilitated the development of CSRVs by using advanced omics, genome editing, epistasis, RNAi, and genetic engineering. This helps in accelerating the identification of key genetic regulators as genes and transcription factors underlying stress tolerance and, further, introducing genetic elements into desirable rice varieties (Abdallah et al. 2014).
- The exploitation of heterosis/hybrid vigor is a multigenic complex trait for hybrid development. The genetic dissection of these traits by GAB provides new insights into the molecular and physiological basis of traits such as magnitude and rate of vegetative growth, pollen fertility, flowering time, and yield (Birchler et al. 2006; Bar-Zvi et al. 2017; Fujimoto et al. 2018), thereby helping in developing super rice hybrids, with high heterosis for traits that influence multiple stresses.

1.5 Introgression Breeding for Climate-Smart Inbred and Hybrid Rice Varieties

Ensuring food security and sustainable rice production under climatic threats is the need of the hour. Hence, immediate action is required for the development of rice varieties that can have climate resilience. Immense genetic variation exists in the primary gene pool itself that could provide a solution for securing global food

Fig. 1.6 Generation of climate-smart rice hybrids (CSRH)

security (Fig. 1.6). The identification and deployment of genomic regions associated with traits that influence GCA will facilitate the development of CSRVs. Remarkable progress has been made to improve tolerance of abiotic stresses triggered by GCA in mega-rice cultivars (Hasan et al. 2015; Singh et al. 2016). However, locally adapted high-yielding commercial mega-varieties show a sensitive reaction to most biotic and abiotic stresses and have not been able to show a similar yield performance in varied agroecosystems. Hence, a systematic breeding strategy and molecular genetic information are essential to developing rice varieties that can cope with biotic and abiotic stresses. This was achieved by using the available sufficient molecular genetic diversity in the primary gene pool for dissecting complex traits by using high-resolution SNPs retrieved from the high-quality genome sequence and modified backcross breeding strategies (Li 2001; Li and Zhang 2013).

The GSR-BT is a revolutionary breeding methodology that integrates conventional breeding and recent trends in genomics/molecular biology employed for introgression breeding for developing CSRVs (Ali et al. 2017; Feng et al. 2018). The GSR-BT assures simultaneous product development and QTL/gene discovery in an integrated manner. It mainly involves three key steps: (i) identification of genome-wide donor segments for selection and discovering QTLs for allele mining, (ii) introgression of valuable and hidden QTLs/genes/alleles from different donors into highly adaptable elite rice cultivars for developing trait-specific ILs, and (iii) pyramiding traits/QTLs by using ILs possessing single/a few desirable traits/genes by a designed QTL pyramiding (DQP) approach into the desired genetic background of rice cultivars for multiple complex traits (Pang et al. 2017a; Feng et al. 2018). Several reports on deciphering the complex genetic nature of CS traits using GSR-BT proved its efficacy in developing CSRV (Zhu et al. 2015; Pang et al.

2017a; Ali et al. 2017; Feng et al. 2018). Feng et al. (2018) identified nine and five putative QTLs conferring drought tolerance and low nitrogen use efficiency followed by Pang et al. (2017a) and Zhu et al. (2015) for the cold and salinity tolerance. The QTL, *qCT-3-2* detected by Zhu et al. (2015) has been detected repeatedly in all environments. The same locus partially overlapped with another two QTLs *qPSST-3* and *qCT-3-3* detected for booting and seedling stages cold tolerance (Suh et al. 2010; Zhang et al. 2014). Similarly, there are plenty of reports showing QTLs and genes responsible for CS traits, drought (Zhou et al. 2010; Xu et al. 2005), cold (Jiang et al. 2008), and salinity (Wang et al. 2012; Cheng et al. 2012). Additionally, there are several studies ongoing to identify novel genomic regions and exploring the physiological and molecular mechanisms underlying the CS traits by applying GAB using GSR-BT (Table 1.3).

This has emerged as a dependable tool for accelerating breeding cycles for crop improvement, especially for precise introgression of abiotic stress tolerance, heat and cold tolerance, drought tolerance, submergence and salinity tolerance, increased CO_2 concentration, and biotic stress resistance through GAB. The GSR-BT provides many advantages such as fast-tracking of the best combination of haplotypes, high accuracy selection efficiency, identification of superior materials for QTL/gene pyramiding, genomics-assisted recurrent selection scheme (GARSS), and also in generating novel rice varieties for modern breeding programs to facilitate crop improvement (Zhang 2007; Ali et al. 2012, 2017; Chauhan et al. 2015) (Fig. 1.7). The GSR-BT involving an early backcross methodology helps in achieving higher success in the integration of genomic regions with less or no genetic drag derived from the donor parent. The GSR-BT has successfully proved the development of breeding cultivars possessing multiple stress tolerance and high yield potential by using existing genomics approaches (Table 1.4) (Ali et al. 2006, 2016).

The development of climate-smart rice hybrids (CSRH) requires a similar breeding strategy; however, here we will be adopting the parental lines (recipient and donor parents) from known maintainer and restorer pools. This methodology has been well laid out in Fig. 1.7. We exploited 3K resources to develop our heterotic pools and will be fully positioned to exploit the different pools for enhancing heterosis (Ali et al. unpublished). CSR inbred materials could be directly used in two-line hybrid rice technology as the pollen parent. CSRVs with restorer genes were used to develop the first generation of CSRH. However, once both parents are developed, that is, CSR male sterile lines and CSR restorer lines, the second generation of CSRH will be made available. At IRRI, we have developed several CSRH and results from their adaptation trials are now available, and soon these will be shared and licensed to Hybrid Rice Development Consortium (HRDC) members; (Ali et al. unpublished) (Fig. 1.8).

Table 1.3 List of bioinformatics tools and database resources for structural and functional genomics studies

S. No	Genomics databases	Information	References
I	*Structural genomics*		
1	MSU Rice Genome Annotation Project Database	Nipponbare nucleotide sequence information and annotation	https://rice.plantbiology.msu.edu/
2	The Rice Annotation Project (RAP)	Provides an accurate genomic structure and functions of genes and QTLs from the annotation information	https://rapdb.dna.affrc.go.jp/
3	PlantGDB	Mapping of cDNA and EST sequences and interpretations	https://www.plantgdb.org/OsGDB/
4	Oryzabase	Comprehensive of classical rice genetics to advanced genomics	https://shigen.nig.ac.jp/rice/oryzabase/
5	Gramene	Integrative and comparative functional genomics	https://www.gramene.org/
II	*Functional genomics*		
1	Rice Expression Database (RED)	Genes expression profiles, integration and visualizations from RNA-sequence data generated by different rice growth stage	https://expression.ic4r.org/
2	Rice Expression Profile Database (RiceXPro)	Genes expression profiles from microarray analysis	https://ricexpro.dna.affrc.go.jp/
3	RiceVarMap v2.0	Comprehensive studies of genomic variations and its functional annotations from the sequencing of 4,726 rice accessions	https://ricevarmap.ncpgr.cn/v2/
4	TIGR Rice Genome Project BLAST	BLAST server for nucleotide and protein alignments	https://blast.jcvi.org/euk-blast/index.cgi?project=osa1
5	OryGenesDB	Compare the annotations of Gene loci and models from MSU RGAP and RAP-DB	https://orygenesdb.cirad.fr/data.html
6	Rice SNP-Seek Database	3K rice panel of phenotypic traits and their association of SNPs, and InDels	https://snp-seek.irri.org/
7	Q-TARO	View of chromosomal regions assisted reported QTLs and compare to multiple QTLs in category wise	https://qtaro.abr.affrc.go.jp/
8	RPAN: Rice Pan-genome Browser	3,010 accession of genomics and expression profiles	https://cgm.sjtu.edu.cn/3kricedb/

(continued)

Table 1.3 (continued)

S. No	Genomics databases	Information	References
9	Information Commons for Rice (IC4R)	Genome sequence, annotation and integrating with other omics tools	https://ic4r.org/
10	PlantTFDB 4.0	About transcription factors and their associated binding motifs and predictions of 165 species	https://planttfdb.cbi.pku.edu.cn/
11	RiceGE: Rice Functional Genomic Express Database	Graphical representation of annotated genes, SNPs, and PAVs on the chromosomes	https://signal.salk.edu/cgi-bin/RiceGE
12	Rice TE Database	Transposable and repeated elements in the genome sequence	https://www.genome.arizona.edu/cgi-bin/rite/index.cgi
13	Oryza sativa small RNAs	Noncoding RNA molecules, which gives cytoplasm	https://sundarlab.ucdavis.edu/cgi-bin/smrna_browse/rice2/?name=miR168a

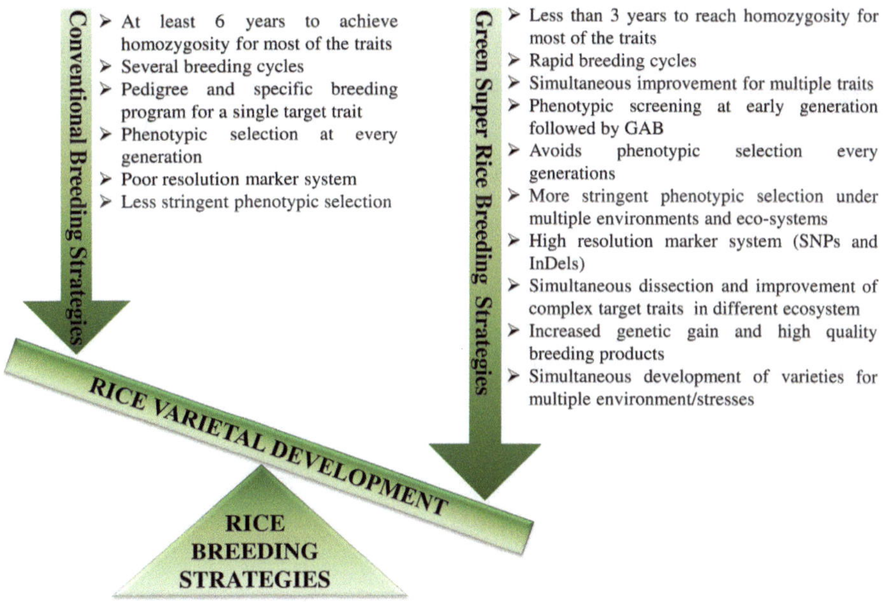

Fig. 1.7 Schematic diagram showing a comparison of conventional and GSR breeding strategies

Table 1.4 The performance of pyramided ILs developed using a DQP pyramiding approach for grain yield across different stresses and seasons

DQ pyramided lines	Average yield across condition (t ha^{-1})				Average yield across conditions (t ha^{-1})
	Irrigated (t ha^{-1})	Low input (t ha^{-1})	75% Nitrogen (t ha^{-1})	Rainfed (t ha^{-1})	
GSR IR2-DQ28-S2-C1	6.57	3.18	5.20	3.15	4.52
GSR IR2-DQ3-D1-R3	6.38	3.20	4.08	3.70	4.34
GSR IR2-DQ3-D1-L1	6.29	3.19	4.27	3.34	4.27
GSR IR2-DQ11-S3-R1	6.36	2.97	3.91	3.13	4.09
GSR IR2-DQ2-S1-S1	6.16	2.82	4.24	3.10	4.08
GSR IR2-DQ24-Y1-S1	5.70	2.81	4.58	3.18	4.07
GSR IR2-DQ45-S3-S1	6.11	2.99	4.30	2.78	4.05
GSR IR2-DQ25-L1-L1	6.35	3.17	4.04	2.58	4.04
GSR IR2-DQ15-C2-S1	5.74	3.04	4.15	3.10	4.01
GSR IR2-DQ25-L1-C1	6.05	2.94	3.88	3.15	4.01
GSR IR2-DQ8-Y2-C1	6.24	2.81	4.30	2.60	3.99
GSR IR2-DQ42-S1-Y1	6.36	3.03	4.04	2.49	3.98
GSR IR2-DQ20-Y1-S1	5.87	3.00	4.25	2.75	3.97
GSR IR2-DQ40-S4-R1	6.13	2.89	4.13	2.69	3.96
GSR IR2-DQ41-L3-R1	6.81	2.66	3.91	2.46	3.96
GSR IR2-DQ44-Y1-L1	5.61	3.06	4.06	3.11	3.96
GSR IR2-DQ25-S2-Y1	6.00	3.12	3.91	2.79	3.96
GSR IR2-DQ43-C3-R1	6.03	2.85	3.94	2.98	3.95
PSB Rc82 ^{LIC}	5.10	2.31	3.35	2.13	3.22
Rc 192^{SC}	3.07	2.59	3.02	2.14	2.70
Rc 222	4.84	2.87	3.48	1.82	3.25
UPLRi7	4.77	2.56	3.39	2.84	3.39

(continued)

Table 1.4 (continued)

DQ pyramided lines	Average yield across condition (t ha^{-1})				Average yield across conditions (t ha^{-1})
	Irrigated (t ha^{-1})	Low input (t ha^{-1})	75% Nitrogen (t ha^{-1})	Rainfed (t ha^{-1})	
Pr > F, G	0.13 (DS) 0.0 (WS)	0.00 (DS) 0.00 (WS)	0.02 (DS) 0.12 (WS)	0.01 (WS)	–
Heritability	0.25 (DS) 0.71 (WS)	0.46 (DS) 0.59 (WS)	0.30 (DS) 0.62 (WS)	0.65 (WS)	–

SC Salinity check; *LIC* Low-input check

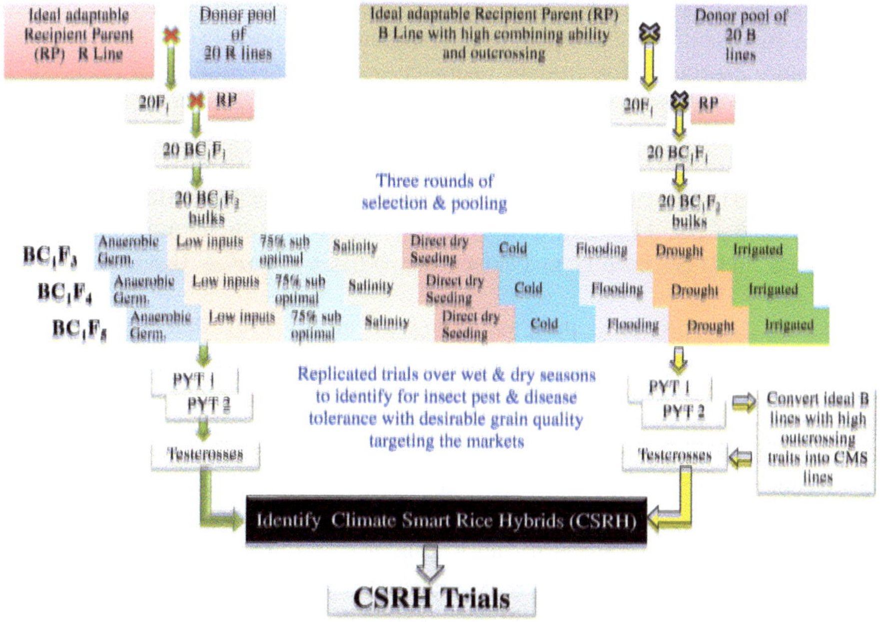

Fig. 1.8 Schematic representation of breeding strategies used to develop climate-smart restorer/pollen parents, cytoplasmic male sterile (CMS) lines, and their maintainer lines

1.6 Designed QTL Pyramiding (DQP) for Efficient Trait Stacking

Most agronomically important traits, including grain yield and abiotic stresses, show complex non-Mendelian inheritance and are highly influenced by the environment. These traits are under the influence of polygenes with minor cumulative effects and hence are detected as QTLs. Understanding the molecular genetics and breeding efficiency of these QTLs is of significant concern to rice breeders. To date, only a few researchers have tried to dissect and use QTLs that confer useful traits such as these QTLs that are affected by QTL × environmental interactions (Q × E), low resolution of the detected QTLs, and epistasis between the QTLs (Bocianowski 2013; Liu et al. 2014; Zhu et al. 2015). Hence, to overcome these predicaments, a systematic breeding strategy, such as the GSR-BT, is necessary for the simultaneous dissection and use of useful QTLs toward breeding CSRVs. This enables us to develop genome-wide trait-specific ILs for dissecting QTLs and to pyramid these QTLs that control useful traits in the desired genetic backgrounds of rice mega-varieties.

A designed QTL pyramiding approach under the GSR breeding strategy involves the identification of superior introgression lines with a set of complex abiotic stress tolerance traits based on the molecular data. Combining the traits in different combinations derived from trait-/gene-specific introgression lines would complement well, especially for combining valuable abiotic and biotic stress tolerances. In the past, we were successful in developing several pyramiding lines in an IR64 recipient background, especially for drought tolerance traits. We identified as many as 48 promising pyramiding lines (Guan et al. 2010), out of which 20 were screened under drought, low inputs, and irrigated conditions. Eventually, this resulted in the identification and release of five varieties: NSIC Rc29 = IR83140-B-36-B (upland), NSIC Rc434 = IR83142-B-19-B (drought lowland), NSIC Rc390 = IR83140-B-28-B (salinity), and GSR 11 = IR83140-B-11-B (salinity) for their cultivation in the Philippines and one recently in Pakistan i.e. NIBGE 3GSR = IR83142-B-8-B-B (drought).

Recently, in the GSR program, Pang et al. (2017a) demonstrated DQP approaches for improving the complex trait of salinity tolerance. A total of 200 F_4 plants were derived from the cross between two ILs, GPDQ3, and GPDQ4, which were developed from the GSR-BT. The simultaneous mapping and improvement for salinity tolerance were carried out using 200 F_4 plants, which further resulted in the detection of three QTLs conferring salinity tolerance. A total of 32 M SNP data retrieved from the tGBS were used for the genotyping of the QTL mapping. The detected QTLs were introgressed, and ILs showed strong salinity tolerance, which proved the efficacy of the detected and introgressed QTLs. Similarly, Feng et al. (2018) detected nine QTLs for drought tolerance (DT) and seven QTLs for low nitrogen tolerance (LNT) by using a DQP population generated by three sets of BC_1F_4 trait-specific ILs derived from HHZ × Teqing, HHZ × CDR22, and HHZ × OM1723 in the genetic background of variety Huanghuazhan (Ali et al. 2017). The QTLs were discovered through a systematic GSR-BT for developing several superior GSR lines combined with improved DQP approaches. Additionally, we were successful in nominating as many as eight DQ pyramiding lines in an HHZ background and another four in a WTR-1 background into the All India Coordinated Rice Improvement Project (AICRIP) trials in different testing categories. This demonstrates the potential of DQP and is a highly recommended breeding strategy to complement complex traits and stack traits to develop CSRVs.

1.7 Climate-Smart Rice Varietal Products and Deployment

In the past three decades, several breeders have targeted increasing yield potential and tolerance of biotic and abiotic stresses. Recently released rice varieties in multilocation trials in China reached a high yield potential of up to 12 t ha^{-1} under more input use of water and fertilizers (Jiang et al. 2016). However, the application of higher doses of fertilizers, continuous irrigation, and overuse of pesticides increases rice production costs to farmers and also depletes soil nutrient capacity, hinders human health, and affects soil health because of severe pest and disease outbreaks (Aktar et al. 2009). Hence, the development of CSRVs using GSR-BT that showed higher performance even with minimal agronomic management under low-input conditions will be a boon for marginal farmers.

To breed such varieties by GSR-BT will help to address the challenges of tolerance of multiple biotic and abiotic stresses and nutrient- and water-use efficiency in different target environments (Zhang 2007; Ali et al. 2013). In 2009, a GSR breeding program began at IRRI, which resulted in the development of 4,973 ILs from three recipient backgrounds (HHZ, WTR-1, and TME50518) using 71 donors. These ILs showed significantly higher grain yield potential than the checks under more than one environment and in further MET in different Asian and African countries. To date, 54 GSR varieties have been deployed worldwide, with nearly 26 IRRI-bred GSR varieties bred from three recipient parental backgrounds: IR64, HHZ, and WTR-1 alone (Table 1.5). A total of 97 GSR cultivars are currently undergoing national varietal testing, and 91 of them are from IRRI-bred GSR lines belonging to IR64 (6), HHZ (72), and WTR-1 (13) backgrounds. GSR varieties showed consistently higher grain yield under low inputs and tolerance of multiple biotic and abiotic stresses (Ali et al. 2012; Chauhan et al. 2015; Yorobe et al. 2016; Li 2017). Hence, the advanced GSR breeding strategy proved its efficacy for improving multiple traits in different target environments by developing CSRVs under ever-changing GCA, which can be easily adopted (Ali et al. 2016, 2017, 2018; Feng et al. 2018).

1.8 Challenges and Future Strategies

The emerging challenges posed by the increasing global population, decreasing arable lands, and climate change exert tremendous pressure on the global production and productivity of rice, on which more than 70.0% of the Asian population depends on food. Ray et al. (2013) estimated that global yield must increase by 1.1–1.3% in the major cereal crops per year to feed the estimated 9.7 billion world population by 2050. Additionally, GCA impacts such as fluctuating global atmospheric temperature, elevated CO_2, and erratic rainfall patterns significantly contributed to the dramatic reduction in rice crop yield (Batley and Edwards 2016).

Table 1.5 List of popular CSRVs developed by the GSR-BT strategy

Country	GSR line	Recurrent parent	Type	Released name	Released system	Year of release
Philippines	IR83140-B-36-B	IR64	IRRI bred	NSIC Rc29 (Katihan 4)	Formal	2014
Philippines	IR83140-B-28-B	IR64	IRRI bred	NSIC Rc390 (Salinas 19)	Formal	2014
Philippines	IR84675-58-4-1-B-B	IR64	IRRI bred	NSIC Rc392 (Salinas 20)	Formal	2014
Philippines	IR83140-B-11-B	IR64	IRRI bred	GSR 11	Informal	2014
Philippines	IR83142-B-19-B	IR64	IRRI bred	NSIC Rc434 (Sahod Ulan 21) or GSR2	Formal	2015
Pakistan	GSR IR1-5-S10-D3-Y2	Huanghuazhan	IRRI bred	NIBGE 1 GSR	Formal	2015
Pakistan	GSR IR1-5-S12-D3-Y2	Huanghuazhan	IRRI bred	NIBGE 2 GSR	Formal	2015
Philippines	GSR IR1-5-S8-D3-SU1	Huanghuazhan	IRRI bred	GSR 5a	Informal	2014
Philippines	GSR IR1-1-Y4-Y1	Huanghuazhan	IRRI bred	GSR 1	Informal	2014
Philippines	GSR IR1-5-S14-S2-Y2	Huanghuazhan	IRRI bred	GSR 5	Informal	2014
Philippines	GSR IR1-12-S2-Y3-Y2	Huanghuazhan	IRRI bred	GSR12A	Informal	2015
Philippines	GSR IR1-8-S6-S3-Y2	Huanghuazhan	IRRI bred	NSIC 2016 Rc480 (GSR8)	Formal	2016
Philippines	GSR IR1-5-S6-S3-D1	Huanghuazhan	IRRI bred	NSIC 2016 Rc436 (Tubigan 37)	Formal	2016
Philippines	GSR IR1-12-D10-S1-D1	Huanghuazhan	IRRI bred	BPI-NSIC-2018-Rc534 or Salinas 29	Formal	2018
Philippines	GSR IR1-21-Y4-Y2-Y1	Huanghuazhan	IRRI bred	BPI-NSIC-2018-Rc514 or Tubigan 45	Formal	2018
Philippines	GSR IR1-3-S13-Y1-S1	Huanghuazhan	IRRI bred	BPI-NSIC-2018-Rc512 or Tubigan 44	Formal	2018
Vietnam	GSR IR1-11-Y10-D3-Y3	Huanghuazhan	IRRI bred	Gia Loc 501	Informal	2016
Vietnam	GSR IR1-5-D1-D1	Huanghuazhan	IRRI bred	Gia Loc 502	Informal	2016
Philippines	GSR IR2-1-L1-L1-L2	Weed Tolerant Rice1	IRRI bred	GSR 21	Informal	2015

(continued)

Table 1.5 (continued)

Country	GSR line	Recurrent parent	Type	Released name	Released system	Year of release
Philippines	GSR IR2-12-Y15-Y2-Y2	Weed Tolerant Rice1	IRRI bred	GSR 22	Informal	2015
Philippines	GSR IR2-7-L4-L1-R2	Weed Tolerant Rice1	IRRI bred	GSR 23	Informal	2015
Philippines	GSR IR2-1-L1-N1-L2	Weed Tolerant Rice1	IRRI bred	GSR 24	Informal	2015
Philippines	GSR IR2-19-R2-SU2-R2	Weed Tolerant Rice1	IRRI bred	BPI-NSIC-2018-Rc556 or Salinas 32	Formal	2018
Philippines	GSR IR2-6-R4-S3-Y2	Weed Tolerant Rice1	IRRI bred	BPI-NSIC-2018-Rc554 or Salinas 31	Formal	2018
Indonesia	GSR IR1-5-D1-D1	Huanghuazhan	IRRI bred	Inpari 46 GSR	Formal	2019
Pakistan	IR-83142-B-8-B-B	IR64	IRRI bred	NIBGE GSR3	Formal	2019

A conventional breeding approach is much slower and involves screening large populations to identify the desired genetic variation and increase the frequency of favorable alleles. Hence, there is an urgent need to integrate various genomics-based technologies with conventional breeding methodologies. This integrated approach not only helps to increase the precision of crop improvement but also accelerates the development of breeding products, which is the need of the hour.

Novel systematic plant breeding designs and strategies and advanced genomics technologies are required to understand the molecular genetic information of useful agronomic and morphophysiological traits associated with QTLs and genes under the worst situations of climate change while ensuring stable yields. By the use of advanced GSR-BT and modern genomics tools such as NGS, tGBS, allele mining, and GS, the GSR-BT began, which resulted in the release of more than 26 CSRVs and several putative QTLs for drought, salinity, and submergence tolerance, weed-competitive ability, low input use efficiency, and biotic stress tolerance, leading to the development of climate-smart rice inbreds and hybrids (Ali et al. 2017; Feng et al. 2018; Wambugu et al. 2018; Zhang et al. 2018).

Furthermore, increasing the efficiency and selection of breeding materials can be achieved by GS that uses GEBV, GBLUP, and KBLUP multivariate analysis models as these help in predicting the performance of their progenies. Hence, GS is used in the development of superior breeding lines among the populations (Bassi et al. 2016). The availability of the 3K genome needs to be fine-tuned in a comprehensive and systematic way for the mining of useful genes/alleles for TOI for creating novel and sustainable rice varieties that can adapt to GCA in future breeding. Deep sequencing of recipient and donor parents in GSR breeding programs will be an excellent opportunity for trait discovery and understanding the genetics of complex traits. This will further trigger the development of superior high-yielding climate-smart inbreds and hybrids. Further integration of different approaches such as GS into breeding schemes, genome sequencing and editing technologies, high-throughput phenotyping technologies, artificial intelligence machine algorithms, and crop simulation models will also allow rapid improvement of molecular breeding programs for developing climate-resilient rice varieties.

1.9 Conclusions

Developing CSRVs is critical to tackling global climate change in a post-genomics era. Delivering and deploying CSRVs is challenging as it should be done in rapid breeding cycles with great precision. This challenge has been addressed by applying the integrated approach of fine-tuned conventional breeding methodologies designed by GSR-BT along with modern genomics tools, NGS, and tGBS, which could help to speed up the breeding of CSRVs. The availability of cost-effective and highly efficient genome sequencing techniques helped in discovering even the rarest favorable alleles through an allele mining approach. GS and GBS are the essential tools that enhance the efficiency and accuracy of plant selection, which in turn increases genetic gain as a response to selection. The evolving techniques of GBS are making it highly useful, especially for the precise identification of the genetic factors that influence the trait through QTL mapping and GWAS analysis tools. The success of GSR-BT might be attributed to innovative introgression breeding and designed QTL pyramiding efforts integrated with genomics and appropriate cross-tolerance screening and selection techniques that tapped the hidden genetic diversity from a large number of donors into adaptable varietal backgrounds such as HHZ, WTR-1, and TME80518. The development and deployment of CSRVs in Asia and Africa from our work are already helping to increase global rice production and productivity with increased farm income.

References

Abberton M, Batley J, Bentley A, Bryant J, Cai H, Cockram J, Costa de Oliveira A, Cseke LJ, Dempewolf H, De Pace C (2016) Global agricultural intensification during climate change: a role for genomics. Plant Biotechnol J 14:1095–1098

Abdallah NA, Moses V, Prakash CS (2014) The impact of possible climate changes on developing countries: the needs for plants tolerant to abiotic stresses. GM Crops Food 5:77–80

Abdelrahman M, Al-Sadi AM, Pour-Aboughadareh A, Burritt DJ, Tran L-SP (2018) Genome editing using CRISPR/Cas9–targeted mutagenesis: an opportunity for yield improvements of crop plants grown under environmental stresses. Plant Physiol Biochem 131:31–36

Aktar W, Sengupta D, Chowdhury A (2009) Impact of pesticides use in agriculture: their benefits and hazards. Interdiscip Toxicol 2:1–12

Alexandrov N, Tai S, Wang W, Mansueto L, Palis K, Fuentes RR, Ulat VJ, Chebotarov D, Zhang G, Li Z (2014) SNP-Seek database of SNPs derived from 3000 rice genomes. Nucl Acids Res 43:D1023–D1027

Ali AJ, Xu JL, Ismail AM, Fu BY, Vijaykumar CHM, Gao YM, Domingo J, Maghirang R, Yu SB, Gregorio G (2006) Hidden diversity for abiotic and biotic stress tolerances in the primary gene pool of rice revealed by a large backcross breeding program. Field Crops Res 97:66–76

Ali J, Aslam UM, Tariq R, Murugaiyan V, Schnable PS, Li D, Marfori-Nazarea CM, Hernandez JE, Arif M, Xu J (2018) Exploiting the genomic diversity of rice (*Oryza sativa* L.): SNP-typing in 11 early-backcross introgression-breeding populations. Front Plant Sci 9

Ali J, Franje NJ, Revilleza JE, Acero B (2016) Breeding for low-input responsive Green Super Rice (GSR) varieties for rainfed lowlands of Asia and Africa. Philipp J Crop Sci

Ali J, Xu J-L, Gao Y-M, Ma X-F, Meng L-J, Wang Y, Pang Y-L, Guan Y-S, Xu M-R, Revilleza JE (2017) Harnessing the hidden genetic diversity for improving multiple abiotic stress tolerance in rice (*Oryza sativa* L.). PLoS ONE 12:e0172515

Ali J, Xu JL, Gao YM, Fontanilla M, Li ZK (2013) Breeding for yield potential and enhanced productivity across different rice ecologies through green super rice (GSR) breeding strategy. Int Dialog Percept Prospect Des Rice 60–68

Ali J, Xu JL, Gao YM, Fontanilla MA, Li ZK (2012) Green super rice (GSR) technology: an innovative breeding strategy—achievements & advances. In: The 12th SABRAO congress on plant breeding. Towards 2025: challenges in a rapidly changing world

Altieri MA, Nicholls CI (2017) The adaptation and mitigation potential of traditional agriculture in a changing climate. Clim Change 140:33–45

Anantha MS, Patel D, Quintana M, Swain P, Dwivedi JL, Torres RO, Verulkar SB, Variar M, Mandal NP, Kumar A (2016) Trait combinations that improve rice yield under drought: Sahbhagi Dhan and new drought-tolerant varieties in South Asia. Crop Sci 56:408–421

Atlin GN, Cairns JE, Das B (2017) Rapid breeding and varietal replacement are critical to adaptation of cropping systems in the developing world to climate change. Glob Food Sec 12:31–37

Bachlava E, Taylor CA, Tang S, Bowers JE, Mandel JR, Burke JM, Knapp SJ (2012) SNP discovery and development of a high-density genotyping array for sunflower. PLoS ONE 7: e29814

Balachiranjeevi CH, Prahalada GD, Mahender A, Mohamad Jamaluddin, Ma. Anna Lynn Sevilla1, Corinne Mira Marfori-Nazarea, Ricky Vinarao, Ali J (2019) Identification of a novel locus, BPH38(t), conferring resistance to brown planthopper (*Nilaparvata lugens Stal.*) using early backcross population in rice (*Oryza sativa* L.). Euphytica 215:185

Bar-Zvi D, Lupo O, Levy AA, Barkai N (2017) Hybrid vigor: the best of both parents, or a genomic clash? Curr Opin Syst Biol 6:22–27

Barakate A, Stephens J (2016) An overview of CRISPR-based tools and their improvements: new opportunities in understanding plant–pathogen interactions for better crop protection. Front Plant Sci 7:765

Bassi FM, Bentley AR, Charmet G, Ortiz R, Crossa J (2016) Breeding schemes for the implementation of genomic selection in wheat (*Triticum* spp.). Plant Sci 242:23–36

Batley J, Edwards D (2016) The application of genomics and bioinformatics to accelerate crop improvement in a changing climate. Curr Opin Plant Biol 30:78–81

Bayer MM, Rapazote-Flores P, Ganal M, Hedley PE, Macaulay M, Plieske J, Ramsay L, Russell J, Shaw PD, Thomas W (2017) Development and evaluation of a barley 50k iSelect SNP array. Front Plant Sci 8:1792

Beckage B, Gross LJ, Lacasse K, Carr E, Metcalf SS, Winter JM, Howe PD, Fefferman N, Franck T, Zia A (2018) Linking models of human behaviour and climate alters projected climate change. Nat Clim Change 8:79

Bhat JA, Ali S, Salgotra RK, Mir ZA, Dutta S, Jadon V, Tyagi A, Mushtaq M, Jain N, Singh PK (2016) Genomic selection in the era of next generation sequencing for complex traits in plant breeding. Front Genet 7:221

Bhatia D, Wing RA, Yu Y, Chougule K, Kudrna D, Lee S, Rang A, Singh K (2018) Genotyping by sequencing of rice interspecific backcross inbred lines identifies QTLs for grain weight and grain length. Euphytica 214:41

Birchler JA, Yao H, Chudalayandi S (2006) Unraveling the genetic basis of hybrid vigor. Proc Natl Acad Sci 103:12957–12958

Bocianowski J (2013) Epistasis interaction of QTL effects as a genetic parameter influencing estimation of the genetic additive effect. Genet Mol Biol 36:93–100

Bortesi L, Fischer R (2015) The CRISPR/Cas9 system for plant genome editing and beyond. Biotechnol Adv 33:41–52

Challinor AJ, Müller C, Asseng S, Deva C, Nicklin KJ, Wallach D, Vanuytrecht E, Whitfield S, Ramirez-Villegas J, Koehler AK (2018) Improving the use of crop models for risk assessment and climate change adaptation. Agric Syst 159:296–306

Chang CC, Lee HL, Hsu SH (2013) Food security: global trends and regional perspective with reference to East Asia. Pacific Rev 26:589–613

Chauhan BS, Opeña J, Ali J (2015) Response of 10 elite "Green Super Rice" genotypes to weed infestation in aerobic rice systems. Plant Prod Sci 18:228–233

Chen H, Xie W, He H, Yu H, Chen W, Li J, Yu R, Yao Y, Zhang W, He Y (2014a) A high-density SNP genotyping array for rice biology and molecular breeding. Mol Plant 7:541–553

Chen W, Gao Y, Xie W, Gong L, Lu K, Wang W, Li Y, Liu X, Zhang H, Dong H (2014b) Genome-wide association analyses provide genetic and biochemical insights into natural variation in rice metabolism. Nat Genet 46:714

Cheng L, Wang Y, Meng L, Hu X, Cui Y, Sun Y, Zhu L, Ali J, Xu J, Li Z (2012) Identification of salt-tolerant QTLs with strong genetic background effect using two sets of reciprocal introgression lines in rice. Genome 55:45–55

Choi JY, Platts AE, Fuller DQ, Wing RA, Purugganan MD (2017) The rice paradox: multiple origins but single domestication in Asian rice. Mol Biol Evol 34:969–979

Civáň P, Brown TA (2017) Origin of rice (*Oryza sativa* L.) domestication genes. Genet Resour Crop Evol 64:1125–1132

Collard BCY, Ismail AM (2013) EIRLSBN: twenty years of achievements in rice breeding. IRRI

Collard BCY, Jahufer MZZ, Brouwer JB, Pang ECK (2005) An introduction to markers, quantitative trait loci (QTL) mapping and marker-assisted selection for crop improvement: the basic concepts. Euphytica 142:169–196

Dar MH, Singh S, Zaidi NW, Shukla S (2012) Sahbhagi Dhan: science's answer to drought problems. STRASA News 5:1–3

Das G, Patra JK, Baek K-H (2017) Insight into MAS: a molecular tool for development of stress resistant and quality of rice through gene stacking. Front Plant Sci 8:985

Das KN, Basumatary A, Ahmed S (2016) Targeted yield precision model assessment for rice-rice crop sequences in farmers' fields in humid, sub-tropical Northeastern India. J Soil Sci Plant Nutr 16:48–59

Davey JW, Hohenlohe PA, Etter PD, Boone JQ, Catchen JM, Blaxter ML (2011) Genome-wide genetic marker discovery and genotyping using next-generation sequencing. Nat Rev Genet 12:499

Deokar AA, Ramsay L, Sharpe AG, Diapari M, Sindhu A, Bett K, Warkentin TD, Tar'an B (2014) Genome wide SNP identification in chickpea for use in development of a high density genetic map and improvement of chickpea reference genome assembly. BMC Genomics 15:708

Descalsota GIL, Swamy BPM, Zaw H, Inabangan-Asilo MA, Amparado A, Mauleon R, Chadha-Mohanty P, Arocena EC, Raghavan C, Leung H (2018) Genome-wide association mapping in a rice MAGIC Plus population detects QTLs and genes useful for biofortification. Front Plant Sci 9:1347

Dimaano NGB, Ali J, Cruz PCS, Baltazar AM, Diaz MGQ, Acero BL, Li Z (2017) Performance of newly developed weed-competitive rice cultivars under lowland and upland weedy conditions. Weed Sci 65:798–817

Dimaano NG, Ali J, Mahender A, Cruz PCS, Baltazar AM, Diaz MGQ, Pang YL, Jr BA, Li Z (2019) Novel quantitative trait loci for weed-competitive ability traits using the early generation of backcross rice populations. (unpublished)

Duitama J, Silva A, Sanabria Y, Cruz DF, Quintero C, Ballen C, Lorieux M, Scheffler B, Farmer A, Torres E (2015) Whole genome sequencing of elite rice cultivars as a comprehensive information resource for marker assisted selection. PLoS ONE 10:e0124617

Endo A, Masafumi M, Kaya H, Toki S (2016) Efficient targeted mutagenesis of rice and tobacco genomes using Cpf1 from *Francisella novicida*. Sci Rep 6:38169

Fahad S, Nie L, Khan FA, Chen Y, Hussain S, Wu C, Xiong D, Jing W, Saud S, Khan FA (2014) Disease resistance in rice and the role of molecular breeding in protecting rice crops against diseases. Biotechnol Lett 36:1407–1420

Fahad S, Bajwa AA, Nazir U, Anjum SA, Farooq A, Zohaib A, Sadia S, Nasim W, Adkins S, Saud S (2017) Crop production under drought and heat stress: plant responses and management options. Front Plant Sci 8:1147

Feltus FA, Wan J, Schulze SR, Estill JC, Jiang N, Paterson AH (2004) An SNP resource for rice genetics and breeding based on subspecies indica and japonica genome alignments. Genome Res 14:1812–1819

Feng B, Chen K, Cui Y, Wu Z, Zheng T, Zhu Y, Ali J, Wang B, Xu J, Zhang W (2018) Genetic dissection and simultaneous improvement of drought and low nitrogen tolerances by designed QTL pyramiding in rice. Front Plant Sci 9:306

Fujimoto R, Uezono K, Ishikura S, Osabe K, Peacock WJ, Dennis ES (2018) Recent research on the mechanism of heterosis is important for crop and vegetable breeding systems. Breed Sci 68:145–158

Fumoto T, Kobayashi K, Li C, Yagi K, Hasegawa T (2008) Revising a process-based biogeochemistry model (DNDC) to simulate methane emission from rice paddy fields under various residue management and fertilizer regimes. Glob Change Biol 14:382–402

Ganal MW, Durstewitz G, Polley A, Bérard A, Buckler ES, Charcosset A, Clarke JD, Graner E-M, Hansen M, Joets J (2011) A large maize (*Zea mays* L.) SNP genotyping array: development and germplasm genotyping, and genetic mapping to compare with the B73 reference genome. PLoS ONE 6:e28334

Garcia-Oliveira AL, Tan L, Fu Y, Sun C (2009) Genetic identification of quantitative trait loci for contents of mineral nutrients in rice grain. J Integr Plant Biol 51:84–92

Garg AK, Kim J-K, Owens TG, Ranwala AP, Do Choi Y, Kochian LV, Wu RJ (2002) Trehalose accumulation in rice plants confers high tolerance levels to different abiotic stresses. Proc Natl Acad Sci 99:15898–15903

Garris A, Tai T, Coburn J, Kresovich S, McCouch SR (2005) Genetic structure and diversity in *Oryza sativa* L. Genetics 169:1631–1638

Gimhani DR, Gregorio GB, Kottearachchi NS, Samarasinghe WLG (2016) SNP-based discovery of salinity-tolerant QTLs in a bi-parental population of rice (*Oryza sativa*). Mol Genet Genomics 291:2081–2099

Goff SA, Ricke D, Lan T-H, Presting G, Wang R, Dunn M, Glazebrook J, Sessions A, Oeller P, Varma H (2002) A draft sequence of the rice genome (*Oryza sativa* L. ssp. *japonica*). Science 296:92–100

Grenier C, Cao T-V, Ospina Y, Quintero C, Châtel MH, Tohme J, Courtois B, Ahmadi N (2015) Accuracy of genomic selection in a rice synthetic population developed for recurrent selection breeding. PLoS ONE 10:e0136594

Gregorio GB (1997) Tagging salinity tolerance genes in rice using amplified fragment length polymorphism (AFLP). PhD. thesis, University of the Philippines, Los Baños

Gu J, Yin X, Zhang C, Wang H, Struik PC (2014) Linking ecophysiological modelling with quantitative genetics to support marker-assisted crop design for improved yields of rice (*Oryza sativa*) under drought stress. Ann Bot 114:499–511

Guan YS, Serraj R, Liu SH, Xu JL, Ali J, Wang WS, Venus E, Zhu LH, Li ZK (2010) Simultaneously improving yield under drought stress and non-stress conditions: a case study of rice (*Oryza sativa* L.). J Exp Bot 61:4145–4156

Guo L, Gao Z, Qian Q (2014) Application of resequencing to rice genomics, functional genomics and evolutionary analysis. Rice 7:4

Hampton J, Conner A, Boelt B, Chastain T, Rolston P (2016) Climate change: seed production and options for adaptation. Agriculture 6:33

Hasan MM, Rafii MY, Ismail MR, Mahmood M, Rahim HA, Alam MA, Ashkani S, Malek MA, Latif MA (2015) Marker-assisted backcrossing: a useful method for rice improvement. Biotechnol Biotechnol Equip 29:237–254

Hatfield JL, Prueger JH (2015) Temperature extremes: effect on plant growth and development. Weather Clim Extrem 10:4–10

He J, Zhao X, Laroche A, Lu Z-X, Liu H, Li Z (2014) Genotyping-by-sequencing (GBS), an ultimate marker-assisted selection (MAS) tool to accelerate plant breeding. Front Plant Sci 5:484

Hu W, Wen M, Han Z, Tan C, Xing Y (2013) Scanning QTLs for grain shape using a whole genome SNP array in rice. J Plant Biochem Physiol 1:104

Huang X, Feng Q, Qian Q, Zhao Q, Wang L, Wang A, Guan J, Fan D, Weng Q, Huang T (2009) High-throughput genotyping by whole-genome resequencing. Genome Res

Huang X, Han B (2014) Natural variations and genome-wide association studies in crop plants. Annu Rev Plant Biol 65:531–551

Huang X, Kurata N, Wang Z-X, Wang A, Zhao Q, Zhao Y, Liu K, Lu H, Li W, Guo Y (2012) A map of rice genome variation reveals the origin of cultivated rice. Nature 490:497

Huang X, Sang T, Zhao Q, Feng Q, Zhao Y, Li C, Zhu C, Lu T, Zhang Z, Li M (2010) Genome-wide association studies of 14 agronomic traits in rice landraces. Nat Genet 42:961

Huq MA, Akter S, Nou IIS, Kim HT, Jung YJ, Kang KK (2016) Identification of functional SNPs in genes and their effects on plant phenotypes. J Plant Biotechnol 43:1–11
Iizumi T, Furuya J, Shen Z, Kim W, Okada M, Fujimori S, Hasegawa T, Nishimori M (2017) Responses of crop yield growth to global temperature and socioeconomic changes. Sci Rep 7:7800
IPCC (2007) Climate change 2007: synthesis report. Contribution of working groups I, II and III to the fourth assessment report of the intergovernmental panel on climate change. In: Core Writing Team, Pachauri RK, Reisinger A (eds) IPCC, Geneva, Switzerland, 104p
Jacob SR, Tyagi V, Agrawal A, Chakrabarty SK, Tyagi RK (2015) Indian plant germplasm on the global platter: an analysis. PLoS ONE 10:e0126634
Jaganathan D, Ramasamy K, Sellamuthu G, Jayabalan S, Venkataraman G (2018) CRISPR for crop improvement: an update review. Front Plant Sci 9:985
Jain RK, Jain S (2000) Transgenic strategies for genetic improvement of Basmati rice. Indian J Exp Biol 38(1):6–17
Jat RK, Sapkota TB, Singh RG, Jat ML, Kumar M, Gupta RK (2014) Seven years of conservation agriculture in a rice–wheat rotation of Eastern Gangetic Plains of South Asia: yield trends and economic profitability. Field Crops Res 164:199–210
Jewel ZA, Ali J, Mahender A, Hernandez J, Pang Y, Li Z (2019) Identification of quantitative trait loci associated with nutrient use efficiency traits using SNP markers in an early backcross population of rice (*Oryza sativa* L.). Int J Mol Sci 20:900
Jiang L, Xun M, Wang J, Wan J (2008) QTL analysis of cold tolerance at seedling stage in rice (Oryza sativa L.) using recombination inbred lines. J Cereal Sci 48:173–179
Jiang P, Xie X, Huang M, Zhou X, Zhang R, Chen J, Wu D, Xia B, Xiong H, Xu F (2016) Potential yield increase of hybrid rice at five locations in southern China. Rice 9:11
Jockson MT (2010) International rice gene bank operations manual. Germplasm Charact Eval 34–40
Kamburova VS, Nikitina E V, Shermatov SE, Buriev ZT, Kumpatla SP, Emani C, Abdurakhmonov IY (2017) Genome editing in plants: an overview of tools and applications. Int J Agron. Article ID 7315351
Kamthan A, Chaudhuri A, Kamthan M, Datta A (2016) Genetically modified (GM) crops: milestones and new advances in crop improvement. Theor Appl Genet 129:1639–1655
Khanday I, Skinner D, Yang B, Mercier R, Sundaresan V (2018) A male-expressed rice embryogenic trigger redirected for asexual propagation through seeds. Nature 19. https://doi.org/10.1038/s41586-018-0785-8
Khatri-Chhetri A, Aggarwal PK, Joshi PK, Vyas S (2017) Farmers' prioritization of climate-smart agriculture (CSA) technologies. Agric Syst 151:184–191
Khush GS (2005) IR varieties and their impact. Int Rice Res Inst
Khush GS (1987) Rice breeding: past, present and future. J Genet 66:195–216
Kole C, Muthamilarasan M, Henry R, Edwards D, Sharma R, Abberton M, Batley J, Bentley A, Blakeney M, Bryant J (2015) Application of genomics-assisted breeding for generation of climate resilient crops: progress and prospects. Front Plant Sci 6:563
Kumar S, Banks TW, Cloutier S (2012) SNP discovery through next-generation sequencing and its applications. Int J Plant Genomics
Kumar N, Kumar N, Shukla A, Shankhdhar SC, Shankhdhar D (2015a) Impact of terminal heat stress on pollen viability and yield attributes of rice (*Oryza sativa* L.). Cereal Res Commun 43:616–626
Kumar V, Singh A, Mithra SVA, Krishnamurthy SL, Parida SK, Jain S, Tiwari KK, Kumar P, Rao AR, Sharma SK (2015b) Genome-wide association mapping of salinity tolerance in rice (*Oryza sativa*). DNA Res 22:133–145
Kurokawa Y, Noda T, Yamagata Y, Angeles-Shim R, Sunohara H, Uehara K, Furuta T, Nagai K, Jena KK, Yasui H (2016) Construction of a versatile SNP array for pyramiding useful genes of rice. Plant Sci 242:131–139

Leung H, Raghavan C, Zhou B, Oliva R, Choi IR, Lacorte V, Jubay ML, Cruz CV, Gregorio G, Singh RK (2015) Allele mining and enhanced genetic recombination for rice breeding. Rice 8:34

Li JY, Wang J, Zeigler RS (2014) The 3,000 rice genomes project: new opportunities and challenges for future rice research. Gigascience 3:8

Li T, Ali J, Marcaida M, Angeles O, Franje NJ, Revilleza JE, Manalo E, Redoña E, Xu J, Li Z (2016) Combining limited multiple environment trials data with crop modeling to identify widely adaptable rice varieties. PLoS ONE 11:e0164456

Li Y, Xiao J, Chen L, Huang X, Cheng Z, Han B, Zhang Q, Wu C (2018) Rice functional genomics research: past decade and future. Mol Plant 11:359–380

Li ZK (2001) QTL mapping in rice: a few critical considerations. In: Rice genetics IV. World scientific, pp 153–171

Li ZK, Zhang F (2013) Rice breeding in the post-genomics era: from concept to practice. Curr Opin Plant Biol 16:261–269

Li Z (2017) Breeding green super rice (GSR) varieties for sustainable rice cultivation 1–22. https://doi.org/10.19103/AS.2016.0003.05

Lipper L, Thornton P, Campbell BM, Baedeker T, Braimoh A, Bwalya M, Caron P, Cattaneo A, Garrity D, Henry K (2014) Climate-smart agriculture for food security. Nat Clim Change 4:1068

Liu L, Fan X-D (2014) CRISPR–Cas system: a powerful tool for genome engineering. Plant Mol Biol 85:209–218

Liu S, Waqas MA, Wang S, Xiong X, Wan Y (2017) Effects of increased levels of atmospheric CO_2 and high temperatures on rice growth and quality. PLoS ONE 12:e0187724

Liu Y, Chen L, Fu D, Lou Q, Mei H, Xiong L, Li M, Xu X, Mei X, Luo L (2014) Dissection of additive, epistatic effect and QTL× environment interaction of quantitative trait loci for sheath blight resistance in rice. Hereditas 151:28–37

Lobell DB, Sibley A, Ortiz-Monasterio JI (2012) Extreme heat effects on wheat senescence in India. Nat Clim Change 2:186

Londo JP, Chiang YC, Hung KH, Chiang TY, Schaal BA (2006) Phylogeography of Asian wild rice, *Oryza rufipogon*, reveals multiple independent domestications of cultivated rice, *Oryza sativa*. Proc Natl Acad Sci 103:9578–9583

Mackill DJ (1996) Rainfed lowland rice improvement. Int Rice Res Inst

Madan P, Jagadish SVK, Craufurd PQ, Fitzgerald M, Lafarge T, Wheeler TR (2012) Effect of elevated CO_2 and high temperature on seed-set and grain quality of rice. J Exp Bot 63:3843–3852

Mahender A, Ali J, Prahalada GD, Sevilla M.AL, Balachiranjeevi CH, Md Jamal, Maqsood U, Li Z (2019) Genetic dissection of developmental responses of agro-morphological traits under different doses of nutrient fertilizers using high-density SNP markers. PLoS ONE 14 (7): e0220066. https://doi.org/10.1371/journal.pone.0220066

Mandal N, Shukla VD, Variar M, Kumar A (2009) Drought-tolerant variety Sahbhagi Dhan released for cultivation in India. STRASA News 2:11–12

Malzahn A, Lowder L, Qi Y (2017) Plant genome editing with TALEN and CRISPR. Cell Biosci 7:21

Mao X, Zheng Y, Xiao K, Wei Y, Zhu Y, Cai Q, Chen L, Xie H, Zhang J (2018) OsPRX2 contributes to stomatal closure and improves potassium deficiency tolerance in rice. Biochem Biophys Res Commun 495:461–467

Marcaida M, Li T, Angeles O, Evangelista GK, Fontanilla MA, Xu J, Gao Y, Li Z, Ali J (2014) Biomass accumulation and partitioning of newly developed Green Super Rice (GSR) cultivars under drought stress during the reproductive stage. Field Crops Res 162:30–38. https://doi.org/10.1016/j.fcr.2014.03.013

Matthews RB, Kropff MJ, Horie T, Bachelet D (1997) Simulating the impact of climate change on rice production in Asia and evaluating options for adaptation. Agric Syst 54:399–425

McCouch S, Baute GJ, Bradeen J, Bramel P, Bretting PK, Buckler E, Burke JM, Charest D, Cloutier S, Cole G (2013) Agriculture: feeding the future. Nature 499:23

McCouch SR, Zhao K, Wright M, Tung C-W, Ebana K, Thomson M, Reynolds A, Wang D, DeClerck G, Ali ML (2010) Development of genome-wide SNP assays for rice. Breed Sci 60:524–535
McNally KL, Childs KL, Bohnert R, Davidson RM, Zhao K, Ulat VJ, Zeller G, Clark RM, Hoen DR, Bureau TE (2009) Genomewide SNP variation reveals relationships among landraces and modern varieties of rice. Proc Natl Acad Sci 106:12273–12278
Mishra R, Joshi RK, Zhao K (2018) Genome editing in rice: recent advances, challenges, and future implications. Front Plant Sci 9
Mohanty S (2009) Rice and the global financial crisis. Rice Today 8:40
Mottaleb KA, Rejesus RM, Mohanty S, Murty M, Li T, Valera HG, Gumma MK (2012) Ex ante impact assessment of a drought tolerant rice variety in the presence of climate change. In: Agricultural and applied economics association annual meeting, Seattle, Washington, USA
Muthayya S, Sugimoto JD, Montgomery S, Maberly GF (2014) An overview of global rice production, supply, trade, and consumption. Ann N Y Acad Sci 1324:7–14
Nelson GC, Van Der Mensbrugghe D, Ahammad H, Blanc E, Calvin K, Hasegawa T, Havlik P, Heyhoe E, Kyle P, Lotze-Campen H (2014) Agriculture and climate change in global scenarios: why don't the models agree? Agric Econ 45:85–101
Normile D (1997) Yangtze seen as earliest rice site. Science 275:309
Najeeb S, Ali J, Mahender A, Pang YL, Zilhas J, Murugaiyan V, Lakshminarayana RV, Li Z (2019) Identification of main-effect quantitative trait loci (QTLs) for low-temperature stress tolerance germination and early seedling vigor-related traits in rice (*Oryza sativa* L.). Mol Breeding 11032:1090. https://doi.org/10.1007/s11032-019-1090-4
Ott A, Liu S, Schnable JC, Yeh C-T 'Eddy,' Wang K-S, Schnable PS (2017) tGBS® genotyping-by-sequencing enables reliable genotyping of heterozygous loci. Nucl Acids Res 45:e178–e178
Pang Y, Chen K, Wang X, Wang W, Xu J, Ali J, Li Z (2017a) Simultaneous improvement and genetic dissection of salt tolerance of rice (*Oryza sativa* L.) by designed QTL pyramiding. Front Plant Sci 8: https://doi.org/10.3389/fpls.2017.01275
Pang Y, Chen K, Wang X, Xu J, Ali J, Li Z (2017b) Recurrent selection breeding by dominant male sterility for multiple abiotic stresses tolerant rice cultivars. Euphytica 213:268
Pareek CS, Smoczynski R, Tretyn A (2011) Sequencing technologies and genome sequencing. J Appl Genet 52:413–435
Peng S, Huang J, Sheehy JE, Laza RC, Visperas RM, Zhong X, Centeno GS, Khush GS, Cassman KG (2004) Rice yields decline with higher night temperature from global warming. Proc Natl Acad Sci 101:9971–9975
Porter JR, Xie L, Challinor AJ, Cochrane K, Howden SM, Iqbal MM, Lobell DB, Travasso MI, Netra Chhetri NC, Garrett K (2014) Food security and food production systems. In: Field CB et al (eds) Climate change 2014: impacts, adaptation, and vulnerability. Cambridge University Press, pp 485–533
Portères R (1976) African cereals: *Eleusine*, Fonio, Black Fonio, Tefi', *Brachiaria*, *Paspalum*, *Pennisetum*, and African Rice. Orig African Plant Domest 409
Prasanna V (2014) Impact of monsoon rainfall on the total foodgrain yield over India. J Earth Syst Sci 123:1129–1145
Ps S, Sv AM, Prakash C, Mk R, Tiwari R, Mohapatra T, Singh NK (2017) High resolution mapping of QTLs for heat tolerance in rice using a 5K SNP array. Rice 10
Rahman MH, Zhang Y, Zhang K, Rahman MS, Barman HN, Riaz A, Chen Y, Wu W, Zhan X, Cao L (2017) Genetic dissection of the major quantitative trait locus (qSE11), and its validation as the major influence on the rate of stigma exsertion in rce (*Oryza sativa* L.). Front Plant Sci 8:1818
Rasheed A, Hao Y, Xia X, Khan A, Xu Y, Varshney RK, He Z (2017) Crop breeding chips and genotyping platforms: Progress, challenges, and perspectives. Mol Plant 10:1047–1064
Ray DK, Mueller ND, West PC, Foley JA (2013) Yield trends are insufficient to double global crop production by 2050. PLoS ONE 8:e66428

Reuter JA, Spacek DV, Snyder MP (2015) High-throughput sequencing technologies. Mol Cell 58:586–597

Rimbert H, Darrier B, Navarro J, Kitt J, Choulet F, Leveugle M, Duarte J, Rivière N, Eversole K, Le Gouis J (2018) High throughput SNP discovery and genotyping in hexaploid wheat. PLoS ONE 13:e0186329

Rosegrant MW, Koo J, Cenacchi N, Ringler C, Robertson RD, Fisher M, Cox CM, Garrett K, Perez ND, Sabbagh P (2014) Food security in a world of natural resource scarcity: the role of agricultural technologies. Int Food Policy Res Inst

Rötter RP, Tao F, Höhn JG, Palosuo T (2015) Use of crop simulation modelling to aid ideotype design of future cereal cultivars. J Exp Bot 66:3463–3476

Sanghera GS, Wani SH, Hussain W, Singh NB (2011) Engineering cold stress tolerance in crop plants. Curr Genomics 12:30

Sapkota TB, Jat ML, Aryal JP, Jat RK, Khatri-Chhetri A (2015) Climate change adaptation, greenhouse gas mitigation and economic profitability of conservation agriculture: Some examples from cereal systems of Indo-Gangetic Plains. J Integr Agric 14:1524–1533

Sasaki T (2017) Breeding green super rice (GSR) varieties for sustainable rice cultivation. Z. Li, Chinese Academy of Agricultural Sciences, China; and J. Ali, International Rice Research Institute, Philippines. In: Achieving sustainable cultivation of rice, vol 1. Burleigh Dodds Science Publishing, pp 131–152

Scheben A, Batley J, Edwards D (2017) Genotyping-by-sequencing approaches to characterize crop genomes: choosing the right tool for the right application. Plant Biotechnol J 15:149–161

Seck PA, Diagne A, Mohanty S, Wopereis MCS (2012) Crops that feed the world 7: rice. Food Secur 4:7–24

Semenov MA, Stratonovitch P (2013) Designing high-yielding wheat ideotypes for a changing climate. Food Energy Secur 2:185–196

Serraj R, McNally KL, Slamet-Loedin I, Kohli A, Haefele SM, Atlin G, Kumar A (2011) Drought resistance improvement in rice: an integrated genetic and resource management strategy. Plant Prod Sci 14:1–14

Shan Q, Wang Y, Li J, Zhang Y, Chen K, Liang Z, Zhang K, Liu J, Xi JJ, Qiu J-L (2013) Targeted genome modification of crop plants using a CRISPR-Cas system. Nat Biotechnol 31:686

Shen C, Que Z, Xia Y, Tang N, Li D, He R, Cao M (2017) Knockout of the annexin gene *OsAnn3* via CRISPR/Cas9-mediated genome editing decreased cold tolerance in rice. J Plant Biol 60:539–547

Shen YJ, Jiang H, Jin JP, Zhang ZB, Xi B, He YY, Wang G, Wang C, Qian L, Li X (2004) Development of genome-wide DNA polymorphism database for map-based cloning of rice genes. Plant Physiol 135:1198–1205

Singh B, Singh N, Mishra S, Tripathi K, Singh BP, Rai V, Singh AK, Singh NK (2018) Morphological and molecular data reveal three distinct populations of Indian wild rice *Oryza rufipogon* Griff. species complex. Front Plant Sci 9:123

Singh N, Jayaswal PK, Panda K, Mandal P, Kumar V, Singh B, Mishra S, Singh Y, Singh R, Rai V (2015) Single-copy gene based 50 K SNP chip for genetic studies and molecular breeding in rice. Sci Rep 5:11600

Singh R, Singh Y, Xalaxo S, Verulkar S, Yadav N, Singh S, Singh N, Prasad KSN, Kondayya K, Rao PVR (2016) From QTL to variety: harnessing the benefits of QTLs for drought, flood and salt tolerance in mega rice varieties of India through a multi-institutional network. Plant Sci 242:278–287

Sonah H, Bastien M, Iquira E, Tardivel A, Légaré G, Boyle B, Normandeau É, Laroche J, Larose S, Jean M (2013) An improved genotyping by sequencing (GBS) approach offering increased versatility and efficiency of SNP discovery and genotyping. PLoS ONE 8:e54603

Song G, Jia M, Chen K, Kong X, Khattak B, Xie C, Li A, Mao L (2016) CRISPR/Cas9: a powerful tool for crop genome editing. Crop J 4:75–82

Song Q, Hyten DL, Jia G, Quigley CV, Fickus EW, Nelson RL, Cregan PB (2013) Development and evaluation of SoySNP50K, a high-density genotyping array for soybean. PLoS ONE 8: e54985

Su C-F, Wang Y-C, Hsieh T-H, Lu C-A, Tseng T-H, Yu S-M (2010) A novel MYBS3-dependent pathway confers cold tolerance in rice. Plant Physiol 153:145–158

Suh JP, Jeung JU, Lee JI, Choi YH, Yea JD, Virk PS, Mackill DJ, Jena KK (2010) Identification and analysis of QTLs controlling cold tolerance at the reproductive stage and validation of effective QTLs in cold-tolerant genotypes of rice (Oryza sativa L.). Theor Appl Genet 120:985–995

Sun H, Huang Q-M, Su J (2005) Highly effective expression of glutamine synthetase genes GS1 and GS2 in transgenic rice plants increases nitrogen-deficiency tolerance. Zhi wu sheng li yu fen zi sheng wu xue xue bao. J Plant Physiol Mol Biol 31:492–498

Sun Y, Tong C, He S, Wang K, Chen L (2018) Identification of nitrogen, phosphorus, and potassium deficiencies based on temporal dynamics of leaf morphology and color. Sustainability 10:762

Taghavi SM, Mendoza TC, Acero B, Li T, Siddiq SA, Yorobe J, Li Z, Ali J (2017) Carbon dioxide equivalent emissions of newly developed rice varieties. J Agric Sci 9:107–123

Takagi H, Abe A, Yoshida K, Kosugi S, Natsume S, Mitsuoka C, Uemura A, Utsushi H, Tamiru M, Takuno S (2013) QTL-seq: rapid mapping of quantitative trait loci in rice by whole genome resequencing of DNA from two bulked populations. Plant J 74:174–183

Tan C, Han Z, Yu H, Zhan W, Xie W, Chen X, Zhao H, Zhou F, Xing Y (2013) QTL scanning for rice yield using a whole genome SNP array. J Genet Genomics 40:629–638

Tang W, Wu T, Ye J, Sun J, Jiang Y, Yu J, Tang J, Chen G, Wang C, Wan J (2016) SNP-based analysis of genetic diversity reveals important alleles associated with seed size in rice. BMC Plant Biol 16:93

The Core Writing Team, Pachauri RK, Meyer LA (2014) IPCC, 2014: climate change 2014: synthesis report. In: Contribution of working groups I. II III to fifth assessment report of the intergovernmental panel on climate change, IPCC, Geneva, Switzerland

Thomson MJ, Singh N, Dwiyanti MS, Wang DR, Wright MH, Perez FA, DeClerck G, Chin JH, Malitic-Layaoen GA, Juanillas VM (2017) Large-scale deployment of a rice 6 K SNP array for genetics and breeding applications. Rice 10:40

Tilman D, Balzer C, Hill J, Befort BL (2011) Global food demand and the sustainable intensification of agriculture. Proc Natl Acad Sci 108:20260–20264

Tong S, Berry HL, Ebi K, Bambrick H, Hu W, Green D, Hanna E, Wang Z, Butler CD (2016) Climate change, food, water and population health in China. Bull World Health Organ 94:759

Torkamaneh D, Laroche J, Belzile F (2016) Genome-wide SNP calling from genotyping by sequencing (GBS) data: a comparison of seven pipelines and two sequencing technologies. PLoS ONE 11:e0161333

Tung C-W, Zhao K, Wright MH, Ali ML, Jung J, Kimball J, Tyagi W, Thomson MJ, McNally K, Leung H (2010) Development of a research platform for dissecting phenotype–genotype associations in rice (*Oryza* spp.). Rice 3:205–217

Tyagi AK, Khurana JP (2003) Plant molecular biology and biotechnology research in the post-recombinant DNA era. In: Biotechnology in India I. Springer, pp 91–121

Tyagi AK, Khurana JP, Khurana P, Raghuvanshi S, Gaur A, Kapur A, Gupta V, Kumar D, Ravi V, Vij S (2004) Structural and functional analysis of rice genome. J Genet 83:79–99

Varshney RK, Singh VK, Kumar A, Powell W, Sorrells ME (2018) Can genomics deliver climate-change ready crops? Curr Opin Plant Biol 45:1692

Varshney RK, Terauchi R, McCouch SR (2014) Harvesting the promising fruits of genomics: applying genome sequencing technologies to crop breeding. PLoS Biol 12:e1001883

Vermeulen SJ, Campbell BM, Ingram JSI (2012) Climate change and food systems. Annu Rev Environ Resour 37:195–222

Wambugu PW, Ndjiondjop M-N, Henry RJ (2018) Role of genomics in promoting the utilization of plant genetic resources in genebanks. Brief Funct Genomics 17:198–206

Wang C, Liu Q, Shen Y, Hua Y, Wang J, Lin J, Wu M, Sun T, Cheng Z, Mercier R (2019) Clonal seeds from hybrid rice by simultaneous genome engineering of meiosis and fertilization genes. Nat Biotechnol 1

Wang F, Wang C, Liu P, Lei C, Hao W, Gao Y, Liu Y-G, Zhao K (2016) Enhanced rice blast resistance by CRISPR/Cas9-targeted mutagenesis of the ERF transcription factor gene OsERF922. PLoS ONE 11:e0154027

Wang H, Zhang MS, Guo R, Shi DC, Liu B, Lin XY, Yang CW (2012) Effects of salt stress on ion balance and nitrogen metabolism of old and young leaves in rice (Oryza sativa L.) BMC Plant Biol 12(2012):194

Wang W, Mauleon R, Hu Z, Chebotarov D, Tai S, Wu Z, Li M, Zheng T, Fuentes RR, Zhang F (2018) Genomic variation in 3,010 diverse accessions of Asian cultivated rice. Nature 557:43

Wang X, Pang Y, Wang C, Chen K, Zhu Y, Shen C, Ali J, Xu J, Li Z (2017a) New candidate genes affecting rice grain appearance and milling quality detected by genome-wide and gene-based association analyses. Front Plant Sci 7:1–11

Wang X, Pang Y, Zhang J, Wu Z, Chen K, Ali J, Ye G, Xu J, Li Z (2017b) Genome-wide and gene-based association mapping for rice eating and cooking characteristics and protein content. Sci Rep 7:1–10

Wang Y, Zang J, Sun Y, Ali J, Xu J, Li Z (2013) Background-independent quantitative trait loci for drought tolerance identified using advanced backcross introgression lines in rice. Crop Sci 53(2):430. https://doi.org/10.2135/cropsci2012.06.0361

Wei X, Qiao W, Chen Y, Wang R, Cao L, Zhang W, Yuan N, Li Z, Zeng H, Yang Q (2012) Domestication and geographic origin of *Oryza sativa* in China: insights from multilocus analysis of nucleotide variation of *O. sativa* and *O. rufipogon*. Mol Ecol 21:5073–5087

Wickland DP, Battu G, Hudson KA, Diers BW, Hudson ME (2017) A comparison of genotyping-by-sequencing analysis methods on low-coverage crop datasets shows advantages of a new workflow GB-eaSy. BMC Bioinformatics 18:586

Xie K, Yang Y (2013) RNA-guided genome editing in plants using a CRISPR–Cas system. Mol Plant 6:1975–1983

Xie W, Wang G, Yuan M, Yao W, Lyu K, Zhao H, Yang M, Li P, Zhang X, Yuan J (2015) Breeding signatures of rice improvement revealed by a genomic variation map from a large germplasm collection. Proc Natl Acad Sci 112:E5411–E5419

Xu JL, Lafitte HR, Gao YM, Fu BY, Torres R, Li ZK (2005) QTLs for drought escape and tolerance identified in a set of random introgression lines of rice. Theor Appl Genet 111:1642–1650

Xu Q, Yuan X, Wang S, Feng Y, Yu H, Wang Y, Yang Y, Li X (2016) The genetic diversity and structure of indica rice in China as detected by single nucleotide polymorphism analysis. BMC Genet 17:53

Xu C, Ren Y, Jian Y, Guo Z, Zhang Y, Xie C, Fu J, Wang H, Wang G, Xu Y (2017a) Development of a maize 55 K SNP array with improved genome coverage for molecular breeding. Mol Breed 37:20

Xu Y, Li P, Yang Z, Xu C (2017b) Genetic mapping of quantitative trait loci in crops. Crop J 5:175–184

Yang X, Xia X, Zhang Z, Nong B, Zeng Y, Xiong F, Wu Y, Gao J, Deng G, Li D (2017) QTL mapping by whole genome re-sequencing and analysis of candidate genes for nitrogen use efficiency in rice. Front Plant Sci 8:1634

Ye C, Tenorio FA, Argayoso MA, Laza MA, Koh H-J, Redoña ED, Jagadish KSV, Gregorio GB (2015) Identifying and confirming quantitative trait loci associated with heat tolerance at flowering stage in different rice populations. BMC Genet 16:41

Yorobe JM, Ali J, Pede VO, Rejesus RM, Velarde OP, Wang H (2016) Yield and income effects of rice varieties with tolerance of multiple abiotic stresses: the case of green super rice (GSR) and flooding in the Philippines. Agric Econ (United Kingdom) 47:261–271

Yun W, Jinping Z, Yong S, Jauhar A, Jianlong X, Zhikang L (2012) Identification of genetic overlaps for salt and drought tolerance using simple sequence repeat markers on an advanced backcross population in rice. Crop Sci 52:1583–1592

Yu J, Hu S, Wang J, Wong GK-S, Li S, Liu B, Deng Y, Dai L, Zhou Y, Zhang X (2002) A draft sequence of the rice genome (*Oryza sativa* L. ssp. *indica*). Science 296:79–92

Yu J, Wang J, Lin W, Li S, Li H, Zhou J, Ni P, Dong W, Hu S, Zeng C (2005) The genomes of *Oryza sativa*: a history of duplications. PLoS Biol 3:e38

Zhang H, Liu S, Chen G, Liu X, Xuan N, Yang Y, Liu W, Ding H, Yao F (2018) Genetic dissection of QTLs and differentiation analysis of alleles for heading date genes in rice. PLoS ONE 13:e0190491

Zhang H, Zhang J, Wei P, Zhang B, Gou F, Feng Z, Mao Y, Yang L, Zhang H, Xu N (2014) The CRISPR/Cas9 system produces specific and homozygous targeted gene editing in rice in one generation. Plant Biotechnol J 12:797–807

Zhang Q (2007) Strategies for developing Green Super Rice. Proc Natl Acad Sci 104:16402–16409

Zhao C, Liu B, Piao S, Wang X, Lobell DB, Huang Y, Huang M, Yao Y, Bassu S, Ciais P (2017) Temperature increase reduces global yields of major crops in four independent estimates. Proc Natl Acad Sci 114:9326–9331

Zhou L, Zeng Y, Zheng W, Tang B, Yang S, Zhang H, Li J, Li Z (2010) Fine mapping a QTL qCTB7 for cold tolerance at the booting stage on rice chromosome 7 using a near-isogenic line. Theor Appl Genet 121:895–905

Zhou J, Peng Z, Long J, Sosso D, Liu B, Eom J, Huang S, Liu S, Vera Cruz C, Frommer WB (2015) Gene targeting by the TAL effector PthXo2 reveals cryptic resistance gene for bacterial blight of rice. Plant J 82:632–643

Zhu H, Liu Z, Fu X, Dai Z, Wang S, Zhang G, Zeng R, Liu G (2015) Detection and characterization of epistasis between QTLs on plant height in rice using single segment substitution lines. Breed Sci 65:192–200

Chapter 2
Advanced Genomics and Breeding Tools to Accelerate the Development of Climate Resilient Wheat

Yuanfeng Hao, Awais Rasheed, Robert Jackson, Yonggui Xiao, Yong Zhang, Xianchun Xia and Zhonghu He

Abstract Knowledge-based breeding to develop high-yielding wheat cultivars is the key to keep pace with increasing food demand, not only in optimal but also in stressed conditions. Resilience to climate extremes and variability has become one of the most important crop breeding targets. Genomics will play an important role to uncover the basis of adaptability to heat, drought, salinity and other abiotic stresses, and disease resistances in wheat. Deeper understanding of the physiological and genetic bases of drought and heat resistance is crucial for maintaining and improving breeding program efficiency. The high-quality wheat reference genome sequence is recently decoded and new genotyping tools are being developed based on the most updated genomics information to be used in practical breeding programs. In this chapter, we focused on the (i) quantitative trait loci (QTLs) analysis related to drought, heat, salinity tolerance, and diseases resistance in wheat, (ii) functional genes discovered for important breeding traits and development of markers for use in breeding, (iii) role of wheat genetic resources to enhance the genetic diversity and expansion of alleles for important genes, (iv) improved genotyping and phenotyping approaches to understand the genetic basis of wheat production traits, and (iv) future strategies to accelerate the rate of genetic gain in a changing climate.

Keywords Climate changes · Functional markers · Genetic resources · Genomics · QTL · Wheat breeding

---

Y. Hao · A. Rasheed (✉) · Y. Xiao · Y. Zhang · X. Xia · Z. He (✉)
Institute of Crop Sciences, Chinese Academy of Agricultural Sciences (CAAS), Beijing, China
e-mail: a.rasheed@cgiar.org

Z. He
e-mail: hezhonghu02@caas.cn

A. Rasheed · Z. He
International Maize and Wheat Improvement Center (CIMMYT), C/O CAAS, Beijing, China

R. Jackson
NIAB, Cambridge, UK

C. Kole (ed.), *Genomic Designing of Climate-Smart Cereal Crops*,
https://doi.org/10.1007/978-3-319-93381-8_2

2.1 Introduction

Sustainable crop production is an important component to ensure food security. Wheat (*Triticum aestivum* L.) is one of the most important food crops and its production could be severely affected by the climate anomalies and extremes. Future agricultural production will face multiple challenges from global climate changes. These include increased CO_2, temperature, drought episodes, and occurrence of new and virulent races of pathogens and pests. For wheat, the estimated global loss will be about 6% for each Celsius degree increase. Through adjustment of planting and harvest times, expansion of croplands to more permissive areas, wheat production could be increased to some extent, but the continuous lift in yield will heavily rely on the integration of advanced genomics and easy-to-use phenotypic tools (Fig. 2.1). Winter wheat, for example, commonly faces many challenges during the entire life cycle from the early seedling establishment (Fig. 2.1I) to the end of harvest (Fig. 2.1V). Development and dissemination of climate resilient wheat cultivars to meet increasing food demand is urgently needed.

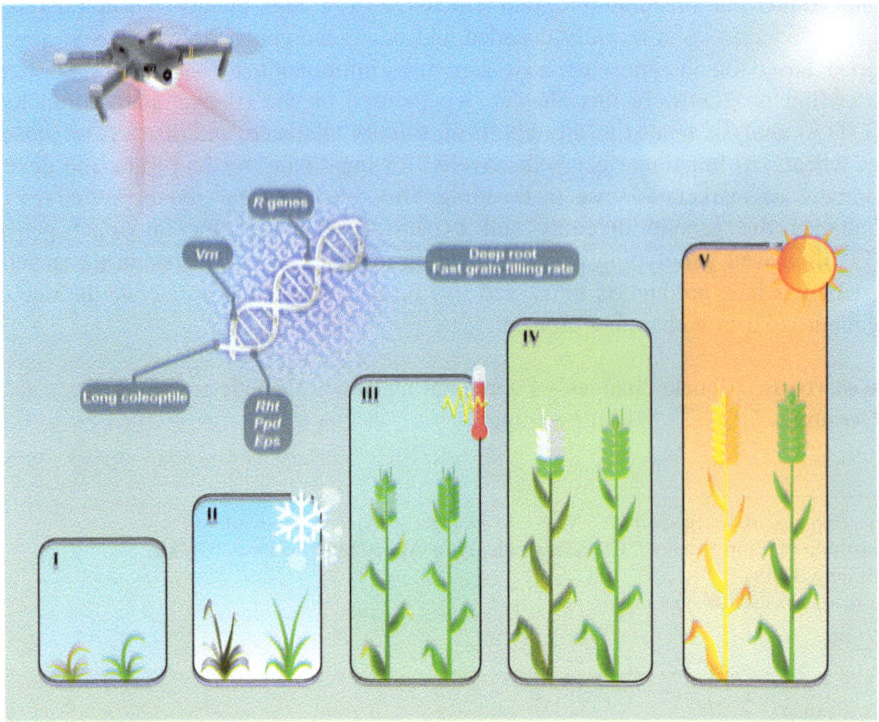

Fig. 2.1 Integrated genomics and phenomics approaches are key to develop climate resilient wheat cultivars. The climate-smart traits include seedling vigor/drought adaptability (**I**), winter hardiness (**II**), resistance to late spring frost from extreme weather patterns (**III**), disease resistance (**IV**), and terminal heat stress tolerance (**V**)

2.2 Climate Vulnerabilities and Wheat Production

Wheat is the most widely cultivated crop in the world and contributes about 21% of the global food production. In the past five decades, there was not much change for wheat area harvested with the value of about 220 million hectares (ha), but significant increases were observed for both production and yield (Fig. 2.2). The world wheat production has increased about 1.5 times from 294.3 million tons in 1967–749.5 million tons in 2016. Average yield as a result boosted about 2.0 tons per ha from about 1.4 in 1966 to the current 3.4 (Fig. 2.2).

Notably, the yield steadily increased to about 0.5 tons per 12–13 years, which was equal to yearly increment of yield about 40 kg per ha (Fig. 2.3). However, this increment probably is not enough to keep pace with the world population growth and higher food requirement for rapid urbanization (Godfray et al. 2010; Tester and Langridge 2010). It is estimated that the world population will reach 9.7 billion in 2050 under the medium variant scenario. More than two-thirds of the people may be living in urban areas compared to the current 54 percent (FAO 2017). Urbanization has been accompanied by a shift in dietary patterns on the food system typically resulting in more food requirements. To meet the demands, wheat production in certain areas such as sub-Saharan Africa and South Asia would need to more than double by 2050 (Ray et al. 2013), while in the rest of the world the projected increase would be about one-third above current levels (FAO 2017). However, global climate change, which is predicted to get worse in the future with extreme temperature, altered rainfall pattern (leading to droughts and floods), extreme weather events, and changing patterns of pathogens and pests, will

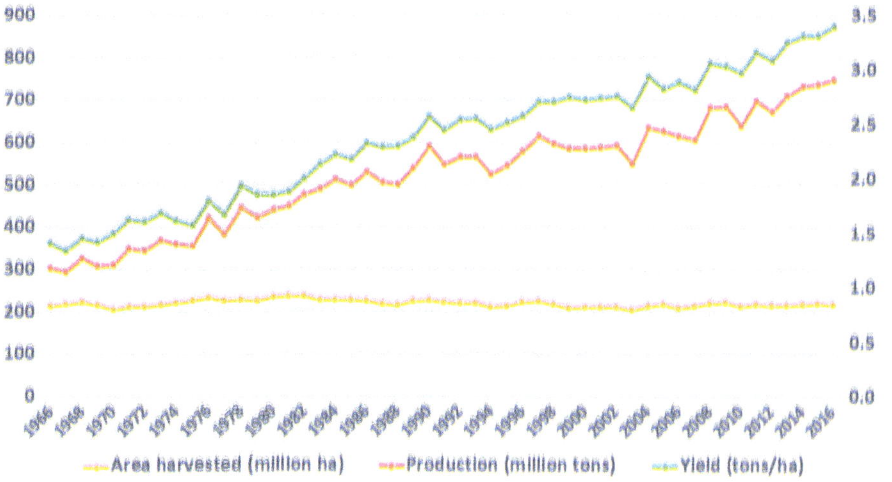

Fig. 2.2 World wheat production, area harvested and yield in 1966–2016 source from FAO (www.fao.org/faostat/, updated on September 15, 2018)

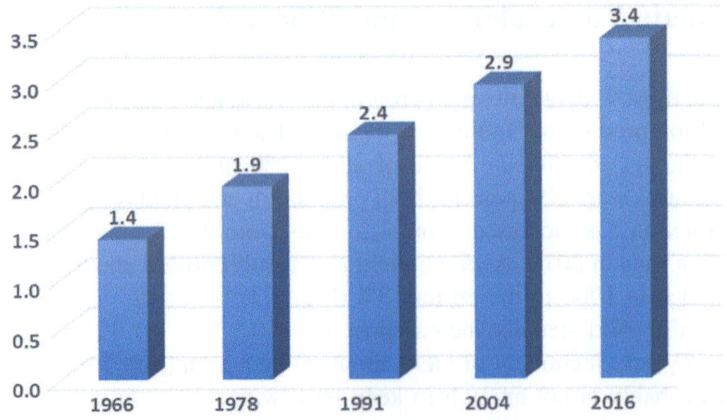

Fig. 2.3 World average wheat yield increased about 0.5 ton per ha in a 12–13-year interval

adversely affect the wheat production to some extent and make the future wheat increase more challenging (Abberton et al. 2016).

2.2.1 Climate Across Wheat Mega-Environments

The International Maize and Wheat Improvement Center (CIMMYT) develops improved wheat germplasm for use worldwide particularly in developing countries, where wheat was grown on about 110 million ha (Lantican et al. 2005). To address the needs of these diverse wheat-growing areas, in 1988, CIMMYT's Strategic Plan proposed the term mega-environment (ME) to subdivide global wheat domains to target germplasm development (Rajaram et al. 1995). A ME is defined as a broad, not necessarily contiguous area by similar biotic and abiotic stresses, cropping system requirements, consumer preferences, and volume of production. By 1993, the 12 MEs have been well defined, which includes six for spring wheat (ME1–ME6), three for facultative wheat (ME7–ME9), and three for winter wheat (ME10–ME12) (Braun et al. 1996). The detailed information as described below includes representative sites and the weather information, major constraints and planting areas. Most of the information refers to the Wheat Atlas (http://wheatatlas.org/) and the publications (Rajaram et al. 1995; Braun et al. 1996; Ortiz et al. 2008b; He et al. 2013) with updates.

Spring Wheat

ME1: Autumn sown (spring sown rare), white grain, low rainfall, irrigated, average minimum temperature for coolest quarter 11 °C > T $\geq$ 3 °C. Major constraints: leaf rust, stem rust, stripe rust, lodging. Representative sites: Gangetic Valley, India; Indus Valley, Pakistan; Nile Valley, Egypt; Yaqui Valley, Mexico; 32 million ha coverage.

ME2A: Spring sown, largely red grain (except Ethiopia), highland summer rain (elevation $\geq$ 1400 m), wettest quarter precipitation $\geq$ 250 mm, average minimum temperature for wettest quarter 16 °C > T $\geq$ 3 °C. Major constraints: barley yellow dwarf virus, stripe rust, head scab, Septoria, tan spot. Representative sites: Kulumsa, Ethiopia; Toluca, Mexico; 10 million ha coverage.

ME2B: Autumn sown, largely red grain, lowland winter rain (elev. < 1400 m), coolest quarter precipitation $\geq$ 150 mm, average minimum temperature for coolest quarter 16 °C > T $\geq$ 3 °C. Major constraints: barley yellow dwarf virus, stripe rust, head scab, Septoria, tan spot. Representative sites: Izmir, Turkey; Pergamino, Argentina.

ME3: Same as general ME2 (A and B merged)–main difference soil pH < 5.2, largely red grain (except Himalayas). Additional constraints: acid soil (Aluminum, Manganese toxicity). Representative sites: Passo Fundo, Brazil; Mpika, Zambia; 1.7 million ha coverage.

ME4A: Autumn sown, largely white grain, Mediterranean climate, post-flowering moisture deficits, late season frosts, low rainfall, wettest quarter precipitation 100–400 mm, average minimum temperature for coolest quarter 11 °C > T $\geq$ 3 °C. Major constraints: drought, heat, rust, common bunt, Septoria, tan spot, root diseases. Representative sites: Aleppo, Syria; Settat, Morocco; 10 million ha coverage.

ME4B: Autumn sown, largely red grain, winter drought or Southern Cone-type rainfall, pre-flowering moisture deficits, low rainfall, wettest quarter precipitation 100–400 mm, average minimum temperature for coolest quarter 11 °C > T $\geq$ 3 °C. Major constraints: drought, rust, common bunt, Septoria, *Fusarium*, tan spot, root diseases. Representative sites: Marcos Juárez, Argentina; 5.8 million ha coverage.

ME4C: Autumn sown (after monsoons), only white grain, receding moisture levels after sowing, low rainfall, wettest quarter precipitation 100–400 mm, average minimum temperature for coolest quarter 16 °C > T $\geq$ 3 °C. Major constraints: drought, rust, common bunt, Septoria, tan spot, root diseases. Representative sites: Dharwad, India; 5.8 million ha coverage.

ME5: Autumn sown, tropical high rainfall, and/or irrigated, average minimum temperature for coolest quarter 16 °C > T > 11 °C. Major constraints: heat, spot blotch, leaf and stem rust. Subdivided into high humidity areas ME5A (representative sites Joydebpur, Bangladesh and Paraguay, 3.9 million ha coverage), and low humidity areas ME5B (Kano, Nigeria and Wad Medani Sudan, 3.2 million ha coverage).

ME6: Spring sown, high latitude (>45° N or S)—photosensitive cultivars, average minimum temperature for coolest quarter T < −13 °C, average minimum temperature for warmest quarter T $\geq$ 9 °C. Major constraints: rust, root diseases, tan spot. Subdivided into high rainfall areas ME6A (Harbin, Heilongjiang, China) and semiarid areas ME6B (Astana, Kazakhstan) with a total coverage of 5.4 million ha.

Facultative Wheat

ME7: Autumn sown, irrigated, average minimum temperature for coolest quarter 3 °C > T ≥ −2 °C. Major constraints: cold, stripe rust, mildew. Representative Sites: Zhengzhou, Henan, China; 9 million ha coverage. Sites currently listed as ME7 sometimes do not meet this criterion, although almost all these sites fit the ME10 criteria.

ME8: Autumn sown, high rainfall/irrigated, annual precipitation ≥ 500 mm, average minimum temperature for coolest quarter 6 °C > T ≥ 1 °C. Major constraints: cold, stripe rust, mildew, Septoria, root rots. Representative Sites: Temuco, Chile; Corvallis, Oregon, USA; 0.7 million ha coverage. Perhaps annual precipitation ≥ 800 mm for USA to have distinct ME8/9 sites.

ME9: Autumn sown, low rainfall, annual precipitation <500 mm, average minimum temperature for coolest quarter 3 °C > T ≥ −2 °C. Major constraints: cold, drought, stripe rust, root rots. Representative Sites: Diyarbakir, Turkey; Vernon, Texas, USA; 6.8 million ha coverage. Perhaps annual precipitation <800 mm for USA to have distinct ME8/9 sites.

Winter Wheat

ME10: Autumn sown, irrigated, average minimum temperature for coolest quarter −2 °C > T ≥ −13 °C. Major constraints: winter kill, rust, mildew. Representative Sites: Beijing, China; 4.6 million ha coverage.

ME11: Autumn sown, high rainfall/irrigated, annual precipitation ≥ 500 mm, average minimum temperature for coolest quarter 1 °C > T ≥ −13 °C. Major constraints: winter kill, rust, Septoria, mildew. Representative Sites: Cambridge, UK; Krasnodar, Russia. Perhaps annual precipitation ≥ 800 mm for USA to have distinct ME11/12 sites.

ME12: Autumn sown, low rainfall, annual precipitation <500 mm, average minimum temperature for coolest quarter 1 °C > T ≥ −13 °C. Major constraints: winter kill, drought, stripe rust, bunts, root rots. Representative Sites: Ft. Collins, Colorado; Manhattan, Kansas, USA; 7.9 million ha coverage. Perhaps annual precipitation <800 mm for USA to have distinct ME11/12 sites.

2.2.2 *Effects of Global Warming and Climate Change on Wheat Production*

Human activities from pollution to overpopulation are driving up the Earth's temperature and fundamentally changing the world. The main cause is a phenomenon known as the greenhouse effect. Gases in the atmosphere such as water vapor, carbon dioxide, methane, nitrous oxide, and chlorofluorocarbons let the sun's light in but keep some of the heat from escaping like the glass walls of a greenhouse. The more greenhouse gases in the atmosphere, the more heat gets trapped strengthening the greenhouse effect and increasing the Earth's temperature. Human activities like the burning of fossil fuels have increased the amount of CO_2 from the past around 280 ppm (parts per million) in the preindustrial era to the current 408 ppm (data from NASA/GISS). The number is not only the highest level record in history but also rising continuously at about 2 ppm more per year to a new higher level (data from NOAA). As one of the consequences, the global temperature has increased more than 1 °C since 1880. Seventeen of the 18 warmest years in the 136-year record all have occurred since 2001. The year 2016 ranks as the warmest on record. Drought is another consequence from climate changes. Severe droughts in wheat-growing seasons have been observed in Australia, Russia, USA, China, and EU countries over the last few years (Lobell et al. 2011). Wheat is one of the typical crops that prefers relatively cool temperatures and sensitive to both heat and drought in particular in early booting and flowering developmental stage (Barber et al. 2017; Stratonovitch and Semenov 2015). Here, we will use three levels of data including history trends, current observations, and future predictions so as to provide a whole picture on how deeply the climate change has and will affect the wheat production.

Global wheat production decline from climate change in history

Lobell et al. (2011) have developed a database of yield response models to evaluate the impact of the recent climate trends on major crop yields for the period 1980–2008. At the global scale, wheat will lose about 5.5% (4.9% loss from heat and 0.6% from drought) relative to what would have been achieved without the climate trends when the positive effect from elevated CO_2 was not considered. While countries showed different responses for the climate changes, Russia has lost almost 15% of wheat production mostly attributing to the increased temperature and extreme weather, whereas the United States showed no effect because of the lack of significant climate trends. China and India, the two largest wheat producers, showed 2.4% and 5.5% of yield loss, respectively, due to the impact of the temperature and precipitation trends.

Wheat yield losses in the current field

Ciudad Obregon (27° 24′ N, 109° 56′ W; 39 masl), Mexico, is one of the major experimental stations for CIMMYT spring wheat. In 2014–2015, growing season (referred to as 2015), the day maximum temperature did not change much during January to April (the four key months for wheat development and maturity for spring wheat) in comparison to the mean maximum temperature of 1993–2014, but the night minimum temperature increased around 3.4 °C. This change has had a significant negative impact on the yield performance of most wheats. Using the same control cultivar, Borlaug 100, in yield trial in CIMMYT experimental plots as an example, the average yield in 2013 and 2014 was about 9.0–10.0 tons per ha in normal seasons, which reduced to 6.2–7.0 tons in 2015, about 30% decline observed in warmer seasons. This decline was also observed in farmers' fields in the

Yaqui valley across Ciudad Obregon where average yield decreased from the 7.4 tons/ha in 2013 to 5.2 tons/ha in 2015 (Ravi Singh and Hans Braun, personal communications). The yield loss was about 10% for each 1 °C increase in growing-season minimum temperature, which was consistent with the observation on rice (Peng et al. 2004).

Significant wheat yield losses were also observed in 2018 in Gaoyi (37° 36′ N, 114° 36′ E), Hebei province, China. Gaoyi located, in the northern China, one of the research experimental stations for the Institute of Crop Sciences, Chinese Academy of Agricultural Sciences (CAAS). The same high-yielding cultivar Liangxing 99 was used as control cultivar across yield trials in the last 5 years. The average yield of Liangxin 99 was about 7.3 tons/ha in 2018, about 30% decline relative to the mean yield of 10.5 tons/ha over the 3 previous years. When historical weather data (source: https://www.tianqi.com/) was used for analysis, we did not find much difference for the mean day maximum and night minimum temperatures in March and April over 2016–2018, but observed extreme weather pattern in 2018 (Fig. 2.4). In the wheat booting stage (late March), the day maximum temperature was above 30 °C, and suddenly fell down to 2–7 °C in the following days (the so-called late spring frost in Fig. 2.1III). In the middle and late of April (anthesis stage), the similar fluctuations of temperature were also observed (Fig. 2.4), which significantly affected the fertility of floret and therefore the grain number, and finally decrease in yield (Fig. 2.1III). At grain filling stage, a heat wave (maximum temperature over 33 °C in 5 consecutive days) occurred around 7–10 days earlier in 2018 than previous years, resulting in early senescence (Fig. 2.1V) and causing a decrease in grain weight of about 25% (our field observation).

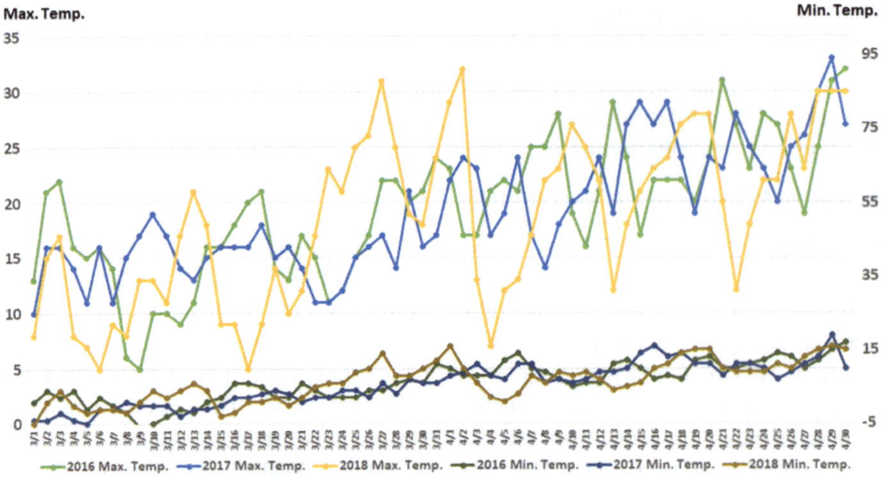

Fig. 2.4 Extreme weather pattern (in yellow) observed in 2018 in Gaoyi, Hebei province, China. Upper, maximum temperature; Lower, minimum temperature

Rising temperature will reduce future wheat production

Climate change has dramatically affected the wheat production in the past (Lobell and Gourdji 2012). Continued emission of greenhouse gases will increase Earth's temperature by 2.0 °C and 3.7 °C in the middle and the end of the twenty-first century, respectively, according to the Intergovernmental Panel on Climate Change (IPCC) RCP8.5 scenario (IPCC 2013). Most of wheat-growing areas are expected to become warmer with reduced precipitation (Ortiz et al. 2008b). It is estimated that global wheat yield is projected to decline between 4.1% and 6.4% with a 1 °C global temperature increase, and warmer regions are likely to suffer more yield losses than cooler regions (Liu et al. 2016). Additionally, heat wave and other adverse weather conditions will increase in frequency, intensity, and duration with the rising temperature (Tebaldi et al. 2006; Rahmstorf and Coumou 2011), which will add further threat for wheat production (Trnka et al. 2014).

For the 12 MEs delineated by CIMMYT, most of the spring wheat-growing areas probably with the only exception of ME6 (high latitude) will suffer significant yield loss from the increasing heat (He et al. 2013). Using the wheat major producing areas in the plains of the Indo-Ganges as an example, as much as 51% of its area, currently the most favorable and high-yielding environment (ME1) will likely reclassified as a heat stressed, irrigated, short season, and low-yielding production environment (ME5) as a result of climate change (Ortiz et al. 2008b). This area included most of northern and eastern India, the eastern parts of Pakistan, virtually all of Bangladesh, and southern plains of Nepal, where approximately 90 million tons of wheat grains were produced (about 12% of global production). Contrarily, the high latitude ME6 is probably the only area which will fully benefit from the climate change. This ME comprises the cool temperature regions of North America and northern Eurasia with the coolest quarter minimum temperature above −13 °C and the warmest quarter minimum temperature below 9 °C as mentioned earlier. Presently, North American wheat can be grown up to 55° N, but under future climate change scenario in particular the rising temperature, the ME6 may shift northward up to 65° N as a major expansion of potential wheat-growing areas. However, this expansion will face many difficulties, such as suitable soils, land competition from forestry or protected areas, and infrastructure factors, which finally increase the uncertainty. For the facultative and winter wheat environments (ME7–ME12), both positive and negative impacts could be predicted from the climate changes while again more uncertainty remains compared to the spring wheat. For example, warmer winters will reduce the severity of winterkill and benefit yield, while warmer spring and summer will drive shorter life cycles, resulting in less seasonal photosynthesis, shorter reproductive phase, and thus lower yield (Ainsworth and Ort 2010; He et al. 2013).

2.3 Major Constraints for Wheat Production

2.3.1 Abiotic Stresses

Apart from temperature, the other important abiotic stresses due to the consequence of climate change are drought and salinity. Wheat yield is estimated to be reduced by 0.15–0.20 tons ha^{-1} for each 10 mm water deficiency (Gate 1995). Drought severely affects vegetative and reproductive growth of wheat, however some growth stages are more sensitive to drought than others. Almost 65 million ha of wheat area was affected by drought stress in 2013. The unpredicted climate change scenario will likely increase the episodes and intensity of drought stress. The morphological and physiological traits affected by drought include root, shoot, tiller number, spike length, grain numbers per spike, 1000-grain weight, awn length, peduncle length, relative water content, membrane stability, and photosynthetic activity. Therefore, it is very important to determine the drought-sensitive traits in the target environment in parallel to grain yield.

Soil salinity and sodicity are also a major wheat production limiting factors globally after drought stress. A soil is considered saline when the electrical conductivity of the saturated paste extract is above 4 dS/m (equivalent to about 40 mMNaCl). A soil is considered sodic when divalent cations are significantly replaced by Na^+ (Rengasamy 2010). According to an estimate, 230 million ha of irrigated land and 1,500 million ha of dryland agriculture are salt-affected (FAO 2017). Chartres and Noble (2015) reported that almost 11% of the world's irrigated land is affected by salinization which constitutes about 100 Mha of soil. It is of high concern that salt-affected land and its unceasing expansion are highest in developing and most populated countries like India (7 Mha; Vashev et al. 2010), Bangladesh (1 Mha; Hossain 2010), and Pakistan (6 Mha; Qureshi et al. 2008; Vashev et al. 2010) which is a great threat to sustainable agriculture.

2.3.2 Wheat Diseases

The life of wheat faces many challenges from diseases including fungal, bacterial, and viral diseases (Prescott et al. 1986). The three rust diseases (stripe rust, leaf rust, and stem rust), powdery mildew, and *Fusarium* head blight (FHB) are major constraints for wheat production in a worldwide perspective (Singh et al. 2016).

Wheat stripe (yellow) rust, caused by the fungal pathogen *Puccinia striiformis* f. sp. *tritic in* (*Pst*), has evolved to be the largest biotic limitation to global wheat production (Schwessinger 2017). New pathogen genotypes are more aggressive and able to infect previously resistant wheat cultivars, leading to rapid pathogen migration across continents (Wellings 2007; Hovmøller et al. 2016; Schwessinger 2017). In the USA, stripe rust has been an occasional problem in the southern and central wheat regions from 1941 until 1999, but the disease has been consistently severe in the region since 2000 (Chen et al. 2002; Line 2002; Chen 2013, 2014). The change mainly attributed to the major shift of *Pst* isolate. The new isolates are better adapted and more aggressive at warmer temperatures than the old isolates (Milus et al. 2006, 2009; Chen 2013). In Australia, this cool temperature pathogen was previously thought to be unadaptable to the country's environments, but the disease was finally detected in 1979 most likely transmitted through human activities (Wellings 2007). New exotic pathogens became more adapted in drier and warmer conditions in Australia indicated the major shift of *Pst* pathotype in the context of climate changes (Wellings 2007). Up to 2000, most of the predominant races of *Pst* in Europe were typical to the NW-European genetic group (Schwessinger 2017), while in 2011, two new races termed Warrior and Kranich, were detected on both wheat and triticale in many European countries (Sorensen et al. 2014; Hovmøller et al. 2016). Compared to old ones, the two new races can cause more diseases on adult plants of wheat, and produce more conidiospores in wheat tissues under warmer temperature (Hovmøller et al. 2016). In China, the unusual prevalence of *Pst* occurred in a very early stage of wheat development in January, 2017 (Yao et al. 2019), which is mainly due to warmer winter, appropriate moisture from unexpected raining, and lack of host resistance. Collectively, partially due to climate changes, more aggressive *Pst* isolates are becoming predominant and adapt well in warmer wheat planting areas, which represents forthcoming great challenges for wheat production.

Different from stripe rust, which is favorable in a cooler climate with optimum temperature 2–15 °C, leaf rust (brown rust) and stem rust (black rust) develop well in warmer temperature conditions at 10–30 °C and 15–35 °C, respectively. With the global temperature rising, the two wheat diseases have the potential to expand their prevalent areas and cause more wheat production losses (Juroszek and von Tiedemann 2013). Leaf rust, caused by *Puciniatri ticina* (*Pt*), is the most common and widely distributed wheat disease (Kolmer 2005). Losses from this disease are usually less damaging than those from stripe rust and stem rust (Li et al. 2014). Yield losses are mostly due to decreasing kernel number per spike and reducing kernel weight (Huerta-Espino et al. 2011).

Stem rust, caused by the fungus *Pucciniagraminis* f. sp. *triticin* (*Pgt*), has severely devastated wheat production historically and can cause up to 100% loss of yield (Saunders et al. 2019). However, the disease has been under satisfactory control since 1950s attribute to the wide adoption of resistant cultivars (*Sr31* carriers in 1BL.1RS) and the removal of alternative host barberry. Stem rust reemerged in the form of the Ug99 in Uganda in 1998. At least eight races belonging to the Ug99 lineage were identified in several eastern and southern African highlands (Singh et al. 2008, 2011, 2015). Most of wheat cultivars grown were highly susceptible to these new races (Singh et al. 2011). Gene discoveries and development of wheat cultivars for Ug99 resistance were carried out subsequently under the umbrella of the Borlaug Global Rust Initiative (http://www.globalrust.org). Using Mexico–Kenya shuttle breeding scheme including two generations per year, CIMMYT in 2010 began releasing advanced breeding lines from stem rust resistant donors for Ug99 indicating significant progresses achieved (Singh et al. 2015). The reemergence of stem rust diseases exists not only in Africa and Asia but also in Europe. In 2013, Germany experienced its first major outbreak of stem rust disease in decades, and in 2016, both bread and durum wheat were ravaged by stem rust in Sicily, making the largest European outbreak for many years (Lewis et al. 2018). One new *Pst* race named UK-01 identified in the UK suddenly draws public attentions since more than 80% of current UK wheat cultivars are highly susceptible to this strain. Climate changes over the past 25 years suggest conducive conditions for infection of the disease (Lewis et al. 2018; Saunders et al. 2019) indicating potential threat of the disease to wheat productions.

Powdery mildew, caused by *Blumeria graminis* f. sp. *tritici* (*Bgt*), is another important disease in wheat-growing regions with temperate and maritime climates (Cowger et al. 2012). *Bgt* is a cool-season pathogen favored by temperatures in 10–20 °C range. In recent decades, this disease has become important in some warmer and drier areas because of intensive production with higher plant densities, nitrogen fertilizers, and water supply (Cowger et al. 2012; Singh et al. 2016). According to prediction models of climate change on disease, the importance of powdery mildew in wheat in the UK, Northern of Germany, Finland, and Sweden will increase (Juroszek and von Tiedemann 2013).

All the wheat diseases mentioned above are caused by biotrophic fungi, which are obligate parasites that attach only living plants. While, the diseases of *Fusarium* head blight (FHB) and wheat blast (WB) are necrotrophic fungi, which are facultative parasites surviving on both living and dead tissues (Singh et al. 2016). FHB, also called scab, is primarily incited by *Fusarium graminearum* Schwabe. FHB epidemics have been reported in most of the wheat production areas in the world. The increased rotation of maize–wheat, conservative cultivation and global climate changes, the three major factors have made FHB epidemics more frequent and severe, even in regions where FHB has not been reported (McMullen et al. 2012; Bai et al. 2018; Zhu et al. 2019). For example, FHB was historically prevalent only in the Middle and Lower Yangtze Valleys and part of Northeastern spring wheat regions, but moved to Yellow and Huai River Valleys, the major wheat producing areas, in recent years in China (Zhu et al. 2019). In the state of Georgia, USA, FHB was not a problem in the past, but epidemics occurred in 3 consecutive years of 2014–2016 causing major losses of wheat production (https://swvt.uga.edu/winter-crops/wheat.html). In the context of climate change, FHB incidence during wheat anthesis is projected to become more severe, especially in southern UK (Juroszek and von Tiedemann 2013). Therefore, improvement of FHB host resistance should have a high priority in the future in order to reduce mycotoxin contamination of grain and to ensure food security (Shah et al. 2018).

Wheat blast represents potential threat to wheat production under the climate change scenario (Cruz et al. 2016a). It was first emerged in the state of Parana in Brazil in the mid-1980s. The pathogen has spread to Bolivia, Paraguay, and Argentina (Urashima et al. 1994). Most severe blast epidemics were characterized by continuous rainfall with warm temperature 18–25 °C during flowering stage, followed by sunny, hot, and humid weather. The disease appeared in Bangladesh for the first time in Asia during February 2016 (Callaway 2016; Islam et al. 2016), which quickly drawing wide public concerns on its potential threat to major wheat producing regions including India and the south of China. Preliminary analysis revealed that central India, Bangladesh, Ethiopia, Eurasia, and North American are areas of risk (Duveiller et al. 2011). In the USA, the state of Louisiana, Mississippi, and Florida are in high risk of the disease outbreaks under the prediction models of climate change (Cruz et al. 2016a).

2.4 Identification of Climate-Smart Traits and Genes

2.4.1 Modulating Major Adaptability Genes for Climate Adaptability

Flowering time is one of the most important developmental traits for wheat adaptability and yield stability in target environments. Three genes controlling vernalization (*Vrn*), photoperiod response (*Ppd*), and early flowering (*Eps*) are known to be major determinants of flowering time optimization (Kamran et al. 2014). The manipulation of these genes is very important in the stress escape mechanism where early maturity is achieved to avoid terminal drought and heat stresses. Cultivars such as newly released Zhongmai 578 (Zhongmai 255/Jimai 22, Zhongmai 255 = Yumai 49/Sunstate) in China have very early maturity, about 5–7 days earlier than control, and fast grain filling rate. They produced a high thousand-kernel weight (TKW) and high grain yield even in unfavorable years such as in 2017–2018, which was a growing season with high abiotic stress. Normal cultivars lost 30% of their TKW, while Zhongmai 578 only slightly decrease by 4%. Yield of Zhongmai 578 in stressed years is comparable to the values in favorable years fully indicating the importance of earliness, fast grain filling rate, and TKW stability for climate resilient wheat development.

It has been known that *VRN1*, *Ppd1*, and *Elf3* (*Eps*) have the extended roles beyond flowering time (Kamran et al. 2014). *VRN1* loci are known to increase number of spikelets (Hailu and Merker 2008), final leaf number (Hay and Kirby 1991), and winter hardiness (Fig. 2.1II). It was demonstrated that winter-type *VRN1* alleles significantly increased the heading time and decreased grain yield (GY) under heat stress conditions in bread wheat (Ogbonnaya et al. 2017). Similarly, *Ppd1*'s roles are not confined to photoperiod sensitivity, they also control modifications leading to spikes with more elaborated arrangements and increase number of grains producing spikelets (Boden et al. 2015). The *PRR73* is a paralog of *Ppd-D1* in bread wheat, and accessions having Hap-I at *PRR73-A1* and Hap-II at *PRR73-B1* were earlier in heading and taller under long day conditions than accessions having contrasting haplotype (Diaz et al. 2012).

Plant height is an important trait largely controlled by *Rht-B1* and *Rht-D1* genes (Fig. 2.1III, IV). The important alleles *Rht-B1b* and *Rht-D1b* significantly reduce PH by 14–17%, decrease lodging, and increase harvest index (Rasheed et al. 2016).

Previous studies have shown very broad roles of *Rht* genes having pleiotropic effect on anther extrusion, a major trait in hybrid wheat production (Würschum et al. 2018), resistance against *Fusarium* head blight (Gosman et al. 2009), insect pest resistance (Emebiri et al. 2017), and grain quality. Therefore, modulating plant height by selecting appropriate Rht alleles according to target environment is not only important for pure-line breeding but can also assist in hybrid wheat breeding where tallness of male is required for effective production of hybrids (Würschum et al. 2018).

The GA-sensitive severe dwarf *Rht12* was more heat tolerant ($t_5 = 29.4 \pm 0.72$) than the similarly statured GA-insensitive *Rht-D1c*. The GA-sensitive, semidwarfing *Rht8* conferred greater drought tolerance in one experiment. It is possible that sensitivity of wheat fertility to heat and drought stress is partly due to the widespread use of the semidwarfing alleles that underpinned much of the wheat *Green Revolution*. Rapid increases in wheat yields in major growing areas of the world from the 1960s to the 1990s were associated with reducing wheat statures to less than 1.0 m, increasing harvest index, and reducing lodging risk in fertile and humid conditions (Gooding 2009).

By the 1990s, 80% of registered wheat cultivars contained a semidwarfing *Rht* allele, and in over 90% of these cases, the allele was either *Rht-B1b* or *Rht-D1b* (Worland et al. 1998). The *Rht-1* alleles encode for DELLA proteins (Peng et al. 1999), which repress gibberellic acid (GA)-responsive growth (Murase et al. 2008; Achard and Genschik 2009). Both *Rht-B1b* and *Rht-D1b* are from the Japanese wheat "Norin 10". The two genes can reduce sensitivity to endogenous gibberellins (GA), reduce final crop height by about 15% in UK field conditions (Gooding et al. 2012), and contain single nucleotide substitutions causing premature stop codons in the N-terminal coding region, possibly leading to more repressive forms of DELLA (Pearce et al. 2011). More potent alleles, which gave severe dwarfism, are available at both loci. *Rht-Blc* from "Tom thumb" produces a predicted 30 amino acid insertion within DELLA (Pearce et al. 2011), and a 50% reduction in height. *Rht-D1c* from "Ai-bian" (Borner et al. 1997) overexpresses *Rht-D1b* through increased gene copy number (Pearce et al. 2011) and reduces height by about 55%. Despite of their widespread use, early reviews suggested that both *Rht-B1b* and *Rht-D1b* may confer increased sensitivity to drought and heat stresses (Gale and Youssefian 1985).

In particular, it is known that GA-signaling is critical for the normal development and functioning of reproductive tissues (Mutasa-Gottgens and Hedden 2009). The *Rht* gene interfering with GA-signaling, which may explain preliminary work (Law et al. 1981) found *Rht-B1b*, *Rht-D1b*, and *Rht-B1c* to be associated with marked reductions in fertility in response to increased temperatures (e.g., 30/27 °C day/night for 18 h days) during booting. Further circumstantial evidence is that Norin 10 alleles have reduced selection frequencies in areas of southern Europe, where there is an increased likelihood of high-temperature events during meiosis (Worland 1986). Sip et al. (2011) note a recent decline in the use of *Rht-B1b* and *Rht-D1b* in areas further north, in central Europe.

2.4.2 Drought and Salinity Tolerance Related Genes

Yield reduction from abiotic stresses is mainly due to the reduction in photosynthetic rate as consequences of metabolic limitation, stomatal closure, and oxidative damage to chloroplast (Farooq et al. 2014). Seedling traits like germination and growth including long coleoptile are one of the most critical stages for drought stress (Fig. 2.1I). Drought stress usually delays seed germination and suppresses growth because moisture content in soil is the main source of seed germination (Mickky and Aldesuquy 2017). This in turn affects root/shoot ratio, dry root weight, biomass, and grain yield of wheat (Ahmadi et al. 2012). The choice of *Rht* gene such as *Rht8* instead of *Rht-B1b* and *Rht-D1b* benefits the development of long coleoptile, good seedling emergence, and consequence yield in deep sowing conditions (Rebetzke et al. 2007b). In addition to the major *Rht* genes, several QTLs on chromosome 1A, 2B, 2D, 4A, 5A, 5D, and 6B were also identified to be associated with coleoptile length (Rebetzke et al. 2007a; Rebetzke et al. 2014). The one on chromosome arm 1AS has recently been characterized as a major locus named *Lcol-A1* (Bovill et al. 2019). The locus was associated with increased emergence from depth. Two flanking markers IWB58229 and IWA710 were developed and could be used for selection of long coleoptile wheats (Bovill et al. 2019).

Cell membrane stability by electrolyte leakage is one of the most important physiological and biochemical traits of seeds. The stable cell membrane with minimum leakage is the physiological marker for drought tolerance. This characteristic known as cell membrane thermostability or relative cell injury percentage (RCI%) is used as a standard to identify drought-tolerant genotypes (Ahmad et al. 2015). At flowering, drought stress significantly reduces pollination which ultimately lead to relatively fewer number of grains per spike (Akram 2011). At the onset of drought, plants cease photosynthesis but respiration continues. The stored carbohydrates are then used by plants until the resumption of photosynthesis (Hammad and Ali 2014). If the photosynthesis remains ceased for longer period of times, shortage of carbohydrates occurs that sustains parental respiration but florets get deprived of it. This ultimately affects the grain setting and grain yield (Ruan et al. 2010).

Drought tolerance is a polygenic mechanism and very complex to understand both physiological and molecular basis of drought adaptability. Drought usually comes with heat particularly in terminal stage. After several years of screening in terminal heat-prone environments in China, cultivars such as Zhongmai 895 and its sister line Zhongmai 875 (Zhoumai 16/Liken 4) showed exceptional heat and drought tolerance in late growth stage. Their thousand-kernel weights were about 5–10 g higher than normal controls in both favorable and stressed environments. The stay-green character and the deep roots we have observed are probably part of reasons for their high-yielding performance in stressed environments (Fig. 2.1V). Other cultivars such as Zhongmai 23, Shiluan 02-1, Liangxing 99, Zhongmai 30 (GY13028), Zhongmai 31 (GY14038) and their derivatives are also tolerant to terminal heat and drought stresses based on our long-term observations.

Various types of enzymes and genes are involved in the regulation of drought response that include late embryo abundant proteins (LEA), vacuolar acid invertase, Abscisic acid (ABA) genes expression, Glutathione-S-transferase (GST), invertase genes (INV), dehydration responsive element binding proteins (DREB), ethylene responsive factors (ERF), zinc-finger proteins (ZFP), and WRKY (Kulkarni et al. 2017). Invertase (β-D-fructofuranosidefructanohydrolase) is an enzyme engaged in the transportation of sucrose from source to sink for plant growth and development (Krishnan et al. 1985). There are three types of invertase enzymes classified on the basis of subcellular localization, solubility, and isoelectric point. These enzymes include cell wall invertase, vacuolar invertase, and cytoplasmic invertase (Roitsch and Gonzalez 2004). The cell wall invertase enzyme plays an important role in pollen development and plant metabolism. The storage of carbohydrates becomes different under drought conditions as compared to well-watered conditions (Ji et al. 2010).

Salinity tolerance in plants is a complex polygenic trait with both dominant and additive effects being important for inheritance (Yamaguchi and Blumwald 2005; Mujeeb-Kazi et al. 2019). The exclusion of Na^+ from leaves to avoid toxic concentrations is the main mechanism conferring salinity tolerance in wheat (Munns et al. 2006). In bread wheat, this is conferred by the chromosome 4D locus; *Kna1*, which is responsible for K^+:Na^+ discrimination and controls K^+ and Na^+ accumulation in shoot (Gorham et al. 1987; Dubcovsky et al. 1996). The candidate gene is *TaHKT1;5-D*, a Na^+ transporter, which removes Na^+ from the transpiration stream to the leaves and increases the ratio of K^+:Na^+ (Byrt et al. 2014). Durum wheat which lacks the D-genome and any homologues of *Kna1* on the A- and B-genome has a lower salt tolerance than bread wheat (Huang et al. 2008). *Triticum monococcum*, the earliest cultivated wheat, has a homologue counterpart of *Kna1*, *Nax2*, present in the A genome. It was shown that the *NAx2* candidate gene was *TmHKT1;5-A* (Byrt et al. 2007). QTL analysis of *T. monococcum* identified another locus, *Nax1*, responsible for Na^+ exclusion from leaves (Lindsay et al. 2004). Fine mapping showed *TmHKT1;4-A2*, another Na^+ transporter, was the responsible gene (Huang et al. 2006). *Nax1* and *Nax2* are only present in *T. monococcum* and its wild ancestor but not *T. urartu*, the progenitor of durum and bread wheat suggesting potential new breeding targets for salinity resistance (Huang et al. 2008). Previously, several important reviews and perspectives have been published on salinity tolerance mechanisms focusing on crops including wheat (Munns et al. 2006; Roy et al. 2014). The cost of salinity tolerance mechanisms (Munns and Gilliham 2015), the physiological basis of salinity tolerance in crops (Munns et al. 2016; Negrao et al. 2017) and molecular breeding for salinity tolerance (Ashraf and Foolad 2013) were comprehensively discussed.

2.4.3 Disease-Resistance Genes

Host resistance, chemicals, cultural practices, and biological control are major strategies for managing wheat diseases. Among them, host resistance using *R* genes is the most economical and environmentally friendly approach to control the diseases (Singh et al. 2016). Mainly due to the hexaploid nature of common wheat, the 17-Gb genome size, and more than 85% repetitive DNA, the map-based cloning of resistance genes was difficult (Keller et al. 2018). There are at least two types of resistance genes. One is race specific, or major genes usually conferring resistance from the seedling to adult plant stages, and the other is durable resistance, which is often susceptible in seedling but resistant or partially resistant in adult. The durable resistance is often controlled by quantitative trait loci (QTLs), which have small to intermediate effects, and the accumulation of multiple QTLs can lead to a high level of adult plant resistance (Singh et al. 2016).

For breeders, durable resistance genes with pleiotropic multi-pathogen resistance are more attractive than seedling resistance genes. Three genes *Lr34/Yr18/Sr57/Pm38/Ltn1*, *Lr46/Yr29/Sr58/Pm39/Ltn2*, and *Lr67/Yr46/Sr55/Pm46/Ltn3* were identified on wheat chromosome arms 7DS, 1BL, and 4DL, respectively, and all were co-segregating with a morphological marker named leaf tip necrosis (*Ltn*) in flag leaves (Lagudah 2011). *Lr34/Yr18/Sr57/Pm38* provides an important source of partial resistance in adult plants and is more effective when deployed with other adult plant resistance genes. After more than three decades effort, the gene was successfully cloned in 2009 (Krattinger et al. 2009) by map-based approach since the gene was first recognized in 1977 (Dyck 1977). *Lr34* spans 11,805 base pairs, consists of 24 exons, and encodes pleiotropic drug resistance (PDR)-like ATP-binding cassette (ABC) transporter protein. The pleiotropic drug resistance transporters share a common structure with two cytosolic nucleotide-binding domains and two hydrophobic transmembrane domains (Krattinger et al. 2009). Latest study identified ABA as a substrate of the LR34 resistant protein. LR34res-mediated ABA redistribution played a key role on the transcriptional response and physiology of *Lr34res*-expressing plants (Krattinger et al. 2019). Despite its extensive use in breeding for more than 50 years, no increase in virulence toward *Lr34* has been described indicating its durability. Since its isolation, the gene has been transferred into barley, rice, durum, sorghum, and maize. The transgenic lines demonstrated good resistance to barley leaf rust and powdery mildew (Risk et al. 2013), rice blast (Krattinger et al. 2016), durum wheat leaf rust, stripe rust and powdery mildew (Rinaldo et al. 2017), sorghum anthracnose and rust (Schnippenkoetter et al. 2017), and maize common rust and northern corn leaf blight (Sucher et al. 2017). These very positive results fully supported the broad-spectrum resistance nature of the gene not only in wheat but across all major cereal crop species (Keller et al. 2018).

Another cloned APR gene with pleiotropic effect was *Lr67/Yr46/Sr55/Pm46* located on chromosome 4DL (Hiebert et al. 2010). *Lr67* present in landrace PI250413 and then RL6077 through five cycles of backcrossing to Thatcher. Later, the gene was found conferring resistance to multiple pathogens such as stripe rust (Herrera-Foessel et al. 2011), stem rust, and powdery mildew (Herrera-Foessel et al. 2014). The *Lr67* resistance gene encodes a predicted hexose transporter (Moore et al. 2015). Both the resistant versions of *Lr34* and *Lr67* coding protein only differed by two critical amino acid changes from their susceptible protein versions (Krattinger et al. 2009; Moore et al. 2015). The LR67sus showed a high affinity to glucose in yeast uptake experiments, whereas glucose uptake was abolished in LR67res the resistant hexose transporter through a dominant-negative interference mechanism by forming inactive heteromultimeric protein complexes (Moore et al. 2015). The *Lr67res* allele is present in older, tall wheat cultivars rather than modern semi-dwarf cultivars released by CIMMYT, which may partially attribute to the intense selection for *Rht-D1b* gene and the repulsion linkage between the two genes (Moore et al. 2015). Exome sequence data from 267 barley landraces and wild accessions was screened and neither of the *Lr67res* was detected. Like the resistance gene *Lr34*, transgenic *Lr67res* barley exhibited early senescence and higher pathogenesis-related gene expression and showed seedling resistance to the barley leaf rust and powdery mildew (Milne et al. 2019).

Apart from the two multi-pathogen resistance genes, at least 22 *R* genes have been isolated by map-based or rapid gene cloning approaches in common wheat (Keller et al. 2018; Bettgenhaeuser and Krattinger 2019). These genes included four leaf rust resistance genes (*Lr1*, *Lr10*, *Lr21* and *Lr22a*), six stripe rust resistance genes (*Yr5*, *Yr7*, *YrSP*, *Yr10*, *Yr15*, and *Yr36*), seven stem rust resistance genes (*Sr13*, *Sr22*, *Sr33*, *Sr35*, *Sr45*, *Sr46*, and *Sr50*), and five powdery mildew resistance genes (*Pm2*, *Pm3*, *Pm8*, *Pm2,1* and *Pm60*). All the genes encode typical intracellular immune receptors of the nucleotide-binding and leucine-rich repeat (NLR) family with the only exceptions of *Yr15* and *Yr36*, which encode proteins with kinase domain (Fu et al. 2009; Klymiuk et al. 2018).

The *Lr10* and *Lr21* were the first disease-resistance genes cloned in wheat in 2003 at the time wheat genome sequence information was limited and map-based cloning was difficult. *Lr10* is a single-copy gene on chromosome arm 1AS (Feuillet et al. 2003). *Lr21* is mapped to a gene-rich region at the distal end of chromosome arm 1DS (Huang et al. 2003). *Lr1* located at the distal end of chromosome arm 5DL was cloned by two independent groups in 2007 (Cloutier et al. 2007; Qiu et al. 2007). Different from the previous three leaf rust resistance genes, *Lr22a* was recently rapidly isolated in hexaploid wheat on chromosome 2DS using targeted chromosome-based cloning via long-range assembly (TACCA) strategy. TACCA combines genome-complexity reduction by chromosome flow sorting with Chicago long-range linkage to assemble complex genomes (Thind et al. 2017).

In addition to *Lr34* and *Lr67*, *Yr36* was another partial, broad-spectrum resistance gene originally identified in wild emmer wheat. *Yr36* on chromosome arm 6BS encodes a WHEAT KINASE START1 (WKS1) protein with an N-terminal non-RD kinase domain in connection with a START lipid-binding domain at the C-terminus and confers race nonspecific resistance against stripe rust disease (Fu et al. 2009). WKS1 interacts with the thylakoid-associated ascorbate peroxidase (tAPX), and ascorbate peroxidases are involved in the detoxification of peroxides.

The interaction of WKS1 and tAPX might reduce the cell's ability to detoxify reactive oxygen species and lead to cell death. As the response is much slower than a typical hypersensitive cell death response, *Yr36*-containing wheat plants will allow the limited pathogen growth and restricted sporulation in a partial resistance manner (Gou et al. 2015). *Yr10*, a seedling or all-stage resistance gene, was characterized as a typical CC-NBS-LRR resistance gene on chromosome arm 1BS (Liu et al. 2014), while *Yr15* on the same chromosome 1BS, a broad-spectrum resistance gene derived from wild emmer encodes a putative kinase-pseudokinase protein named WTK1 (wheat tandem kinase 1) (Klymiuk et al. 2018). The three stripe rust resistance genes *Yr5*, *Yr7*, and *YrSP* on chromosome 2BL were simultaneously cloned from hexaploid wheat using MutRenSeq (a rapid gene cloning strategy combining chemical mutagenesis with exome capture, and sequencing for typical NLR genes), and all encode NLR protein with a noncanonical N-terminal zinc-finger BED domain (Marchal et al. 2018).

Map-based cloning of stem rust resistance gene has become increasingly important for wheat breeding since the new emerging virulent *Pgt* races designated Ug99 in 1998. The wheat gene *Sr33* on chromosome 1D introgressed from *Aegilopstauschii* and *Sr35* on chromosome arm $3A^{m}L$ from *Triticum monococcum* were subsequently cloned (Periyannan et al. 2013; Saintenac et al. 2013). *Sr33* is orthologous to the barley *Mla* gene (Periyannan et al. 2013). *Sr35* confers near immunity to Ug99 and related races (Saintenac et al. 2013). In 2015, rye-derived stem rust resistance gene *Sr50* (previously known as *SrR*) was isolated, which is also a homologous gene to *Mla* (Mago et al. 2015). Another gene *Sr13* on chromosome 6AL was cloned from durum wheat, which showed better resistance at high temperatures (Zhang et al. 2017a). Different from the four genes isolated by map-based cloning strategy, *Sr22* on chromosome 7A and *Sr45* on 1D were cloned by MutRenSeq (Steuernagel et al. 2016), and *Sr46* was isolated using AgRenSeq (association genetics with R gene enrichment sequencing) (Arora et al. 2019). Both strategies are useful to rapidly isolate NLR-encoding genes.

Pm2 was also isolated using rapid gene cloning strategy dubbed MutChromSeq. The method can reduce the complexity of genome based on flow sorting and sequencing of mutant chromosomes to identify causal SNPs (Sánchez-Martín et al. 2016). *Pm3* was among the earliest genes cloned in hexaploid wheat (Yahiaoui et al. 2004). The gene is located on chromosome arm 1AS in a similar region of *Lr10*. Later, up to 17 alleles at *Pm3* locus have been identified mainly by PCR amplification approach (Srichumpa et al. 2005; Yahiaoui et al. 2006; Bhullar et al. 2009). Using homology-based cloning and subsequent physical and genetic mapping, *Pm8* in 1RS is isolated as a rye orthologue of the *Pm3* gene (Hurni et al. 2013). Another gene *Pm21* in a *Haynaldiavillosa*-wheat 6VS.6AL translocation was initially cloned as a serine/threonine kinase gene (Cao et al. 2011) but later as a typical CC-NBS-LRR gene indicating the complexity of the gene's nature (He et al. 2018a; Xing et al. 2018). With the release of reference sequence of AA genome, a powdery mildew resistance gene named *Pm60* was rapidly cloned from *Triticum urartu* on 7U chromosome (Zou et al. 2018) indicating a huge game change for gene cloning in new genomic era.

For necrotrophic fungi caused disease such as *Fusarium* head blight (FHB), only seven genes *Fhb1* to *Fhb7* were formally designated (Bai et al. 2018). *Fhb1* was previously cloned as a pore-forming toxin-like (*PFT*) gene through positional mapping, mutation analysis, gene silencing, and transgenic overexpression (Rawat et al. 2016), however, it was subsequently found to be present in many FHB-susceptible cultivars putting the role of the gene in dispute (He et al. 2018b; Zhu et al. 2018). In two recent studies, two research groups identified a histidine-rich calcium-binding (*TaHRC* or *His*) gene adjacent to *PFT* from Ning 7840 and Wangshuibai, respectively, as a new *Fhb1* candidate (Li et al. 2019; Su et al. 2019). Despite agreement on *His* candidacy, they reached contrasting conclusions regarding the gene's function. Therefore, the underlying genetic basis of *Fhb1* resistance remains unclear.

Compared to FHB, the source of wheat blast resistance was even less reported. The currently major resistance source is an alien chromosome segment from *Aegilopsventricosa*-wheat 2NS-2AS translocation (Cruz and Valent 2017). Tested in growth chamber, greenhouse and field environments, the cultivars with 2NS had 50.4–72.3% less head blast than those without 2NS among more than 400 cultivars including both spring and winter types (Cruz et al. 2016b). Our test also revealed that 2NS had better blast resistance in the field. More than 300 Chinese leading cultivars or elite lines were tested for wheat blast in Bangladesh and Bolivia, and more than 90% of cultivars were fully susceptible. Among them, one cultivar Zhongmai 175, released by CAAS, showed the best resistance to the disease across environments (our unpublished data). The cultivar possessed 2NS translocation inherited from a French parent Fr81-4. It was widely planted in Northern Winter Wheat Regions in China with accumulating planting area more than 3 million hectares.

2.4.4 Underlying QTL for Climate-Smart Traits

Most of the QTL studies for drought and heat tolerance have been conducted in wheat using linkage mapping. QTL for various traits at vegetative and reproductive stages were identified with varying phenotypic effects (Table 2.1). Initially, wheat substitution lines in Langdon background were used to map genomic regions associated with heat tolerance (Sun and Quick 1991). Later on, wheat homeologous group 3 was identified to be associated with heat tolerance in wheat cv. Hope (Xu et al. 1996). QTLs for heat tolerance during the grain filling stage have been found on several chromosomal regions including 1A, 1B, 2B, 3B, 5A, and 6D (Mason et al. 2010). Talukder et al. (2014) identified five QTLs for heat tolerance in wheat distributed on chromosomes 1B, 1D, 2B, 6A, and 7A. In addition, QTLs on chromosomes 2A, 3A, 4A, 6A, 6B, and 7A were identified to be associated with yield components and physiological traits, such as stay green and senescence of wheat (Vijayalakshmi et al. 2010). There is only one QTL for heat tolerance which was fine mapped within $\sim$1 Mbp interval with 22 genes on chromosome 3B

Table 2.1 A summary of QTL studies in wheat for drought, heat, and salinity tolerance

Stress	Mapping strategy	Traits	Treatment	QTL	Mapping population and genotyping	Reference
Heat tolerance	Linkage mapping	Yield components	3 day 38 °C daytime heat stress	14	Halberd/Karl 92 RILs	Mason et al. (2010)
	Linkage mapping	Yield components, canopy temperature	High temp environment	3	NW1014/HUW468	Vijayalakshmi et al. (2010)
	Linkage mapping	Physiological traits	Controlled conditions	5	Ventor/Karl 92	Talukder et al. (2014)
	Linkage mapping	Flag leaf cuticular wax	Heat treatment of 38 ° C	2	Halberd/Karl 92 RILs	Mondal et al. (2015)
	GWAS	Physiological and morphological traits	Late and normal sowing	27	188 Spring wheat from ICARDA	Ogbonnaya et al. (2017)
	Linkage mapping	HSI of yield components	High temp environment	3	NW1014/HUW468	Paliwal et al. (2012)
	Linkage mapping	Physiological and morphological	Field conditions	12 additive and 17 epistatic QTL	Hanxuan 10/Lumai 14 DH population	Li et al. (2013)
	Linkage mapping	Morphophysiological	Late and normal sowing	8	RAC875/Kukri DH	Bennett et al. (2012)
	GWAS	Spike ethylene	Heat stress field conditions	39 associations	130 spring wheats from CIMMYT	Valluru et al. (2017)
	GWAS	Seedling traits	Controlled conditions		200 hard red winter wheat	Maulana et al. (2018)
Drought tolerance						Touzy et al. (2019)
	GWAS	Agronomic and quality traits		20	120 elite wheats	Tadesse et al. (2015)
	Linkage mapping	ABA	ABA treatment	1	F2 of CS/CS (Hope5A)	Iehisa et al. (2014)
		Agronomic traits		13	SeriM82/Babax RILs	

(continued)

Table 2.1 (continued)

Stress	Mapping strategy	Traits	Treatment	QTL	Mapping population and genotyping	Reference
	Linkage mapping		Water-limited conditions			Tahmasebi et al. (2016)
	Linkage mapping	Photosynthetic-related traits	Water-limited conditions	225	Chuan 35050/Shannong	Xu et al. (2017)
	Linkage mapping	Agronomic traits	Water-limited conditions	37	WL711/C306 RILs	Kadam et al. (2012)
	Linkage mapping	Growth rate and WUE	Water-limited conditions	20	Drysdale/Gladius RILs	Parent et al. (2015)
	Linkage mapping	Stem water-soluble carbohydrate	Water-limited conditions	48 additive and 62 epistatic	Hanxuan 10/Lumai 14	Yang et al. (2007)
	Linkage mapping	Stomatal density and size	Rainfed environment	4	RAC875/Kukri DHs	Shahinnia et al. (2016)
	GWAS	Agronomic traits	Rainfed environment	44	150 spring wheat	Ain et al. (2015)
	Linkage mapping	Agronomic traits	Water-limited conditions	13	WL711/C306	Shukla et al. (2015)
	Linkage mapping	Agronomic traits	Water-limited conditions	4	Dharwar Dry/Sitta RILs	Kirigwi et al. (2007)
	Linkage mapping	Agronomic traits	Water-limited conditions	104	Seri/Babax RILs	Pinto et al. (2010)
	Linkage mapping	Yield	24 environments	2	Chinese Spring/SQ1 DH	Quarrie et al. (2006)
	GWAS	Agronomic traits	Water-limited conditions	62 MTAs	93 accessions	Mwadzingeni et al. (2017)
	GWAS	Agronomic traits	Water-limited conditions	121 MTAs	208 durum wheats	Sukumaran et al. (2018)

(continued)

Table 2.1 (continued)

Stress	Mapping strategy	Traits	Treatment	QTL	Mapping population and genotyping	Reference
	Linkage mapping	Physiological and morphological	Water-limited controlled conditions	10	C306/HUW206	Kumar et al. (2012)
	Linkage mapping	20 traits	16 environments	163	RAC875/Kukri DH	Bennett et al. (2012)
	Linkage mapping	Stem reserves mobilization		3	W7984/Opata 85	Saleem et al. (2006)
	Linkage mapping	Agronomic traits	10 rainfed environments	16	Kofa/Svevo	Maccaferri et al. (2008)
Salt tolerance	Linkage mapping	Seedling biomass, tillers, chlorophyll, Na and K concentration		40	Berkut/Krichauff	Genc et al. (2010)
	Linkage mapping	Seedling morphology, Na and K conc.		36	Xiaoyan 54/Jing 411	Xu et al. (2012)
	Linkage mapping	Yield traits		Several QTL	Chinese Spring/SQ1	Quarrie et al. (2005)
	Linkage mapping	K/Na ratio		Kna1 locus	4D/4B recombinant Chr lines	Dubcovsky et al. (1996)
	Linkage mapping	Na conc.		Several QTL	Aus29639/Yitpi	Ogbonnaya et al. (2008)
	Linkage mapping	Na conc.		Nax1 locus	Line 149/Tamaroi	Lindsay et al. (2004)

(continued)

Table 2.1 (continued)

Stress	Mapping strategy	Traits	Treatment	QTL	Mapping population and genotyping	Reference
	GWAS	STI, dead leaves		12	Durum wheat	Turki et al. (2015)
	Linkage mapping	Morphological traits		Several QTL	Berkut/Krichauff	Genc et al. (2013)
	Linkage mapping	Yield and yield components		98	Superhead#2/Roshan	Azadi et al. (2015)
	Linkage mapping	Morphological traits		3	Bread wheat accessions	Sardouie-Nasab et al. (2014)
	Linkage mapping	Germination and seedling traits		47	Opata85/W7984	Ma et al. (2007)
	Linkage mapping	Seedling traits, Na and K		34	Chuan 35050/Shannong 483	Xu et al. (2013)
	GWAS	Salt tolerance and mineral conc		49	WTSD91/WN-64	Hussain et al. (2017)
	Linkage mapping	Morphological traits		27	Roshan/Falat (seri82)	Masoudi et al. (2015)
	Linkage mapping	Chloride accumulation		14	Berkut/Krichauff	Genc et al. (2014)
	Linkage mapping	Yield and yield components		36	Opata85/W7984	De Leon et al. (2010)
	GWAS	Seedling traits		37	Natural population	Oyiga et al. (2018)
	Linkage mapping	Na^+ conc.		5	Jandaroi/AUS-14740	Shamaya et al. (2017)

(Thomelin et al. 2016). Mason et al. (2010) also identified a significant QTL on chromosome 3B which was associated with yield components.

Acuna-Galindo et al. (2014) did meta-analysis for all the QTL studies related to heat tolerance and identified the consensus QTLs on chromosomes 1B, 2B, 2D, 4A, 4D, 5A, and 7A. GWAS was successfully used to identify the genetic basis of ethylene production in spike under heat stress in field conditions in CIMMYT spring wheat germplasm (Valluru et al. 2017). This was an important study to discover the genes and germplasm with reduced ethylene effects on grain yield under heat stress conditions. In another GWAS experiment, spring wheat cultivars and advanced lines were subjected to heat stress in several environments by late-sowing and 27 loci were identified to be associated with several physiological and agronomic traits (Ogbonnaya et al. 2017). Similarly, seedling traits were evaluated under controlled heat stress conditions in 200 red winter wheat cultivars and advanced lines (Maulana et al. 2018). Several marker–trait association were identified for leaf chlorophyll content (2B, 2D, 4A, and 4B), shoot length (3B and 7D), number of leaves per seedling (3A, 3B, 4B, 5A, 5B, and 7B), and seedling recovery (2A, 2B, 2D, 3A, 7A, and 7B) using 90 K SNP array.

Most QTL studies for drought resistance have been performed using a biparental spring wheat population. Recently, QTLs for drought tolerance in wheat have been reviewed thoroughly and identified more than 50 original studies on drought tolerance QTL mapping (Gupta et al. 2017). GWAS has also become popular in recent years, due to their direct applicability to existing material, high mapping resolution, and the use of more diverse germplasm (Rasheed and Xia 2019). It is quite challenging to identify stable QTLs underpinning drought adaptability due to the dynamic nature of drought stress in different geographic regions. The multigenic nature of drought tolerance and a large number of genes and biochemical network involved in tolerance add a further layer of complexity. Therefore, integration of all QTL knowledge for drought adaptability to unified framework is very challenging and progress is very slow. Apart from linkage mapping, GWAS is also used to identify genomic regions associated with grain yield and/or related traits in water stress events using high-density molecular markers (Edae et al. 2014; Mwadzingeni et al. 2017; Sukumaran et al. 2015). However, these studies are very limited. It is encouraging that most of the studies were performed under irrigated and water-limited field conditions rather than controlled environments. Most of the drought-tolerant QTL studies were conducted in rainfed or water-limited field conditions to measure the effect of drought on key agronomic or yield related traits (Gupta et al. 2017). However, the QTLs underlying biochemical traits are very important to understand the mechanism underpinning drought adaptability. A large effect QTL on chromosome 5A and seven small effect QTLs were identified ABA content in wheat under drought stress (Iehisa et al. 2014). The 5A QTL is likely to be pleiotropic for carbon isotope discrimination, chlorophyll content, and flag leaf rolling (Peleg et al. 2009). The introgression of QTL for developing drought-tolerant cultivars is not reported.

Many QTLs were characterized for different wheat disease resistances (Hao et al. 2011, 2012, 2013, 2015; Sapkota et al. 2019; Zhang et al. 2011). QTL studies for the three rusts, powdery mildew, and FHB resistance have been carefully reviewed in several literature (Bai et al. 2018; Buerstmayr et al. 2009; Cowger et al. 2012; Li et al. 2014; Liu et al. 2009; Rosewarne et al. 2013), and we will not include this content here.

2.5 Genetic Resources for Climate-Smart Traits and Diversity Analysis

More than 150 different species have been reported in the *Triticeae* tribe, but very few are cultivated including bread wheat, durum wheat, einkorn wheat, rye, and barley. More than 500,000 accessions of these species are stored in 23 genebanks worldwide (Ortiz et al. 2008a). The ploidy among *Triticeae* members is from diploid (2n = 2*x*) to decaploid (2n = 10*x*). Wheat gene pools are classified as primary, secondary, and tertiary based on the hybridization success with bread wheat. However, bread wheat has greater flexibility to cytogenetics applications and has shown to accommodate almost all the *Triticeae* species through wide hybridization. All the immediate progenitors of bread wheat that can be directly hybridized with or without embryo rescue are included in primary gene pool. These species have been most extensively used in pre-breeding and specifically are accessions of *Triticum monococcum*, *Aegilops tauschii*, and *T. dicoccoides*. There has been a tremendous success in using *Aegilops tauschii*, D-genome donor to bread wheat, and its derived amphiploids knowns as synthetic hexaploid wheat (SHW) produced at CIMMYT and elsewhere (Ogbonnaya et al. 2013). SHW and their advanced derivatives have greatly improved yield and yield components, micronutrient content, disease resistance, and the processing quality of elite cultivars (Ogbonnaya et al. 2013). Apart from SHW, there are A genome amphiploids derived from *T. urartu*, the AABB tetraploids (*T. dicoccum*, *T. dicoccoides*, *T. polonicum*, and *T. carthlicum*) and less used AABBDD hexaploids combinations involving *T. spelta* and *T. sphaerococcum*. The untapped genetic diversity and allelic variations in the primary gene pool is an excellent resource to improve bread wheat against abiotic and biotic stresses (Mujeeb-Kazi et al. 2013). Historically, several leading cultivars in Europe are derived from unknown introgressions from *T. dicoccoides* (Gardner et al. 2016).

The secondary and tertiary gene pools species become self-sterile upon hybridization with bread wheat and embryo rescue is mandatory. The most extensive species used in bread wheat improvement are *Secale cereal*, *Agropyron elongatum*, and *Haynaldiavillosa.* The advanced derivatives of cereal rye known as 1BL.1RS translocation were grown on millions of hectares worldwide. The other successful examples of exploiting secondary and tertiary gene pool species include leaf rust and eyespot resistance was transferred from *Ae. umbellulata* and *Ae. ventricosa*, respectively (Garcia-Olmedo et al. 1977). The translocation from *Ae. ventricosa* was recently identified as 20 Mbp in size through whole-genome re-sequencing of four elite cultivars (Pozniak et al. 2017). A mega wheat cultivar,

Xiaoyan 6, in China is an alleged derivative of a cross between wheat and the 70-chromosome grass species *Thinopyrum ponticum* and its derivatives were cultivated over millions of hectares in China. Some other very promising candidates of secondary and tertiary gene pools are *Haynaldiavillosa*, *Th. elongatum*, and *Thinopyrum bessarabicum* which have the advantage of being diploids and preferred for wide hybridization programs.

It is widely documented that genetic diversity of modern wheat had declined considerably like other crops. Reduction in diversity is largely due to the polyploid speciation, domestication, and intensive selection during modern wheat breeding (Liu et al. 2019). The favorable introgressions from wild relatives would not only enhance resilience to biotic and abiotic stresses but will also enhance the genetic diversity of cultivated wheat (Rasheed et al. 2018a). It has been shown that unknown favorable introgression from wild emmer is present in several northern European cultivars, and cultivar Robigus is a direct progenitor of more than one-third of UK cultivars (Gardner et al. 2016). Similarly, the success of the 1BL.1RS translocation in wheat cultivars globally, *Ae. ventricosa* introgressions, and alleged *Th. ponticum* introgressions in wheat cultivars from China indicated the beneficial impact of wild relatives in farmers' fields. Among the primary gene pool, synthetic-derived wheat are a major source of new genetic variation for improving stress adaptation and grain yield potential in bread wheat. More than 62 cultivars have been reported to be synthetic-derivatives (Börner et al. 2015), however, that number has been significantly increased to the present time. The selection of favorable introgressions from synthetic wheat is still challenging due to the absence of molecular markers designed based on the polymorphism between bread wheat and synthetic wheat. However, it is expected that further studies, including re-sequencing of synthetic hexaploid wheat, will help to design new markers to characterize population-level genetic variation in synthetic-derivative and its association with relevant phenotypic traits (Afzal et al. 2019; Rasheed et al. 2018a, b). We recently reviewed the impact of genomics advancements in gene discovery and gene deployment from wheat wild relatives (Rasheed et al. 2018a), therefore these aspects are not described here.

Genebanks are rich-source genetic resources for crop improvement and new allelic variations for the functional genes. The role of genebanks is not only limited to the preservation and physical access to genetic resources but becoming more specified in providing electronic resources related to germplasm collections. The genebank accessions are important sources of alternative favorable alleles of important genes and storehouse of rare superior alleles that retained unselected during domestication and modern breeding. The advancements in breeding strategies and genomics have transformed genebanks into bio-digital resource centers that will facilitate the selection of useful genetic variation and its use in breeding programs and provide easy access to past crop diversity (Milner et al. 2019). It is proposed to link catalogs of germplasm and inquiries into biological mechanisms of plant performance as a long-term joint research goal of genebanks, plant geneticists, and breeders (Milner et al. 2019). This would be possible by sequencing or genotyping the complete genebanks or entire collections of specific gene pools

(Moore 2015; Riaz et al. 2016; Sehgal et al. 2015). The term "digital genebanks" from such huge integrated efforts is important to unravel population structure, detect signatures of selection, develop core collections to avoid redundancy, and map QTL for important phenotypic traits.

2.6 Breeding Chips and Genotyping Platforms to Meet Demand

It is a high priority research area to establish high-throughput (HTP) genotyping platforms without compromising robustness and cost. Huge genome sequencing and gene discovery research is taking place and it is paramount to translate this knowledge for applied breeding. We provided an overview of different HTP genotyping platforms currently available in different crop species. It was generally agreed that data generation is not a problem, but decreasing the cost (per data point or per sample) is a major factor in establishing a HTP genotyping platform to support crop breeding. The genotyping needs to differ for different crop breeding objectives and usually classified into two categories: (i) whole-genome genotyping for genomic prediction, diversity, and association mapping studies and (ii) genotyping few loci for selection breeding, gene introgression, quality control, and hybrid testing.

Whole-genome genotyping

Single nucleotide polymorphisms (SNPs) have become markers of choice due to the abundance in genome and ability to identify them in HTP formats. In most cases, whole-genome genotyping platforms achieve HTP by comprising flexibility. Conventional PCR-based markers such as simple sequence repeats (SSRs) and sequence-tagged sites (STS) are highly flexible to freely choose for any number of samples. However, modern genotyping platforms are nonflexible.

Genotyping-by-sequencing (GBS)

Next generation sequencing (NGS) is becoming a preferred genome-wide genotyping method due to the dramatic decrease in cost, ease in library preparation, and benefit of de novo SNP discovery. Such genotyping methods can be used for various crops irrespective of genome size, genetic background, and ploidy. Scheban et al. (2016) reviewed almost 15 different genotyping-by-sequencing (GBS) techniques that have been used in crop plants. The traditional GBS protocol used a single restriction enzyme digestion followed by adapter ligation, PCR, and sequencing (Elshire et al. 2011). However, a two-enzyme modification was introduced in wheat and barley (Poland et al. 2012). The other promising GBS techniques include skim sequencing and exome capture, where the former is a low-coverage cost-effective way of identifying genetic variation and haplotypes in populations. While exome capture re-sequences the predesigned probes within gene-coding regions (Jordan et al. 2015). Initial exome capture platform was designed for durum wheat to re-sequence 3,497 genes (Saintenac et al. 2011). Later, exome capture was used in bread wheat to sequence 124,201 high confidence genes (Jordan et al. 2015). Another exome capture platform was used to target 57 Mb coding sequence in 43 wheat genotypes and wild species (Winfield et al. 2016). Krasileva et al. (2017) developed another exome capture platform of 84 Mb which covered 82,511 transcripts and the TILLING library of Cadenza and Kronos were characterized. After Elshire GBS, DArT-seq is one of the most widely used platforms in wheat research and breeding. DArTseq platform was selected to genotype genetic resources from CIMMYT under the Seeds of Discovery (SeeD) project. Similarly, genetic diversity studies in durum wheat (Baloch et al. 2017) and bread wheat (Riaz et al. 2017; Vikram et al. 2016) landraces have been conducted using DArTseq approach. Another modification to conventional GBS known as specific locus amplified fragment sequencing (SLAF-seq) was used in species of distant wheat relatives *Agropyron* (Zhang et al. 2015) and *Thinopyrum* (Chen et al. 2013) for genome-specific SNP marker discovery.

These studies indicate the effectiveness of NGS for genetic diversity, gene mapping, and genomic prediction studies. The disadvantages of NGS are the complexity in allele calling when a reference genome sequence is not available, high level of missing data, and lack of efficient SNP imputation tools (Scheben et al. 2016). However, NGS is becoming a popular genome-wide genotyping method despite these disadvantages and significant developments in sequencing chemistry, reductions in the cost of long-read sequencing platforms and availability of pan-genome sequence of wheat will accelerate its adoption in wheat breeding programs.

Array-based genotyping in wheat

SNP chips or arrays are cost-effective and HTP SNP genotyping method and several SNP arrays are available for genotyping wheat. SNP arrays are multiplex and highly robust for high-density genotyping which have highly accurate allele calling compared to NGS. SNP arrays are cost-effective in terms of per data point when genotyping a large number of SNPs and samples (Rasheed et al. 2017). The inflexibility and the overall cost to genotype large population are quite high which make them inaccessible for most crop genetics and breeding programs.

In wheat, the first SNP array of 9 K density was introduced which included almost 9,000 gene-associated SNPs and used it to genotype 2,994 hexaploid wheat accessions (Cavanagh et al. 2013). The array was later updated to 90 K SNP density to genotype 3,380 wheat accessions (Wang et al. 2014a). SNPs in the 9 and 90 K arrays were biased in selection because only a few cultivars and landraces were used to select SNPs in these arrays. This problem was fixed in subsequent SNP arrays, e.g., Winfield et al. (2016) designed a 820 K Affymetrix Axiom SNP array from resequencing exomes of 43 bread wheat and wild species accessions and 475 hexaploid wheat and wheat relative accessions were genotyped. Furthermore, a proportion from the 820 K array was used to develop a breeder-oriented Axiom 35 K SNP (Allen et al. 2017). Another problem was that the intergenic fraction of the wheat genome, which accounts for 98–99% of the total, was poorly selected in SNP arrays. To address this issue, Rimbert et al. (2018) designed a 280 K SNP array based on the 3 million genome-wide SNPs selected from the resequencing data of eight important wheat cultivars. A 660 K SNP has been used to identify QTLs for bread-making quality and kernel number (Jin et al. 2016), and a high-density linkage map of *Agropyron cristatum* was constructed (Zhou et al. 2018). The features of this SNP array and criteria for selection of SNP markers are not known yet. We have recently developed a 50 and 15 K SNP arrays based on the most qualified SNPs selected from the wheat 35, 90, and 660 K SNP arrays. Around 135 and 150 functional markers, and 700 and 1,000 SNPs tightly linked with known QTLs are included in the 50 and 15 K SNP arrays, respectively.

High-throughput single-plex genotyping in wheat

Despite significant progress in developing NGS and SNP array platforms in wheat for genome-wide genotyping, it was still challenging to develop a HTP platform to use single markers in wheat breeding programs. The genotyping requirements for such breeding program were highlighted in six factors: (a) HTP level; how many data points can be generated in a selection cycle, (ii) ease of use, (iii) data quality (sensitivity, reliability, reproducibility, and accuracy), (iv) flexibility (genotyping few samples with many SNPs or many samples with few SNPs), (v) assay development requirements, and (vi) cost per sample or data point. Semagn et al. (2014) reviewed that KASP (Kompetitive Allele Specific PCR) is an acceptable global benchmark technology for such genotyping requirements in terms of both cost-effectiveness and high throughput (Semagn et al. 2014). To support wheat breeding programs, several groups joined hands to convert gel-based functional markers to HTP KASP markers (http://www.cerealsdb.uk.net/). Rasheed et al. (2016) reported the conversion of 72 functional markers into KASP format after validation in a bread wheat diversity panel. This effort was continued in our group and we currently have more than 150 KASP markers for almost 100 functional genes (Rasheed et al. Unpublished data). Several other QTLs linked to important traits and SNPs in functional genes have been converted to KASP markers; examples include *Lr16* (Kassa et al. 2017), *Lr67* (Moore et al. 2015), *Lr23* (Chhetri et al. 2017), and *Yr26* (Wu et al. 2018), as well as many functional markers, including preharvest sprouting gene *TaPHS1* (Liu et al. 2013), earliness per se gene *Eps1* (Zikhali et al. 2016), and *NAM-A1* (Cormier et al. 2015).

The higher cost of KASP is still an issue because the master mix is a commercial proprietary from LGC and until recently there were no other competitors. One open-source uniplex genotyping method was introduced and was termed semi-thermal asymmetric reverse PCR (STARP) (Long et al. 2016). Similarly, Amplifluor was another SNP genotyping approach that can be used with any commercial mastermix. These two approaches will likely reduce the cost per data point (Jatayev et al. 2017). More recently, other commercial alternatives of KASP assays were introduced, including PACE® master mix from 3CR Biosciences (www.3crbio.com) and rhAmp from Integrated DNA Technologies® (https://www.idtdna.com/pages/products/qpcr-and-pcr/genotyping/rhamp-snp-genotyping).

2.7 Transgenesis and Gene Editing for Climate-Smart Traits

Transgenic research has shown great promise in the lab experiment, however its application to improve the crop productivity have shown little promise. In wheat, no transgenic cultivar has been approved yet for commercial cultivation. Araus et al. (2019) critically evaluated the transgenic research in wheat and implications to develop drought-tolerant wheat cultivars. There is a slight chance to improve complex environmental adaptability using transformation technologies. Most of the wheat transgenic plants have failed to translate the benefits observed in controlled environments to field conditions (Passioura 2012). Transgenic research is generally concentrated on the survival of plants suffering from severe water stress, which is rarely an important trait in grain crops such as wheat. Moreover, in many cases, survival has simply been explained by reduced plant size and concomitant slower water uptake compared with the wild-type cultivar (Morran et al. 2011), with the transgenic plants growing in pots or containers instead of the field. This was certainly the case for two members of the wheat homeodomain-leucine zipper I (HD-Zip I) family of transcription factors, which regulate development after plants are exposed to environmental stimuli and stresses. After they were introduced as transgenes into wheat, the plants showed improved resistance to drought and frost, but they exhibited undesirable phenotypic characteristics such as reduced size, biomass, and yield (Kovalchuk et al. 2016; Yang et al. 2018).

Targeted genome editing nucleases, such as zinc-finger nucleases (ZFNs) and transcription activator-like effector nucleases (TALENs), are excellent tools for gene manipulation to validate the functions of gene and developing new traits in plants. The clustered regularly interspersed short palindromic repeats (CRISPR)/Cas system is a more precise nuclease-based approach for efficient and versatile gene editing. In this system, only the 20-nt targeting sequence within the single-guide RNA (sgRNA) needs to be changed to target different genes. The simplicity of the cloning strategy and the few limitations on potential target sites make the CRISPR/Cas system very appealing. Simultaneously editing of three copies of *Mlo* and *TaEDR1* genes in hexaploid wheat conferring resistance to powdery mildew have been reported (Wang et al. 2014b; Zhang et al. 2017b). Shan et al. (2014) described a stepwise protocol for the selection of target sites, as well as the design, construction, verification, and use of sgRNAs for sequence-specific CRISPR/Cas-mediated mutagenesis and gene targeting in rice and wheat. The recent reports of CRISPR/Cas-mediated gene editing of grain size and weight gene, *TaGW7*, identified mutants altered B- and D-homoeologues and increased grain width but reduced grain length (Wang et al. 2019). Similarly, gene editing of *TaGW2* increased 16.3% thousand-grain weight in wheat (Wang et al. 2018). These studies clearly showed that CRISPR-Cas9 based gene editing can help to create new allelic variants targeted by domestication selection and select targets for engineering new gene variants for crop improvement.

2.8 Phenotypic Tools to Measure Climate-Smart Traits

While the genomic tools available for the development of climate-smart crops have advanced rapidly over the past 30 years, the ability to accurately phenotyping these crops in a field environment has not. This has created a "phenotyping bottleneck" whereby large amounts of genomics data are generated for a new cultivar relative to the phenotypic data (Furbank and Tester 2011). As a result, the past decade has seen the rapid development of research into plant phenomics, the characterization of the full set of phenotypes of a given species, in both the academic and commercial sectors (Coppens et al. 2017). Tools created for the study of plant phenomics will play an essential role in the alleviation of the phenotypic bottleneck and the development of climate-smart wheat.

Traditional phenotyping approaches

During domestication of modern wheat, key traits (e.g., flowering time) for specific climates were developed using very simple phenotyping; ease of harvest, yield, consumption quality, and replanting. Modern breeding practices look at a range of more defined key traits but still rely on traditional phenotyping methods. These can be simple physical measurements of growth (e.g., height), time series visual measurements of development (e.g., flowering time), or visual numerical scores or indices (e.g., level of infection). In some cases, destructive measurements of the plant are required (e.g. biomass, end-use quality). Done at the correct scale and replication, these trait assessments are reliable and are the benchmark for phenotypes used in breeders selections.

As discussed in previous sections, genotypic approaches are now commonly applied within breeding programs. The issue with genotype-only selections is the effect of the environment on the desired trait within the field, referred to as the genotype-by-environment interaction. With a changing climate, it is likely that the effect on wheat will be cultivar dependent. This will require the tailoring of selective breeding to changing regional patterns (Trnka et al. 2014). To produce climate-smart wheat, the effects of climate uncertainty on plant processes and productivity across environments need to be accurately defined. This requires the development of accurate, reliable, and objective high-throughput field systems for phenotyping.

High-throughput phenotyping platforms

There are numerous phenotyping platforms in use all with varying suitability for different scales of research and environments. Platforms can broadly be classified into three different types; ground level (e.g., phenocarts), aerial (e.g., drone or planes), or space (e.g., satellites). These types of platforms can carry numerous sensors to analyze numerous traits at varying levels of accuracy and spatial resolution (Araus and Cairns 2014). Before selecting the correct platform, it is essential to consider maximum payload, positioning precision, the field of view, and distance above the crop.

Phenotyping sensors

While it is important to use the correct platform for each trial, it is the sensors mounted on the platform which will be essential to the phenotyping process. Most sensors can be deployed on all types of platforms, but the quality and resolution of the data can vary dramatically. The simplest sensors are the visible light camera (400–700 nm wavelengths, Table 2.2) which functions in a similar way to a digital camera. These cameras (often termed RGB sensors) produce two-dimensional color images and can be deployed on most platforms. Taking numerous images of an individual or group of plants from various angles allows the production of three-dimensional images. Images capture using RGB sensors can be used to measure basic developmental traits (e.g., height and crop cover) from which more complicated traits (e.g., lodging and leaf are indexes) can be calculated (Bendig et al. 2014). The combination of data from RGB and near-infrared (780–2500 nm, Table 2.2) cameras (or dedicated multispectral cameras) can be used to determine vegetation indices (VIs) (Yang et al. 2017). All sensors measure light wavelength function in the same way. One or more bands (e.g., 400–700 nm) capture the reflectance of a surface (e.g., crop canopy) and convert this into a numerical value. VIs are derived from these reflectance values. VIs are more effective than RGB sensors at the segmentation of green healthy plant material, soil, and stressed plant material, resulting in a more accurate prediction of crop cover and plant stress than RGB cameras (Yang et al. 2017). Recently, researchers have been deploying hyperspectral sensors that measure continuous and contiguous ranges of wavelengths (Table 2.2). On aerial based platforms, the data provided by hyperspectral cameras provides more information on traits than multispectral cameras. Hyperspectral cameras can reliably measure nitrogen content and biomass, chlorophyll content, water content, and other photosynthetic parameters (Yang et al. 2017). In addition, novel field-based methods, such as leaf carotenoid content (Zarco-Tejada et al. 2013) and chlorophyll density (Uto et al. 2013). Ultimately light-based sensors are going to assist in the development of climate-smart crops in a variety of ways. The high-throughput nature of the tools will allow for larger trials in more climate diverse environments, helping to identify wheat cultivars with stable yields in various climates. New in field phenotyping techniques developed using hyperspectral cameras will allow for identification and measurement of traits essential to climate-smart wheat. Agronomically, light sensors will allow for the proper management of wheat under varying climate conditions. Additionally, measurement of agronomic performance could mitigate the genotype-by-environment effect seen in trials.

Other sensors will also play an important role in the development of climate-smart wheat. Thermal sensor in particular will be essential to measuring the effect of heat and drought on a crop. To date, thermal sensors have been used to measure water stress (Gonzalez-Dugo et al. 2013), disease (Nilsson 1991), stomatal conductance, and transpiration rate (Baluja et al. 2012). At present, the major limitation of thermal sensors on high-throughput systems is the required calibration and correction for ambient temperature, wind speed, and solar radiation (Sugiura

Table 2.2 Sensors deployed in field phenotyping (adapted from Yang et al. 2017 and Atkinson et al. 2018)

Sensor	Spectral bands	Wavelength range	Potential applications	Advantages	Disadvantages
Digital camera	Red, Green, Blue	–	Leaf color, plant height, lodging, canopy cover, fraction of intercepted radiation, LAI, 3D structure, leaf angle distribution	Low cost, light weight, convenient operation, simple data processing	Low radiometric resolution, lack of proper calibration
Multispectral camera	Red, Green, NIR	490–920 nm	See above and leaf nitrogen content, yield, chlorophyll, biomass, weed emergence	Low cost, flexibility	Less bands, low spectral resolution and discontinuous spectrum
Hyperspectral imager	125–324	380–1000	See above and net photosynthesis, nitrogen, chlorophyll, disease detection	More bands, high spectral resolution, ability of imaging	High cost, cumbersome data processing, sensitive to weather
Thermal imager	–	7.5–13 μm	Canopy temperature, stomatal conductance, water potential	Indirect determination of crop growth status under abiotic and biotic stress	Sensitive to the weather, frequent calibration, difficult to eliminate the influence of soil background
LIDAR	NIR	–	Plant height, biomass	Rich point cloud information, Effective acquisition of high precision horizontal and vertical vegetation canopy structure parameters	High cost, Large data processing
SAR	–	–	Crop identification, crop acreage monitoring, key crop trait estimation, and yield prediction	Collects data even in cloudy weather	High cost, Large data processing, Limited to satetllites therefore can only be used for large plot work

et al. 2007; Deery et al. 2014). Ultimately, this limits the effectiveness of comparing crops over large areas or time.

Machine learning

Machine learning algorithms are promising statistical tools that assist in the analysis of complex data sets and have started to be widely used in many research fields including plant phenotyping and genomics/proteomics (Zhong et al. 2009; Iyer-Pascuzzi et al. 2010). For high-throughput data-driven plant phenotyping, machine learning provides effective trait extraction in image analysis and processing. Discovery of good features for the application is essential to an effective use of learning approaches (Iyer-Pascuzzi et al. 2010). Machine learning approaches can overcome problems on small, complex phenotypic data sets characterized by a low signal-to-noise ratio, and can be used for unbiased trait selection and group classification (Zhao et al. 2016). Building on artificial neural networks, the deep machine learning approaches make greater discriminative and predictive power and shows the success offered by such techniques when applied to the challenging problem of image-based plant phenotyping (Pound et al. 2017). The application of machine learning in wheat have demonstrates state-of-the-art results (>95% accuracy) for spike, root, and shoot feature identification and localization (Pound et al. 2017). Generally, machine learning is widely expected to make plant phenotypical data analyses more effective, robust, and comprehensive (Pound et al. 2017).

Drought and heat tolerance are major determinants for final yield in stressed environments. Delay leaf-senescence during critical periods can extend effective leaf function. Current remote sensing practices measure the amount of green or yellow pixels of crop canopy as a potential reflection of the leaf's physiological changes. However, this approach may not work well on the discrimination of stressed and non-stressed plants. Machine learning approaches could use to find the pigment changes in the wheat canopy, which mainly related to chlorophyll degradations in leaves, stems, and spikes. This would provide dynamic data set to reasonably assess the senescence process. Further machine learning on analysis of field-derived complex agronomic phenotypes under climatic change is expected.

2.9 Future Perspectives

Wheat genetics and breeding research are under paradigm shift due to the recent availability of reference genome sequence of wheat cultivar Chinese Spring (Appels et al. 2018) and the wide application of HTP genotypic and phenotpic platforms. It is likely that de novo sequence assembly from a few other key cultivars will be available very soon. These resources together with other advancements in wheat genomics have the potential to accelerate the rate of genetic progress by improving the selection accuracy and efficiency in breeding programs (Rasheed and Xia 2019). In addition to these advancements, the exponential progress in plant phenotyping using imaging techniques, rapid gene cloning methods based on the mutation screening and NGS, increase in the number of gene discovery studies, and above all the promising gene editing techniques have potential to deliver future climate resilience cultivars if integrated consciously.

Acknowledgements We acknowledge financial support from the National Key Research and Development Program of China (2016YFE0108600 and 2016YFD0101802), National Natural Science Foundation of China for International Collaborations (31761143006), and Agricultural Science and Technology Innovation Program of CAAS.

References

Abberton M, Batley J, Bentley A, Bryant J, Cai H, Cockram J, Costa de Oliveira A, Cseke LJ, Dempewolf H, De Pace C, Edwards D, Gepts P, Greenland A, Hall AE, Henry R, Hori K, Howe GT, Hughes S, Humphreys M, Lightfoot D, Marshall A, Mayes S, Nguyen HT, Ogbonnaya FC, Ortiz R, Paterson AH, Tuberosa R, Valliyodan B, Varshney RK, Yano M (2016) Global agricultural intensification during climate change: a role for genomics. Plant Biotechnol J 14:1095–1098

Achard P, Genschik P (2009) Releasing the brakes of plant growth: how GAs shutdown DELLA proteins. J Exp Bot 60:1085–1092

Acuna-Galindo MA, Mason RE, Subrahmanyam NK, Hays D (2014) Meta-analysis of wheat QTL regions associated with adaptation to drought and heat stress. Crop Sci 55:477–492

Afzal F, Li H, Gul A, Subhani A, Ali A, Mujeeb-Kazi A, Ogbonnaya F, Trethowan R, Xia X, He Z, Rasheed A (2019) Genome-wide analyses reveal footprints of divergent selection and drought adaptive traits in synthetic-derived wheats. G3: Genes, Genomes, Genetics 9:1957

Ahmad I, Khaliq I, Mahmood N, Khan N, Secretariat EF (2015) Morphological and physiological criteria for drought tolerance at seedling stage in wheat. J Anim Plant Sci 25:1041–1048

Ahmadi G, Akbarabadi A, Kahrizi D, Rezaizad A, Gheytouli M (2012) Study of drought tolerance of bread wheat (*Triticum aestivum* L.) genotypes in seedling stage. Biharean Biologist 6:77–80

Ain Q-U, Rasheed A, Anwar A, Mahmood T, Mahmood T, Imtiaz M, He Z, Xia X, Quraishi U (2015) Genome-wide association for grain yield under rainfed conditions in historical wheat cultivars from Pakistan. Front Plant Sci 6

Ainsworth EA, Ort DR (2010) How do we improve crop production in a warming world? Plant Physiol 154:526–530

Akram M (2011) Growth and yield components of wheat under water stress of different growth stages. Bangladesh J Agri Res 36:455–468

Allen AM, Winfield MO, Burridge AJ, Downie RC, Benbow HR, Barker GL, Wilkinson PA, Coghill J, Waterfall C, Davassi A, Scopes G, Pirani A, Webster T, Brew F, Bloor C, Griffiths S, Bentley AR, Alda M, Jack P, Phillips AL, Edwards KJ (2017) Characterization of a Wheat Breeders' Array suitable for high-throughput SNP genotyping of global accessions of hexaploid bread wheat (*Triticum aestivum*). Plant Biotechnol J 15:390–401

Araus JL, Serret MD, Lopes MS (2019) Transgenic solutions to increase yield and stability in wheat: shining hope or flash in the pan? J Exp Bot 70:1419–1424

ArausJL, Cairns JE (2014) Field high-throughput phenotyping: the new crop breeding frontier. Trends Plant Sci 19:52–61

Arora S, Steuernagel B, Gaurav K, Chandramohan S, Long Y, Matny O, Johnson R, Enk J, Periyannan S, Singh N, Asyraf Md Hatta M, Athiyannan N, Cheema J, Yu G, Kangara N, Ghosh S, Szabo LJ, Poland J, Bariana H, Jones JDG, Bentley AR, Ayliffe M, Olson E, Xu SS, Steffenson BJ, Lagudah E, Wulff BBH (2019) Resistance gene cloning from a wild crop relative by sequence capture and association genetics. Nat Biotechnol 37:139–143

Ashraf M, Foolad MR (2013) Crop breeding for salt tolerance in the era of molecular markers and marker-assisted selection. Plant Breed 132:10–20

Atkinson JA, Jackson RJ, Bentley AR, Ober E, Wells DM (2018) Field phenotyping for the future. In: Roberts JA (ed) Annual Plant Reviews online. Wiley

Azadi A, Mardi M, Hervan EM, Mohammadi SA, Moradi F, Tabatabaee MT, Pirseyedi SM, Ebrahimi M, Fayaz F, Kazemi M (2015) QTL mapping of yield and yield components under normal and salt-stress conditions in bread wheat (*Triticum aestivum* L.). Plant Mol Biol Rep 33:102–120

Bai G, Su Z, Cai J (2018) Wheat resistance to *Fusarium* head blight. Can J Plant Path 40:336–346

Baloch FS, Alsaleh A, Shahid MQ, Ciftci V, L ESdM, Aasim M, Nadeem MA, Aktas H, Ozkan H, Hatipoglu R (2017) A whole genome DArTseq and SNP analysis for genetic diversity assessment in durum wheat from central fertile crescent. PLoS One 12:e0167821

Baluja J, Diago MP, Balda P, Zorer R, Meggio F, Morales F et al (2012) Assessment of vineyard water status variability by thermal and multispectral imagery using an unmanned aerial vehicle (UAV). Irrig Sci 30:511–522

Barber HM, Lukac M, Simmonds J, Semenov MA, Gooding MJ (2017) Temporally and genetically discrete periods of wheat sensitivity to high temperature. Front Plant Sci 8

Bendig J, Bolten A, Bennertz S, Broscheit J, Eichfuss S, Bareth G (2014) Estimating biomass of barley using crop surface models (CSMs) derived from UAV-based RGB imaging. Remote Sens 6:10395–10412

Bennett D, Reynolds M, Mullan D, Izanloo A, Kuchel H, Langridge P, Schnurbusch T (2012) Detection of two major grain yield QTL in bread wheat (*Triticum aestivum* L.) under heat, drought and high yield potential environments. Theor Appl Genet 125:1473–1485

Bettgenhaeuser J, Krattinger SG (2019) Rapid gene cloning in cereals. Theor Appl Genet 132:699–711

Bhullar NK, Street K, Mackay M, Yahiaoui N, Keller B (2009) Unlocking wheat genetic resources for the molecular identification of previously undescribed functional alleles at the *Pm3* resistance locus. Proc Nat Acad Sci USA106:9519–9524

Boden SA, Cavanagh C, Cullis BR, Ramm K, Greenwood J, Finnegan EJ, Trevaskis B, Swain SM (2015) *Ppd-1* is a key regulator of inflorescence architecture and paired spikelet development in wheat. Nat Plant 14016

Borner A, Roder M, Korzun V (1997) Comparative molecular mapping of GA insensitive Rht loci on chromosomes 4B and 4D of common wheat (*Triticumaestivum* L.). Theor Appl Genet 95:1133–1137

Börner A, Ogbonnaya FC, Röder MS, Rasheed A, Periyannan S, Lagudah ES (2015) *Aegilops tauschii* introgressions in wheat. In: Molnár-Láng M, Ceoloni C, Doležel J (eds) Alien Introgression in Wheat. Springer International, Switzerland

Bovill WD, Hyles J, Zwart AB, Ford BA, Perera G, Phongkham T, Brooks BJ, Rebetzke GJ, Hayden MJ, Hunt JR, Spielmeyer W (2019) Increase in coleoptile length and establishment by *Lcol-A1*, a genetic locus with major effect in wheat. BMC Plant Biol 19:332

Braun HJ, Rajaram S, Ginkel MV (1996) CIMMYT's approach to breeding for wide adaptation. Euphytica 92:175–183

Buerstmayr H, Ban T, Anderson JA (2009) QTL mapping and marker assisted selection for *Fusarium* head blight resistance in wheat: a review. Plant Breed 128:1–26

Byrt CS, Platten JD, Spielmeyer W, James RA, Lagudah ES, Dennis ES, Munns R (2007) *HKT1;5*-like cation transporters linked to Na+ exclusion loci in wheat, *Nax2* and *Kna1*. Plant Physiol 143:1918–1928

Byrt CS, Xu B, Krishnan M, Lightfoot DJ, Athman A, Jacobs AK, Watson-Haigh NS, Plett D, Munns R, Tester M, Gilliham M (2014) The Na+ transporter, *TaHKT1;5-D*, limits shoot Na + accumulation in bread wheat. Plant J 80:516–526. Callaway E (2016) Devastating wheat fungus appears in Asia for first time. Nature 532:421–422

Callaway E (2016) Devastating wheat fungus appears in Asia for first time. Nature 532:421–422

Cao A, Xing L, Wang X, Yang X, Wang W, Sun Y, Qian C, Ni J, Chen Y, Liu D, Wang X, Chen P (2011) Serine/threonine kinase gene *Stpk-V*, a key member of powdery mildew resistance gene *Pm21*, confers powdery mildew resistance in wheat. Proc Natl Acad Sci USA 108:7727–7732

Cavanagh CR, Chao SM, Wang SC, Huang BE et al (2013) Genome-wide comparative diversity uncovers multiple targets of selection for improvement in hexaploid wheat landraces and cultivars. Proc Natl Acad Sci USA 110:8057–8062

Chartres CJ, Noble A (2015) Sustainable intensification: overcoming land and water constraints on food production. Food Secur 7:235–245

Chen X (2013) High-temperature adult-plant resistance, key for sustainable control of stripe rust. Amme J Plant Sci 4:608–627

Chen XM (2014) Integration of cultivar resistance and fungicide application for control of wheat stripe rust. Can J Plant Pathol 36:311–326

Chen X, Moore M, Milus EA, Long DL, Line RF, Marshall D, Jackson L (2002) Wheat stripe rust epidemics and races of *Puccinia striiformis* f. sp. *tritici* in the United States in 2000. Plant Dis 86:39–46

Chen S, Huang Z, Dai Y, Qin S, Gao Y, Zhang L, Gao Y, Chen J (2013) The development of 7E chromosome-specific molecular markers for *Thinopyrum elongatum* based on SLAF-seq technology. PLoS ONE 8:e65122

Chhetri M, Bariana H, Wong D, Sohail Y, Hayden M, Bansal U (2017) Development of robust molecular markers for marker-assisted selection of leaf rust resistance gene *Lr23* in common and durum wheat breeding programs. Mol Breed 37:21

Cloutier S, McCallum BD, Loutre C, Banks TW, Wicker T, Feuillet C, Keller B, Jordan MC (2007) Leaf rust resistance gene *Lr1*, isolated from bread wheat (*Triticum aestivum* L.) is a member of the large *psr567* gene family. Plant Mol Biol 65:93–106

Coppens F, Wuyts N, Inzé D, Dhondt S (2017) Unlocking the potential of plant phenotyping data through integration and data-driven approaches. Curr Opin Syst Biol 4:58–63

Cormier F, Throude M, Ravel C, Gouis J, Leveugle M, Lafarge S, Exbrayat F, Duranton N, Praud S (2015) Detection of *NAM-A1* natural variants in bread wheat reveals differences in haplotype distribution between a worldwide core collection and european elite germplasm. Agronomy 5:143

Cowger C, Miranda L, Griffey C, Hall M, Murphy JP, Maxwell J (2012) Wheat powdery mildew. In: Sharma I (ed) Disease resistance in wheat. CAB International, Oxfordshire, UK, pp 84–119

Cruz CD, Valent B (2017) Wheat blast disease: danger on the move. Trop Plant Pathol 42:210–222

Cruz CD, Magarey RD, Christie DN, Fowler GA, Fernandes JM, Bockus WW, Valent B, Stack JP (2016a) Climate suitability for *Magnaporthe oryzae Triticum* pathotype in the United States. Plant Dis 100:1979–1987

Cruz CD, Peterson GL, Bockus WW, Kankanala P, Dubcovsky J, Jordan KW, Akhunov E, Chumley F, Baldelomar FD, Valent B (2016b) The 2NS translocation from *Aegilops ventricosa* confers resistance to the *Triticum* patho type of *Magnaporthe oryzae*. Crop Sci 56:990–1000

De Leon JLD, Escoppinichi R, Zavala-Fonseca R, Castellanos T, Roder MS, Mujeeb-Kazi A (2010) Phenotypic and genotypic characterization of salt-tolerant wheat genotypes. Cereal Res Commun 38:15–22

Deery D, Jimenez-Berni J, Jones H, Sirault X, Furbank R (2014) Proximal remote sensing buggies and potential applications for field-based phenotyping. Agronomy 4:349–379

Diaz A, Zikhali M, Turner AS, Isaac P, Laurie DA (2012) Copy number variation affecting the *Photoperiod-B1* and *Vernalization-A1* genes is associated with altered flowering time in wheat (*Triticum aestivum*). PLoS ONE 7:e33234

Dubcovsky J, Maria GS, Epstein E, Luo MC, Dvorak J (1996) Mapping of the K+/Na + discrimination locus Kna1 in wheat. Theor Appl Genet 92:448–454

Duveiller E, Hodson D, Sonder K, Av T (2011) An international perspective on wheat blast. Phytopathology 101:S220

Dyck P (1977) Genetics of leaf rust reaction in three introductions of common wheat. Can J Genet Cytol 19:711–716

Edae EA, Byrne PF, Haley SD, Lopes MS, Reynolds MP (2014) Genome-wide association mapping of yield and yield components of spring wheat under contrasting moisture regimes. Theor Appl Genet 127:791–807

Elshire RJ, Glaubitz JC, Poland JA, Kawamoto K, Buckler E, Mitchell SE (2011) A robust, simple genotyping-by-sequencing (GBS) approach for high diversity species. PLoS ONE 6(5):e19379

Emebiri LC, Tan MK, El-Bouhssini M, Wildman O, Jighly A, Tadesse W, Ogbonnaya FC (2017) QTL mapping identifies a major locus for resistance in wheat to Sunn pest (Eurygaster integriceps) feeding at the vegetative growth stage. Theor Appl Genet 130:309–318

FAO (2017) The future of food and agriculture—trends and challenges. Rome

Farooq M, Hussain M, Siddique KHM (2014) Drought stress in wheat during flowering and grain-filling periods. Crit Rev Plant Sci 33:331–349

Feuillet C, Travella S, Stein N, Albar L, Nublat A, Keller B (2003) Map-based isolation of the leaf rust disease resistance gene *Lr10* from the hexaploid wheat (*Triticum aestivum* L.) genome. Proc Natl Acad Sci USA 100:15253–15258

Fu D, Uauy C, Distelfeld A, Blechl A, Epstein L, Chen X, Sela H, Fahima T, Dubcovsky J (2009) A kinase-START gene confers temperature-dependent resistance to wheat stripe rust. Science 323:1357–1360

Furbank RT, Tester M (2011) Phenomics – technologies to relieve the phenotyping bottleneck. Trends Plant Sci 16:635–644

Gale MD, Youssefian S (1985) Dwarfing genes in wheat. In: Russell GE (ed) Progress in plant breeding 1. Butterworths, London, pp 1–35

Garcia-Olmedo F, Delibes A, Sanchez-Monge R (1977) Transfer of resistance to eyespot disease from *Aegilops ventricosa* to wheat. In: Proceedings of the 8th congress of Eucarpia, pp 91–97

Gardner KA, Lukas M, Mackay IJ (2016) A highly recombined, high-density, eight-founder wheat MAGIC map reveals extensive segregation distortion and genomic locations of introgression segments. Plant Biotechnol J 14:1406–1417

Gate P (1995) Ecophysiologie du blé: de la plante à la culture. Lavoisier, France

Genc Y, Oldach K, Verbyla AP, Lott G, Hassan M, Tester M, Wallwork H, McDonald GK (2010) Sodium exclusion QTL associated with improved seedling growth in bread wheat under salinity stress. Theor Appl Genet 121:877–894

Genc Y, Oldach K, Gogel B, Wallwork H, McDonald GK, Smith AB (2013) Quantitative trait loci for agronomic and physiological traits for a bread wheat population grown in environments with a range of salinity levels. Mol Breed 32:39–59

Genc Y, Taylor J, Rongala J, Oldach K (2014) A major locus for chloride accumulation on chromosome 5A in bread wheat. PLoS ONE 9:e98845

Godfray HCJ, Beddington JR, Crute IR, Haddad L, Lawrence D, Muir JF, Pretty J, Robinson S, Thomas SM, Toulmin C (2010) Food security: the challenge of feeding 9 billion people. Science 327:812–818

Gonzalez-Dugo V, Zarco-Tejada P, Nicolas E, Nortes PA, Alarcon JJ, Intrigliolo DS et al (2013) Using high resolution UAV thermal imagery to assess the variability in the water status of five fruit tree species within a commercial orchard. Precis Agric 14:660–678

Gooding MJ (2009) The wheat crop. In: Khan K, Shewry PR (eds) Wheat: chemistry and technology, 4th edn. AACC International, Minnesota, pp 35–70

Gooding MJ, Addisu M, Uppal RK, Snape JW, Jones HE (2012) Effect of wheat dwarfing genes on nitrogen use efficiency. J AgriSci 150:3–22

Gorham J, Hardy C, Jones RW, Joppa L, Law C (1987) Chromosomal location of a K/Na discrimination character in the D genome of wheat. Theor Appl Genet 74:584–588

Gosman N, Steed A, Hollins T, Bayles R, Jennings P, Nicholson P (2009) Semi-dwarfing *Rht-B1* and *Rht-D1* loci of wheat differ significantly in their influence on resistance to *Fusarium* head blight. Theor Appl Genet 118:695

Gou J-Y, Li K, Wu K, Wang X, Lin H, Cantu D, Uauy C, Dobon-Alonso A, Midorikawa T, Inoue K, Sánchez J, Fu D, Blechl A, Wallington E, Fahima T, Meeta M, Epstein L, Dubcovsky J (2015) Wheat stripe rust resistance protein wks1 reduces the ability of the thylakoid-associated ascorbate peroxidase to detoxify reactive oxygen species. Plant Cell 27:1755–1770

Gupta KP, Balyan SH, Gahlaut V (2017) QTL analysis for drought tolerance in wheat: present status and future possibilities. Agronomy 7, 5. https://doi.org/10.3390/agronomy7010005

Hailu F, Merker A (2008) Variation in gluten strength and yellow pigment in tetraploid wheat germplasm. Genet Res Crop Evol 55:277–285

Hammad SA, Ali OA (2014) Physiological and biochemical studies on drought tolerance of wheat plants by application of amino acids and yeast extract. Ann AgriSci 59:133–145

Hao Y, Chen Z, Wang Y, Bland D, Buck J, Brown-Guedira G, Johnson J (2011) Characterization of a major QTL for adult plant resistance to stripe rust in US soft red winter wheat. Theor Appl Genet 123:1401–1411

Hao Y, Wang Y, Chen Z, Bland D, Li S, Brown-Guedira G, Johnson J (2012) A conserved locus conditioning *Soil-borne wheat mosaic virus* resistance on the long arm of chromosome 5D in common wheat. Mol Breed 30:1453–1464

Hao Y, Cambron SE, Chen Z, Wang Y, Bland DE, Buntin GD, Johnson JW (2013) Characterization of new loci for Hessian fly resistance in common wheat. Theor Appl Genet 126:1067–1076

Hao Y, Parks R, Cowger C, Chen Z, Wang Y, Bland D, Murphy JP, Guedira M, Brown-Guedira G, Johnson J (2015) Molecular characterization of a new powdery mildew resistance gene *Pm54* in soft red winter wheat. Theor Appl Genet 128:465–476

Hay R, Kirby E (1991) Convergence and synchrony, a review of the coordination of development in wheat. Aust J Agri Res 42:661–700

He Z, Joshi AK, Zhang W (2013) Climate vulnerabilities and wheat production. In: Pielke RA (ed) Climate vulnerability. Academic Press, Oxford, pp 57–67

He H, Zhu S, Zhao R, Jiang Z, Ji Y, Ji J, Qiu D, Li H, Bie T (2018a) *Pm21*, encoding a typical CC-NBS-LRR protein, confers broad-spectrum resistance to wheat powdery mildew disease. Mol Plant 11:879–882

He Y, Zhang X, Zhang Y, Ahmad D, Wu L, Jiang P, Ma H (2018b) Molecular characterization and expression of *PFT*, an FHB resistance gene at the *Fhb1* QTL in wheat. Phytopathology 108:730–736

Herrera-Foessel SA, Lagudah ES, Huerta-Espino J, Hayden MJ, Bariana HS, Singh D, Singh RP (2011) New slow-rusting leaf rust and stripe rust resistance genes *Lr67* and *Yr46* in wheat are pleiotropic or closely linked. Theor Appl Genet 122:239–249

Herrera-Foessel S, Singh R, Lillemo M, Huerta-Espino J, Bhavani S, Singh S, Lan C, Calvo-Salazar V, Lagudah E (2014) *Lr67/Yr46* confers adult plant resistance to stem rust and powdery mildew in wheat. Theor Appl Genet 1–9

Hiebert C, Thomas J, McCallum B, Humphreys D, DePauw R, Hayden M, Mago R, Schnippenkoetter W, Spielmeyer W (2010) An introgression on wheat chromosome 4DL in RL6077 (Thatcher*6/PI 250413) confers adult plant resistance to stripe rust and leaf rust (*Lr67*). Theor Appl Genet 121:1083–1091

Hossain M (2010) Global warming induced sea level rise on soil, land and crop production loss in Bangladesh. 19th world congress of soil science, soil solutions for a changing world, Brisbane

Hovmøller MS, Walter S, Bayles RA, Hubbard A, Flath K, Sommerfeldt N, Leconte M, Czembor P, Rodriguez-Algaba J, Thach T, Hansen JG, Lassen P, Justesen AF, Ali S, de Vallavieille-Pope C (2016) Replacement of the European wheat yellow rust population by new races from the centre of diversity in the near-Himalayan region. Plant Pathol 65:402–411

Huang L, Brooks SA, Li W, Fellers JP, Trick HN, Gill BS (2003) Map-based cloning of leaf rust resistance gene *Lr21* from the large and polyploid genome of bread wheat. Genetics 164:655–664

Huang S, Spielmeyer W, Lagudah ES, James RA, Platten JD, Dennis ES, Munns R (2006) A sodium transporter *(HKT7)* is a candidate for *Nax1*, a gene for salt tolerance in durum wheat. Plant Physiol 142:1718–1727

Huang S, Spielmeyer W, Lagudah ES, Munns R (2008) Comparative mapping of *HKT* genes in wheat, barley and rice, key determinants of Na+ transport and salt tolerance. J Exp Bot 59:927–937

Huerta-Espino J, Singh R, Germán S, McCallum B, Park R, Chen W, Bhardwaj S, Goyeau H (2011) Global status of wheat leaf rust caused by *Puccinia triticina*. Euphytica 1–18

Hurni S, Brunner S, Buchmann G, Herren G, Jordan T, Krukowski P, Wicker T, Yahiaoui N, Mago R, Keller B (2013) Rye Pm8 and wheat *Pm3* are orthologous genes and show evolutionary conservation of resistance function against powdery mildew. Plant J 76:957–969

Hussain B, Lucas SJ, Ozturk L, Budak H (2017) Mapping QTLs conferring salt tolerance and micronutrient concentrations at seedling stagein wheat. Sci Rep 7:15662

Iehisa JCM, Matsuura T, Mori IC, Takumi S (2014) Identification of quantitative trait locus for abscisic acid responsiveness on chromosome 5A and association with dehydration tolerance in common wheat seedlings. J Plant Physiol 171:25–34

IPCC (2013) Climate change 2013: the physical science basis. In: Stocker TF, Qin D, Plattner GK, Tignor M, Allen SK, Boschung J, Nauels A, Xia Y, Bex V, Midgley PM (eds.) Contribution of working group i to the fifth assessment report of the intergovernmental panel on climate change. Cambridge University Press, Cambridge, United Kingdom and New York, NY, USA, 1535 pp

Islam MT, Croll D, Gladieux P, Soanes DM, Persoons A, Bhattacharjee P, Hossain MS, Gupta DR, Rahman MM, Mahboob MG, Cook N, Salam MU, Surovy MZ, Sancho VB, Maciel JLN, NhaniJúnior A, Castroagudín VL, Reges JTdA, Ceresini PC, Ravel S, Kellner R, Fournier E, Tharreau D, Lebrun M-H, McDonald BA, Stitt T, Swan D, Talbot NJ, Saunders DGO, Win J, Kamoun S (2016) Emergence of wheat blast in Bangladesh was caused by a South American lineage of *Magnaporthe oryzae*. BMC Biol 14:84

Iyer-Pascuzzi AS, Symonova O, Mileyko Y, Hao Y, Belcher H, Harer J, Belcher H, Harer J, Weitz JS, BenfeyP H (2010) Imaging and analysis platform for automatic phenotyping and trait ranking of plant root systems. Plant Physiol 152:1148–1157

Jatayev S, Kurishbayev A, Zotova L, Khasanova G, Serikbay D, Zhubatkanov A, Botayeva M, Zhumalin A, Turbekova A, Soole K, Langridge P, Shavrukov Y (2017) Advantages of amplifluor-like SNP markers over KASP in plant genotyping. BMC Plant Biol 17:254

Ji X, Shiran B, Wan J, Lewis DC, Jenkins CLD, Condon AG, Richards RA, Dolferus R (2010) Importance of pre-anthesis anther sink strength for maintenance of grain number during reproductive stage water stress in wheat. Plant, Cell Environ 33:926–942

Jin H, Wen W, Liu J, Zhai S, Zhang Y, Yan J, Liu Z, Xia X, He Z (2016) Genome-wide QTL mapping for wheat processing quality parameters in a Gaocheng 8901/Zhoumai 16 recombinant inbred line population. Front Plant Sci 7:1032

Jordan KW, Wang S, Lun Y, Gardiner L-J, MacLachlan R, Hucl P, Wiebe K, Wong D, Forrest KL, Sharpe AG, Sidebottom CH, Hall N, Toomajian C, Close T, Dubcovsky J, Akhunova A, Talbert L, Bansal UK, Bariana HS, Hayden MJ, Pozniak C, Jeddeloh JA, Hall A, Akhunov E (2015) A haplotype map of allohexaploid wheat reveals distinct patterns of selection on homoeologous genomes. Geome Biol 16:48

Juroszek P, von Tiedemann A (2013) Climate change and potential future risks through wheat diseases: a review. Eur J Plant Pathol 136:21–33

Kadam S, Singh K, Shukla S, Goel S, Vikram P, Pawar V, Gaikwad K, Khanna-Chopra R, Singh N (2012) Genomic associations for drought tolerance on the short arm of wheat chromosome 4B. Funct Integr Genom 12:447–464

Kamran A, Iqbal M, Spaner D (2014) Flowering time in wheat (*Triticum aestivum* L.): a key factor for global adaptability. Euphytica 197:1–26

Kassa MT, You FM, Hiebert CW, Pozniak CJ, Fobert PR, Sharpe AG, Menzies JG, Humphreys DG, Rezac Harrison N, Fellers JP, McCallum BD, McCartney CA (2017) Highly predictive SNP markers for efficient selection of the wheat leaf rust resistance gene *Lr16*. BMC Plant Biol 17:45

Keller B, Wicker T, Krattinger SG (2018) Advances in wheat and pathogen genomics: implications for disease control. Annu Rev Phytopathol 56:67–87

Kirigwi F, Van Ginkel M, Brown-Guedira G, Gill B, Paulsen GM, Fritz A (2007) Markers associated with a QTL for grain yield in wheat under drought. Mol Breed 20:401–413

Klymiuk V, Yaniv E, Huang L, Raats D, Fatiukha A, Chen S, Feng L, Frenkel Z, Krugman T, Lidzbarsky G, Chang W, Jääskeläinen MJ, Schudoma C, Paulin L, Laine P, Bariana H, Sela H, Saleem K, Sørensen CK, Hovmøller MS, Distelfeld A, Chalhoub B, Dubcovsky J, Korol AB, Schulman AH, Fahima T (2018) Cloning of the wheat *Yr15* resistance gene sheds light on the plant tandem kinase-pseudokinase family. Nat Commun 9:3735

Kolmer JA (2005) Tracking wheat rust on a continental scale. Curr Opin Plant Biol 8:441–449

Kovalchuk N, Chew W, Sornaraj P et al (2016) The homeodomain transcription factor TaHDZipI-2 from wheat regulates frost tolerance, flowering time and spike development in transgenic barley. New Phytol 211:671–687

Krasileva KV, Vasquez-Gross HA, Howell T, Bailey P, Paraiso F, Clissold L, Simmonds J, Ramirez-Gonzalez RH, Wang X, Borrill P, Fosker C, Ayling S, Phillips AL, Uauy C, Dubcovsky J (2017) Uncovering hidden variation in polyploid wheat. Proc Natl Acad Sci USA 114:E913–E921

Krattinger SG, Lagudah ES, Spielmeyer W, Singh RP, Huerta-Espino J, McFadden H, Bossolini E, Selter LL, Keller B (2009) A putative ABC transporter confers durable resistance to multiple fungal pathogens in wheat. Science 323:1360–1363

Krattinger SG, Sucher J, Selter LL, Chauhan H, Zhou B, Tang M, Upadhyaya NM, Mieulet D, Guiderdoni E, Weidenbach D, Schaffrath U, Lagudah ES, Keller B (2016) The wheat durable, multipathogen resistance gene *Lr34* confers partial blast resistance in rice. Plant Biotechnol J 14:1261–1268

Krattinger SG, Kang J, Bräunlich S, Boni R, Chauhan H, Selter LL, Robinson MD, Schmid MW, Wiederhold E, Hensel G, Kumlehn J, Sucher J, Martinoia E, Keller B (2019) Abscisic acid is a substrate of the ABC transporter encoded by the durable wheat disease resistance gene *Lr34*. New Phytol 223:853–866

Krishnan HB, Blanchette JT, Okita TW (1985) Wheat invertases: characterization of cell wall-bound and soluble forms. Plant Physiol 78:241–245

Kulkarni M, Soolanayakanahally R, Ogawa S, Uga Y, Selvaraj MG, Kagale S (2017) Drought response in wheat: key genes and regulatory mechanisms controlling root system architecture and transpiration efficiency. Front Chem 5:106

Kumar S, Sehgal SK, Kumar U, Prasad PVV, Joshi AK, Gill BS (2012) Genomic characterization of drought tolerance-related traits in spring wheat. Euphytica 186:265–276

Lagudah ES (2011) Molecular genetics of race non-specific rust resistance in wheat. Euphytica 179:81–91

Lantican MA, Dubin HJ, Morris ML (2005) Impacts of international wheat breeding research in the developing world, 1988–2002. CIMMYT, Mexico DF

Law CN, Snape JW, Worland AJ (1981) Reduced fertility of wheat associated with Rht3. Plant breeding institute annual report 1980. Plant Breeding Institute, Cambridge, pp 72–73

Lewis CM, Persoons A, Bebber DP, Kigathi RN, Maintz J, Findlay K, Bueno-Sancho V, Corredor-Moreno P, Harrington SA, Kangara N, Berlin A, García R, Germán SE, Hanzalová A, Hodson DP, Hovmøller MS, Huerta-Espino J, Imtiaz M, Mirza JI, Justesen AF, Niks RE, Omrani A, Patpour M, Pretorius ZA, Roohparvar R, Sela H, Singh RP, Steffenson B, Visser B, Fenwick PM, Thomas J, Wulff BBH, Saunders DGO (2018) Potential for re-emergence of wheat stem rust in the United Kingdom. Commun Biol 1:13

Li S, Chang X, Wang C, Jing R (2013) Mapping QTL for heat tolerance at grain filling stage in common wheat. Sci Agric Sin 46:2119–2129

Li Z, Lan C, He Z, Singh RP, Rosewarne GM, Chen X, Xia X (2014) Overview and application of QTL for adult plant resistance to leaf rust and powdery mildew in wheat. Crop Sci 54:1907–1925

Li G, Zhou J, Jia H, Gao Z, Fan M, Luo Y, Zhao P, Xue S, Li N, Yuan Y, Ma S, Kong Z, Jia L, An X, Jiang G, Liu W, Cao W, Zhang R, Fan J, Xu X, Liu Y, Kong Q, Zheng S, Wang Y, Qin B, Cao S, Ding Y, Shi J, Yan H, Wang X, Ran C, Ma Z (2019) Mutation of a histidine-rich calcium-binding-protein gene in wheat confers resistance to *Fusarium* head blight. Nat Genet 51:1106–1112

Lindsay MP, Lagudah ES, Hare RA, Munns R (2004) A locus for sodium exclusion (*Nax1*), a trait for salt tolerance, mapped in durum wheat. Funct Plant Biol 31:1105–1114

Line RF (2002) Stripe rust of wheat and barley in North America: a retrospective historical review. Annu Rev Phytopathol 40:75–118

Liu S, Hall MD, Griffey CA, McKendry AL (2009) Meta-analysis of QTL associated with Fusarium head blight resistance in wheat. Crop Sci 49:1955–1968

Liu S, Sehgal SK, Li J, Lin M, Trick HN, Yu J, Gill BS, Bai G (2013) Cloning and characterization of a critical regulator for preharvest sprouting in wheat. Genetics 195:263–273

Liu W, Frick M, Huel R, Nykiforuk CL, Wang X, Gaudet DA, Eudes F, Conner RL, Kuzyk A, Chen Q, Kang Z, Laroche A (2014) The stripe rust resistance gene *Yr10* encodes an evolutionary-conserved and unique CC-NBS-LRR sequence in wheat. Mol Plant 7:1740–1755

Liu B, Asseng S, Müller C, Ewert F, Elliott J, Lobell DB et al (2016) Similar estimates of temperature impacts on global wheat yield by three independent methods. Nat Clim Change 6:1130

Liu J, Rasheed A, He Z, Imtiaz M, Arif A, Mahmood T, Ghafoor A, Siddiqui SU, Ilyas MK, Wen W, Gao F, Xie C, Xia X (2019) Genome-wide variation patterns between landraces and cultivars uncover divergent selection during modern wheat breeding. Theor Appl Genet

Lobell D, Gourdji S (2012) The influence of climate change on global crop productivity. Plant Physiol 160:1686–1697

Lobell DB, Schlenker W, Costa-Roberts J (2011) Climate trends and global crop production since 1980. Science 333:616–620

Long YM, Chao WS, Ma GJ, Xu SS, Qi LL (2016) An innovative SNP genotyping method adapting to multiple platforms and throughputs. Theor Appl Genet 130:597–607

Ma LQ, Zhou EF, Huo NX, Zhou RH, Wang GY, Jia JZ (2007) Genetic analysis of salt tolerance in a recombinant inbred population of wheat (*Triticum aestivum* L.). Euphytica 153:109–117

Maccaferri M, Sanguineti MC, Corneti S, Ortega JLA, Salem MB, Bort J, DeAmbrogio E, del Moral LFG, Demontis A, El-Ahmed A (2008) Quantitative trait loci for grain yield and adaptation of durum wheat (*Triticum durum* Desf.) across a wide range of water availability. Genetics 178:489–511

Mago R, Zhang P, Vautrin S, Šimková H, Bansal U, Luo M-C, Rouse M, Karaoglu H, Periyannan S, Kolmer J, Jin Y, Ayliffe MA, Bariana H, Park RF, McIntosh R, Doležel J, Bergès H, Spielmeyer W, Lagudah ES, Ellis JG, Dodds PN (2015) The wheat *Sr50* gene reveals rich diversity at a cereal disease resistance locus. Nat Plant 1:15186

Marchal C, Zhang J, Zhang P, Fenwick P, Steuernagel B, Adamski NM, Boyd L, McIntosh R, Wulff BBH, Berry S, Lagudah E, Uauy C (2018) BED-domain-containing immune receptors confer diverse resistance spectra to yellow rust. Nat Plant 4:662–668

Mason RE, Mondal S, Beecher FW, Pacheco A, Jampala B, Ibrahim AMH, Hays DB (2010) QTL associated with heat susceptibility index in wheat (*Triticum aestivum* L.) under short-term reproductive stage heat stress. Euphytica 174:423–436

Masoudi B, Mardi M, Hervan EM, Bihamta MR, Naghavi MR, Nakhoda B, Amini A (2015) QTL mapping of salt tolerance traits with different effects at the seedling stage of bread wheat. Plant Mol Biol Rep 33:1790–1803

Maulana F, Ayalew H, Anderson JD, Kumssa TT, Huang W, Ma X-F (2018) Genome-wide association mapping of seedling heat tolerance in winter wheat. Front Plant Sci 9:1272

McMullen M, Bergstrom G, De Wolf E, Dill-Macky R, Hershman D, Shaner G, Van Sanford D (2012) A unified effort to fight an enemy of wheat and barley: Fusarium head blight. Plant Dis 96:1712–1728

Mickky BM, Aldesuquy HS (2017) Impact of osmotic stress on seedling growth observations, membrane characteristics and antioxidant defense system of different wheat genotypes. Egyp J Basic Appl Sci 4:47–54

Milne RJ, Dibley KE, Schnippenkoetter W, Mascher M, Lui ACW, Wang L, Lo C, Ashton AR, Ryan PR, Lagudah ES (2019) The wheat *Lr67* gene from the sugar transport protein 13 family confers multipathogen resistance in barley. Plant Physiol 179:1285–1297

Milner SG, Jost M, Taketa S, Mazón ER, Himmelbach A, Oppermann M, Weise S, Knüpffer H, Basterrechea M, König P, Schüler D, Sharma R, Pasam RK, Rutten T, Guo G, Xu D, Zhang J, Herren G, Müller T, Krattinger SG, Keller B, Jiang Y, González MY, Zhao Y, Habekuß A, Färber S, Ordon F, Lange M, Börner A, Graner A, Reif JC, Scholz U, Mascher M, Stein N (2019) Genebank genomics highlights the diversity of a global barley collection. Nat Genet 51:319–326

Milus EA, Seyran E, McNew R (2006) Aggressiveness of *Puccinia striiformis* f. sp. *tritici* isolates in the south-central United States. Plant Dis 90:847–852

Milus EA, Kristensen K, Hovmøller MS (2009) Evidence for increased aggressiveness in a recent widespread strain of *Puccinia striiformis* f. sp. *tritici* causing stripe rust of wheat. Phytopathology 99:89–94

Mondal S, Mason RE, Huggins T, Hays DB (2015) QTL on wheat (*Triticum aestivum* L.) chromosomes 1B, 3D and 5A are associated with constitutive production of leaf cuticular wax and may contribute to lower leaf temperatures under heat stress. Euphytica 201:123–130

Moore G (2015) Strategic pre-breeding for wheat improvement. Nat Plants 1:15018

Moore JW, Herrera-Foessel S, Lan C, Schnippenkoetter W, Ayliffe M, Huerta-Espino J, Lillemo M, Viccars L, Milne R, Periyannan S, Kong X, Spielmeyer W, Talbot M, Bariana H, Patrick JW, Dodds P, Singh R, Lagudah E (2015) A recently evolved hexose transporter variant confers resistance to multiple pathogens in wheat. Nat Genet 47:1494–1498

Morran S, Eini O, Pyvovarenko T, Parent B, Singh R, Ismagul A, Eliby S, Shirley N, Langridge P, Lopato S (2011) Improvement of stress tolerance of wheat and barley by modulation of expression of DREB/CBF factors. Plant Biotechnol J 9:230–249

Mujeeb-Kazi A, Kazi AG, Dundas I, Rasheed A, Ogbonnaya F, Kishii M, Bonnett D, Wang RRC, Xu S, Chen PD, Mahmood T, Bux H, Farrakh S (2013) Genetic diversity for wheat improvement as a conduit to food security. Adv Agron 122:179–257

Mujeeb-Kazi A, Munns R, Rasheed A, Ogbonnaya FC, Ali N, Hollington P, Dundas I, Saeed N, Wang R, Rengasamy P, Saddiq MS, Díaz De León JL, Ashraf M, Rajaram S (2019) Breeding strategies for structuring salinity tolerance in wheat. In: Sparks DL (ed) Advances in agronomy. Academic Press, pp 121–187

Munns R, Gilliham M (2015) Salinity tolerance of crops—what is the cost? New Phytol 208:668–673

Munns R, James RA, Läuchli A (2006) Approaches to increasing the salt tolerance of wheat and other cereals. J Exp Bot 57:1025–1043

Munns R, James RA, Gilliham M, Flowers TJ, Colmer TD (2016) Tissue tolerance, an essential but elusive trait for salt-tolerant crops. Funct Plant Biol 43:1103–1113

Murase K, Hirano Y, Sun TP, Hakoshima T (2008) Gibberellin-induced DELLA recognition by the gibberellin receptor GID1. Nature 456:459–463

Mutasa-GottgensE, and Hedden P (2009) Gibberellin as a factor in floral regulatory networks. J Exp Bot 60, 1979–1989

Mwadzingeni L, Shimelis H, Rees DJG, Tsilo TJ (2017) Genome-wide association analysis of agronomic traits in wheat under drought-stressed and non-stressed conditions. PLoS ONE 12: e0171692

Negrao S, Schmockel SM, Tester M (2017) Evaluating physiological responses of plants to salinity stress. Ann Bot 119:1–11

Nilsson HE (1991) Hand-held radiometry and IR-thermography of plant diseases in field plot experiments. Intl J Remote Sens 12:545–557

Ogbonnaya FC, van Ginkel M, Brettell R (2008) Preface: synthetics for wheat improvement. Proceedings of the 1st synthetic wheat symposium, Sept 2006. Aust J Agr Res 59:389–390

Ogbonnaya FC, Abdalla O, Mujeeb-Kazi A, Alvina GK, Xu SS, Gosman N, Lagudah ES, Bonnett D, Sorrells ME, Tsujimoto H (2013) Synthetic hexaploids: harnessing species of the primary gene pool for wheat improvement. Plant Breed Rev 37:35–122

Ogbonnaya FC, Rasheed A, Okechukwu EC, Jighly A, Makdis F, Wuletaw T, Hagras A, Uguru MI, Agbo CU (2017) Genome-wide association study for agronomic and physiological traits in spring wheat evaluated in a range of heat prone environments. Theor Appl Genet 130:1819–1835

Ortiz R, Braun HJ, Crossa J, Crouch JH et al (2008a) Wheat genetic resources enhancement by the International Maize and Wheat Improvement Center (CIMMYT). Genet Resour Crop Evol 55:1095–1140

Ortiz R, Sayre KD, Govaerts B, Gupta R, Subbarao GV, Ban T, Hodson D, Dixon JM, Iván Ortiz-Monasterio J, Reynolds M (2008b) Climate change: can wheat beat the heat? Agri Agri Ecosyst Environ 126:46–58

Oyiga BC, Sharma RC, Baum M, Ogbonnaya FC, Léon J, Ballvora A (2018) Allelic variations and differential expressions detected at quantitative trait loci for salt stress tolerance in wheat. Plant Cell Environ 41:919–935

Paliwal R, Roder MS, Kumar U, Srivastava JP, Joshi AK (2012) QTL mapping of terminal heat tolerance in hexaploid wheat (*Triticum aestivum* L.). Theor Appl Genet 125:561–575

Parent B, Shahinnia F, Maphosa L, Berger B, Rabie H, Chalmers K, Kovalchuk A, Langridge P, Fleury D (2015) Combining field performance with controlled environment plant imaging to identify the genetic control of growth and transpiration underlying yield response to water-deficit stress in wheat. J Exp Bot 66:5481–5492

Passioura JB (2012) Phenotyping for drought tolerance in grain crops: when is it useful to breeders? Funct Plant Biol 39:851–859

Pearce S, Saville R, Vaughan SP, Chandler PM, Wilhelm EP et al (2011) Molecular characterization of *Rht-1* dwarfing genes in hexaploid wheat. Plant Physiol 157:1820–1831

Peleg ZVI, Fahima T, Krugman T, Abbo S, Yakir DAN, Korol AB, Saranga Y (2009) Genomic dissection of drought resistance in durum wheat x wild emmer wheat recombinant inbreed line population. Plant, Cell Environ 32:758–779

Peng JR, Richards DE, Hartley NM, Murphy GP, Devos KM, Flintham JE, Beales J, Fish LJ, Worland AJ, Pelica F, Sudhakar D, Christou P, Snape JW, Gale MD, Harberd NP (1999) 'Green revolution' genes encode mutant gibberellin response modulators. Nature 400:256–261

Peng S, Huang J, Sheehy JE, Laza RC, Visperas RM, Zhong X, Centeno GS, Khush GS, Cassman KG (2004) Rice yields decline with higher night temperature from global warming. Proc Natl Acad Sci USA 101:9971–9975

Periyannan S, Moore J, Ayliffe M, Bansal U, Wang X, Huang L, Deal K, Luo M, Kong X, Bariana H, Mago R, McIntosh R, Dodds P, Dvorak J, Lagudah E (2013) The gene *Sr33*, an ortholog of barley *Mla* genes, encodes resistance to wheat stem rust race Ug99. Science 341:786–788

Pinto RS, Reynolds MP, Mathews KL, McIntyre CL, Olivares-Villegas JJ, Chapman SC (2010) Heat and drought adaptive QTL in a wheat population designed to minimize confounding agronomic effects. Theor Appl Genet 121:1001–1021

Poland JA, Brown PJ, Sorrells ME, Jannink JL (2012) Development of high-density genetic maps for barley and wheat using a novel two-enzyme genotyping-by-sequencing approach. PLoS ONE 7:e32253

Pound MP, Atkinson JA, Townsend AJ, Wilson MH, Griffiths M, Jackson AS, Bulat A, Tzimiropoulos G, Wells DM, Murchie EH, Pridmore TP, French AP (2017) Deep machine learning provides state-of-the-art performance in image-based plant phenotyping. Giga Sci 6:1–10

Pozniak C, Hucl PJ, Stein N, Poland J, Sharpe A, Koh K, Keller B, Ronen G, Muehlbauer G, Distelfeld A, Mayer K, Budak H, Langridge P (2017) Genome assemblies of elite cultivars provides insights into the wheat pan-genome. In: Proceedings of 13th international wheat genetics symposium, 23–28 Apr, Tulln, Austria, 46

Prescott JM, Burnett PA, Saari EE, Ransom JK, Bowman J, De Milliano WAJ, Singh RP, Bekele GT (1986) Wheat diseases and pests: a guide for field identification. CIMMYT, Mexico

Qiu JW, Schürch AC, Yahiaoui N, Dong LL, Fan HJ, Zhang ZJ, Keller B, Ling HQ (2007) Physical mapping and identification of a candidate for the leaf rust resistance gene *Lr1* of wheat. Theor Appl Genet 115:159–168

Quarrie SA, Steed A, Calestani C, Semikhodskii A, Lebreton C, Chinoy C, Steele N, Pljevljakusic D, Waterman E, Weyen J, Schondelmaier J, Habash DZ, Farmer P, Saker L, Clarkson DT, Abugalieva A, Yessimbekova M, Turuspekov Y, Abugalieva S, Tuberosa R, Sanguineti MC, Hollington PA, Aragues R, Royo A, Dodig D (2005) A high-density genetic map of hexaploid wheat (*Triticum aestivum* L.) from the cross Chinese Spring X SQ1 and its use to compare QTLs for grain yield across a range of environments. Theor Appl Genet 110:865–880

Quarrie SA, Pekic Quarrie S, Radosevic R, Rancic D, Kaminska A, Barnes JD, Leverington M, Ceoloni C, Dodig D (2006) Dissecting a wheat QTL for yield present in a range of environments: from the QTL to candidate genes. J Exp Bot 57:2627–2637

Qureshi AS, McCornick PG, Qadir M, Aslam Z (2008) Managing salinity and waterlogging in the Indus Basin of Pakistan. Agric Water Manag 95:1–10

Rahmstorf S, Coumou D (2011) Increase of extreme events in a warming world. Proc Natl Acad Sci USA 108:17905–17909

Rajaram S, Van Ginkel M, Fischer R (1995) CIMMYT's wheat breeding mega-environments (ME). In: Proceedings of the 8th international wheat genetic symposium, vol 2, China Agricultural Scientech Press, Beijing, China, pp 1101–1106

Rasheed A, Xia X (2019) From markers to genome-based breeding in wheat. Theor Appl Genet 132:767–784

Rasheed A, Wen W, Gao FM, Zhai S, Jin H, Liu JD, Guo Q, Zhang YJ, Dreisigacker S, Xia XC, He ZH (2016) Development and validation of KASP assays for functional genes underpinning key economic traits in wheat. Theor Appl Genet 129:1843–1860

Rasheed A, Hao Y, Xia XC, Khan A, Xu Y, Varshney RK, He ZH (2017) Crop breeding chips and genotyping platforms: progress, challenges and perspectives. Mol Plant 10:1047–1064

Rasheed A, Mujeeb-Kazi A, Ogbonnaya FC, He ZH, Rajaram S (2018a) Wheat genetic resources in the post-genomics era: promise and challenges. Ann Bot 121:603–616

Rasheed A, Ogbonnaya FC, Lagudah E, Appels R, He Z (2018b) The goat grass genome's role in wheat improvement. Nat Plants 4:56–58

Rawat N, Pumphrey MO, Liu S, Zhang X, Tiwari VK, Ando K, Trick HN, Bockus WW, Akhunov E, Anderson JA, Gill BS (2016) Wheat *Fhb1* encodes a chimeric lectin with agglutinin domains and a pore-forming toxin-like domain conferring resistance to Fusarium head blight. Nat Genet 48:1576–1580

Ray DK, Mueller ND, West PC, Foley JA (2013) Yield trends are insufficient to double global crop production by 2050. PLoS ONE 8:e66428

Rebetzke GJ, Ellis MH, Bonnett DG, Richards RA (2007a) Molecular mapping of genes for Coleoptile growth in bread wheat (*Triticum aestivum* L.). Theor Appl Genet 114:1173–1183

Rebetzke GJ, Richards RA, Fettell NA, Long M, Condon AG, Forrester RI, Botwright TL (2007b) Genotypic increases in coleoptile length improves stand establishment, vigour and grain yield of deep-sown wheat. Field Crop Res 100:10–23

Rebetzke GJ, Verbyla AP, Verbyla KL, Morell MK, Cavanagh CR (2014) Use of a large multiparent wheat mapping population in genomic dissection of coleoptile and seedling growth. Plant Biotechnol J 12:219–230

Rengasamy P (2010) Soil processes affecting crop production in salt-affected soils. Funct Plant Biol 37:613–620

Riaz A, Athiyannan N, Periyannan S, Afanasenko O, Mitrofanova O, Aitken EAB, Lagudah E, Hickey LT (2016) Mining Vavilov's treasure chest of wheat diversity for adult plant resistance to *Puccinia triticina*. Plant Dis 101:317–323

Riaz A, Hathorn A, Dinglasan E, Ziems L, Richard C, Singh D, Mitrofanova O, Afanasenko O, Aitken E, Godwin I, Hickey L (2017) Into the vault of the Vavilov wheats: old diversity for new alleles. Genet Resour Crop Evol 64:531–544

Rimbert H, Darrier B, Navarro J, Kitt J, Choulet F et al (2018) High throughput SNP discovery and genotyping in hexaploid wheat. PLoS ONE 13:e0186329

Rinaldo A, Gilbert B, Boni R, Krattinger SG, Singh D, Park RF, Lagudah E, Ayliffe M (2017) The *Lr34* adult plant rust resistance gene provides seedling resistance in durum wheat without senescence. Plant Biotechnol J 15:894–905

Risk JM, Selter LL, Chauhan H, Krattinger SG, Kumlehn J, Hensel G, Viccars LA, Richardson TM, Buesing G, Troller A, Lagudah ES, Keller B (2013) The wheat *Lr34* gene provides resistance against multiple fungal pathogens in barley. Plant Biotechnol J 11:847–854

Roitsch T, Gonzalez MC (2004) Function and regulation of plant invertases: sweet sensations. Trend Plant Sci 9:606–613

Rosewarne GM, Herrera-Foessel SA, Singh RP, Huerta-Espino J, Lan CX, He ZH (2013) Quantitative trait loci of stripe rust resistance in wheat. Theor Appl Genet 1–23

Roy SJ, Negrao S, Tester M (2014) Salt resistant crop plants. Curr Opin Biotechnol 26:115–124

Ruan Y-L, Jin Y, Yang Y-J, Li G-J, Boyer JS (2010) Sugar input, metabolism, and signaling mediated by invertase: roles in development, yield potential, and response to drought and heat. Mol Plant 3:942–955

Saintenac C, Jiang D, Akhunov ED (2011) Targeted analysis of nucleotide and copy number variation by exon capture in allotetraploid wheat genome. Genome Biol 12:R88

Saintenac C, Zhang W, Salcedo A, Rouse MN, Trick HN, Akhunov E, Dubcovsky J (2013) Identification of wheat gene *Sr35* that confers resistance to Ug99 stem rust race group. Science 341:783–786

Saleem U, Khaliq I, Mahmood T, Rafique M (2006) Phenotypic and genotypic correlation coefficients between yield and yield components in wheat. J Agric Res 44:1–6

Sánchez-Martín J, Steuernagel B, Ghosh S, Herren G, Hurni S, Adamski N, Vrána J, Kubaláková M, Krattinger SG, Wicker T, Doležel J, Keller B, Wulff BBH (2016) Rapid gene isolation in barley and wheat by mutant chromosome sequencing. Geome Biol 17:221

Sapkota S, Hao Y, Johnson J, Lopez B, Bland D, Chen Z, Sutton S, Buck J, Youmans J, Mergoum M (2019) Genetic mapping of a major gene for leaf rust resistance in soft red winter wheat cultivar AGS 2000. Mol Breed 39:8

Sardouie-Nasab S, Mohammadi-Nejad G, Nakhoda B (2014) Field screening of salinity tolerance in iranian bread wheat lines. Crop Sci 54:1489–1496

Saunders DGO, Pretorius ZA, Hovmøller MS (2019) Tackling the re-emergence of wheat stem rust in Western Europe. Commun Biol 2:51

Scheben A, Batley J, Edwards D (2016) Genotyping-by-sequencing approaches to characterize crop genomes: choosing the right tool for the right application. Plant Biotechnol J 15:149–161

Schnippenkoetter W, Lo C, Liu G, Dibley K, Chan WL, White J, Milne R, Zwart A, Kwong E, Keller B, Godwin I, Krattinger SG, Lagudah E (2017) The wheat *Lr34* multipathogen resistance gene confers resistance to anthracnose and rust in sorghum. Plant Biotechnol J 15:1387–1396

Schwessinger B (2017) Fundamental wheat stripe rust research in the 21st century. New Phytol 213:1625–1631

Sehgal D, Vikram P, Sansaloni CP, Ortiz C, Pierre CS, Payne T, Ellis M, Amri A, Petroli CD, Wenzl P, Singh S (2015) Exploring and mobilizing the gene bank biodiversity for wheat improvement. PLoS ONE 10:e0132112

Semagn K, Babu R, Hearne S, Olsen M (2014) Single nucleotide polymorphism genotyping using Kompetitive Allele Specific PCR (KASP): overview of the technology and its application in crop improvement. Mol Breed 33:1–14

Shah L, Ali A, Yahya M, Zhu Y, Wang S, Si H, Rahman H, Ma C (2018) Integrated control of Fusarium head blight and deoxynivalenol mycotoxin in wheat. Plant Pathol 67:532–548

Shahinnia F, Le Roy J, Laborde B, Sznajder B, Kalambettu P, Mahjourimajd S, Tilbrook J, Fleury D (2016) Genetic association of stomatal traits and yield in wheat grown in low rainfall environments. BMC Plant Biol 16:150

Shamaya NJ, Shavrukov Y, Langridge P, Roy SJ, Tester M (2017) Genetics of Na+ exclusion and salinity tolerance in Afghani durum wheat landraces. BMC Plant Biol 17:209

Shan Q, Wang Y, Li J, Gao C (2014) Genome editing in rice and wheat using the CRISPR/Cas system. Nat Protoc 9:2395

Shukla S, Singh K, Patil RV, Kadam S, Bharti S, Prasad P, Singh NK, Khanna-Chopra R (2015) Genomic regions associated with grain yield under drought stress in wheat (*Triticum aestivum* L.). Euphytica 203:449–467

Singh RP, Hodson DP, Huerta-Espino J, Jin Y, Njau P, Wanyera R, Herrera-Foessel SA, Ward RW (2008) Will stem rust destroy the world's wheat crop? Adv Agron 98:271–309

Singh RP, Hodson DP, Huerta-Espino J, Jin Y, Bhavani S, Njau P, Herrera-Foessel S, Singh PK, Singh S, Govindan V (2011) The emergence of Ug99 races of the stem rust fungus is a threat to world wheat production. Annu Rev Phytopathol 49:465–481

Singh RP, Hodson DP, Jin Y, Lagudah ES, Ayliffe MA, Bhavani S, Rouse MN, Pretorius ZA, Szabo LJ, Huerta-Espino J, Basnet BR, Lan C, Hovmøller MS (2015) Emergence and spread of new races of wheat stem rust fungus: continued threat to food security and prospects of genetic control. Phytopathology 105:872–884

Singh RP, Singh PK, Rutkoski J, Hodson DP, He X, Jørgensen LN, Hovmøller MS, Huerta-Espino J (2016) Disease impact on wheat yield potential and prospects of genetic control. Annu Rev Phytopathol 54:303–322

Sip V, Chrpova J, Zofajova A, Milec Z, Mihalik D, Pankova K, Snape JW (2011) Evidence of selective changes in winter wheat in middle-European environments reflected by allelic diversity at loci affecting plant height and photoperiodic response. J AgriSci 149:313–326

Sorensen CK, Howmoller MS, Leconte M, Dedryverperson F, Vallavieille CPD (2014) New races of *Puccinia striiformis* found in europe reveal race specificity of long-term effective adult plant resistance in wheat. Phytopathology 104:1042–1051

Srichumpa P, Brunner S, Keller B, Yahiaoui N (2005) Allelic series of four powdery mildew resistance genes at the *Pm3* locus in hexaploid bread wheat. Plant Physiol 139:885–895

Appels R, Eversole K, Feuillet C, Keller B, Rogers J, Stein N, et al (2018) Shifting the limits in wheat research and breeding using a fully annotated reference genome. Science 361:eaar7191

Steuernagel B, Periyannan SK, Hernández-Pinzón I, Witek K, Rouse MN, Yu G, Hatta A, Ayliffe M, Bariana H, Jones JDG, Lagudah ES, Wulff BBH (2016) Rapid cloning of disease-resistance genes in plants using mutagenesis and sequence capture. Nat Biotechnol 34:652

Stratonovitch P, Semenov M (2015) Heat tolerance around flowering in wheat identified as a key trait for increased yield potential in Europe under climate change. J Exp Bot 66:3599–3609

Su Z, Bernardo A, Tian B, Chen H, Wang S, Ma H, Cai S, Liu D, Zhang D, Li T, Trick H, St. Amand P, Yu J, Zhang Z, Bai G (2019) A deletion mutation in *TaHRC* confers *Fhb1* resistance to Fusarium head blight in wheat. Nat Genet 51:1099–1105

Sucher J, Boni R, Yang P, Rogowsky P, Büchner H, Kastner C, Kumlehn J, Krattinger SG, Keller B (2017) The durable wheat disease resistance gene *Lr34* confers common rust and northern corn leaf blight resistance in maize. Plant Biotechnol J 15:489–496

Sugiura R, Noguchi N, Ishii K (2007) Correction of low-altitude thermal images applied to estimating soil water status. BiosysEng 96:301–313

Sukumaran S, dreisigacker S, Lopes MS, Chavez P, Reynolds MP (2015) Genome-wide association study for grain yield and related traits in an elite spring wheat population grown in temperate irrigated environments. Theor Appl Genet128:353–363

Sukumaran S, Reynolds MP, Sansaloni C (2018) Genome-wide association analyses identify QTL hotspots for yield and component traits in durum wheat grown under yield potential, drought, and heat stress environments. Front Plant Sci 9:81

Sun Q, Quick J (1991) Chromosomal locations of genes for heat tolerance in tetraploid wheat. Cereal Res Commun 431–437

Tadesse W, Ogbonnaya FC, Jighly A, Sanchez-Garcia M, Sohail Q, Rajaram S, Baum M (2015) Genome-wide association mapping of yield and grain quality traits in winter wheat genotypes. Plos ONE 10:e0141339. Tahmasebi S, Heidari B, Pakniyat H, McIntyre CL (2016) Mapping QTLs associated with agronomic and physiological traits under terminal drought and heat stress conditions in wheat (*Triticum aestivum* L.). Genome 60:26–45

Talukder S, Babar M, Vijayalakshmi K, Poland J, Prasad P, Bowden R, Fritz A (2014) Mapping QTL for the traits associated with heat tolerance in wheat (*Triticum aestivum* L.). BMC Genet 15:97

Tebaldi C, Hayhoe K, Arblaster J, Meehl G (2006) Going to the extremes: an intercomparison of model-simulated historical and future changes in extreme events. Clim Change 79:233–234

Tester M, Langridge P (2010) Breeding technologies to increase crop production in a changing world. Science 327:818–822

Thind AK, Wicker T, Šimková H, Fossati D, Moullet O, Brabant C, Vrána J, Doležel J, Krattinger SG (2017) Rapid cloning of genes in hexaploid wheat using cultivar-specific long-range chromosome assembly. Nat Biotechnol 35:793

Thomelin P, Bonneau J, Taylor J, Choulet F, Sourdille P, Langridge P (2016) Positional cloning of a QTL, qDHY. 3BL, on chromosome 3BL for drought and heat tolerance in bread wheat. In: Proceedings of the plant and animal genome conference (PAG XXIV), San Diego, CA, USA, p P0850

Touzy G, Rincent R, Bogard M, Lafarge S, Dubreuil P, Mini A, Deswarte J-C, Beauchêne K, Le Gouis J, Praud S (2019) Using environmental clustering to identify specific drought tolerance QTLs in bread wheat (*Triticum aestivum* L.). Theor Appl Genet 132:2859–2880

Trnka M, Rötter RP, Ruiz-Ramos M, Kersebaum KC, Olesen JE, Žalud Z, Semenov MA (2014) Adverse weather conditions for European wheat production will become more frequent with climate change. Nat Clim Change 4:637

Turki N, Shehzad T, Harrabi M, Okuno K (2015) Detection of QTLs associated with salinity tolerance in durum wheat based on association analysis. Euphytica 201:29–41

Urashima AS, Igarashi S, Kato H (1994) Host range, mating type, and fertility of Pyricularia grisea from wheat in Brazil. Plant Dis 77:1211–1216

Uto K, Seki H, Saito G, Kosugi Y (2013) Characterization of rice paddies by a UAV-mounted miniature hyperspectral sensor system. IEEE J Sel Top Appl Earth Obs Remote Sens 6:851–860

Valluru R, Reynolds MP, Davies WJ, Sukumaran S (2017) Phenotypic and genome-wide association analysis of spike ethylene in diverse wheat genotypes under heat stress. New Phytol 214:271–283

Vashev B, Gaiser T, Ghawana T, de Vries A, Stahr K (2010) Biosafor project deliverable 9: cropping potentials for saline areas in India. University of Hohenheim, Hohenheim, Pakistan and Bangladesh

Vijayalakshmi K, Fritz AK, Paulsen GM, Bai G, Pandravada S, Gill BS (2010) Modeling and mapping QTL for senescence-related traits in winter wheat under high temperature. Mol Breed 26:163–175

Vikram P, Franco J, Burgueño-Ferreira J, Li H, Sehgal D, Saint Pierre C, Ortiz C, Sneller C, Tattaris M, Guzman C, Sansaloni CP, Fuentes-Davila G, Reynolds M, Sonders K, Singh P, Payne T, Wenzl P, Sharma A, Bains NS, Singh GP, Crossa J, Singh S (2016) Unlocking the genetic diversity of Creole wheats. Sci Rep 6:23092

Wang SC, Wong DB, Forrest K, Allen A, Chao SM et al (2014a) Characterization of polyploid wheat genomic diversity using a high-density 90 000 single nucleotide polymorphism array. Plant Biotechnol J 12:787–796

Wang Y, Cheng X, Shan Q, Zhang Y, Liu J, Gao C, Qiu J-L (2014b) Simultaneous editing of three homoeoalleles in hexaploid bread wheat confers heritable resistance to powdery mildew. Nat Biotechnol 32:947

Wang W, Simmonds J, Pan Q, Davidson D, He F, Battal A, Akhunova A, Trick HN, Uauy C, Akhunov E (2018) Gene editing and mutagenesis reveal inter-cultivar differences and additivity in the contribution of *TaGW2* homoeologues to grain size and weight in wheat. Theor Appl Genet 131:2463–2475

Wang W, Pan Q, Tian B, He F, Chen Y, Bai G, Akhunova A, Trick HN, Akhunov E (2019) Gene editing of the wheat homologs of TONNEAU1-recruiting motif encoding gene affects grain shape and weight in wheat. Plant J. https://doi.org/10.1111/tpj.14440

Wellings CR (2007) *Puccinia striiformis* in Australia: a review of the incursion, evolution, and adaptation of stripe rust in the period 1979–2006. Aust J Agri Res 58:567–575

Winfield MO, Allen AM, Burridge AJ, Barker GL, Benbow HR, Wilkinson PA, Coghill J, Waterfall C, Davassi A, Scopes G, Pirani A, Webster T, Brew F, Bloor C, King J, West C, Griffiths S, King I, Bentley AR, Edwards KJ (2016) High-density SNP genotyping array for hexaploid wheat and its secondary and tertiary gene pool. Plant Biotechnol J 14:1195–1206

Worland AJ (1986) Gibberellic acid insensitive dwarfing genes in Southern European wheats. Euphytica 35:857–866

Worland AJ, Korzun V, Roder MS, Ganal MW, Law CN (1998) Genetic analysis of the dwarfing gene Rht8 in wheat. Part II. The distribution and adaptive significance of allelic variants at the Rht8 locus of wheat as revealed by microsatellite screening. Theor Appl Genet 96:1110–1120

Wu J, Zeng Q, Wang Q, Liu S, Yu S, Mu J, Huang S, Sela H, Distelfeld A, Huang L, Han D, Kang Z (2018) SNP-based pool genotyping and haplotype analysis accelerate fine-mapping of the wheat genomic region containing stripe rust resistance gene *Yr26*. Theor Appl Genet 131:1481–1496

Würschum T, Liu G, Boeven PHG, Longin CFH, Mirdita V, Kazman E, Zhao Y, Reif JC (2018) Exploiting the Rht portfolio for hybrid wheat breeding. Theor Appl Genet 131:1433–1442

Xing L, Hu P, Liu J, Witek K, Zhou S, Xu J, Zhou W, Gao L, Huang Z, Zhang R, Wang X, Chen P, Wang H, Jones JDG, Karafiátová M, Vrána J, Bartoš J, Doležel J, Tian Y, Wu Y, Cao A (2018) *Pm21* from *Haynaldia villosa* encodes a CC-NBS-LRR protein conferring powdery mildew resistance in wheat. Mol Plant 11:874–878

Xu R, Sun Q, Zhang S (1996) Chromosomal location of genes for heat tolerance as measured by membrane thermostability of common wheat cv. Hope. Hereditas 18:1–3

Xu YF, An DG, Liu DC, Zhang AM, Xu HX, Li B (2012) Mapping QTLs with epistatic effects and QTL x treatment interactions for salt tolerance at seedling stage of wheat. Euphytica 186:233–245

Xu Y, Li S, Li L, Zhang X, Xu H, An D (2013) Mapping QTL s for salt tolerance with additive, epistatic and QTL × treatment interaction effects at seedling stage in wheat. Plant Breed 132:276–283

Xu Y-F, Li S-S, Li L-H, Ma F-F, Fu X-Y, Shi Z-L, Xu H-X, Ma P-T, An D-G (2017) QTL mapping for yield and photosynthetic related traits under different water regimes in wheat. Mol Breed 37:34

Yahiaoui N, Srichumpa P, Dudler R, Keller B (2004) Genome analysis at different ploidy levels allows cloning of the powdery mildew resistance gene *Pm3b* from hexaploid wheat. Plant J 37:528–538

Yahiaoui N, Brunner S, Keller B (2006) Rapid generation of new powdery mildew resistance genes after wheat domestication. Plant J 47:85–98

Yamaguchi T, Blumwald E (2005) Developing salt-tolerant crop plants: challenges and opportunities. Trend Plant Sci 10:615–620

Yang D-L, Jing R-L, Chang X-P, Li W (2007) Identification of quantitative trait loci and environmental interactions for accumulation and remobilization of water-soluble carbohydrates in wheat (*Triticum aestivum* L.) stems. Genetics 176:571

Yang G, Liu J, Zhao C, Li Z, Huang Y, Yu H (2017) Unmanned aerial vehicle remote sensing for field-based crop phenotyping: current status and perspectives. Front Plant Sci 8:1111

Yang Y, Luang S, Harris J, Riboni M, Li Y, Bazanova N, Hrmova M, Haefele S, Kovalchuk N, Lopato S (2018) Overexpression of the class I homeodomain transcription factor *TaHDZipI-5* increases drought and frost tolerance in transgenic wheat. Plant Biotechnol J 16:1227–1240

Yao F, Zhang X, Ye X, Li J, Long L, Yu C, Li J, Wang Y, Wu Y, Wang J, Jiang Q, Li W, Ma J, Wei Y, Zheng Y, Chen G (2019) Characterization of molecular diversity and genome-wide association study of stripe rust resistance at the adult plant stage in Northern Chinese wheat landraces. BMC Genet 20:38

Zarco-Tejada PJ, Guillén-Climent ML, Hernández-Clemente R Catalina A, González MR, Martín P (2013) Estimating leaf carotenoid content in vineyards using high resolution hyperspectral imagery acquired from an unmanned aerial vehicle (UAV). AgriForMeteorol171–172:281–294

Zhang X, Shen X, Hao Y, Cai J, Ohm H, Kong L (2011) A genetic map of *Lophopyrum ponticum* chromosome 7E, harboring resistance genes to Fusarium head blight and leaf rust. Theor Appl Genet 122:263–270

Zhang Y, Zhang J, Huang L, Gao A, Zhang J, Yang X, Liu W, Li X, Li L (2015) A high-density genetic map for P genome of *Agropyron Gaertn*. Based on specific-locus amplified fragment sequencing (SLAF-seq). Planta 242:1335–1347

Zhang W, Chen S, Abate Z, Nirmala J, Rouse MN, Dubcovsky J (2017a) Identification and characterization of *Sr13*, a tetraploid wheat gene that confers resistance to the Ug99 stem rust race group. Proc Natl Acad Sci USA 114:E9483–E9492

Zhang Y, Bai Y, Wu G, Zou S, Chen Y, Gao C, Tang D (2017b) Simultaneous modification of three homoeologs of *TaEDR1* by genome editing enhances powdery mildew resistance in wheat. Plant J 91:714–724

Zhao J, Bodner G, Rewald B (2016) Phenotyping: using machine learning for improved pairwise genotype classification based on root traits. Front Plant Sci 7:1864

Zhong D, Novais J, Grift TE, Bohn M, Han J (2009) Maize root complexity analysis using a support vector machine method. Comput Electron Agri 69:46–50

Zhou S, Zhang J, Che Y, Liu W, Lu Y, Yang X, Li X, Jia J, Liu X, Li L (2018) Construction of *Agropyron Gaertn*. genetic linkage maps using a wheat 660 K SNP array reveals a homoeologous relationship with the wheat genome. Plant Biotechnol J 16:818–827

Zhu Z, Xu D, Cheng S, Gao C, Xia X, Hao Y, He Z (2018) Characterization of *Fusarium* head blight resistance gene *Fhb1* and its putative ancestor in chinese wheat germplasm. Acta Agron Sin 44:473–482

Zhu Z, Hao Y, Mergoum M, Bai G, Humphreys G, Cloutier S, Xia X, He Z (2019) Breeding wheat for resistance to Fusarium head blight in the Global North: China, USA, and Canada. Crop J. https://doi.org/10.1016/jcj201906003

Zikhali M, Wingen LU, Griffiths S (2016) Delimitation of the Earliness *per se* D1 (*Eps-D1*) flowering gene to a subtelomeric chromosomal deletion in bread wheat (*Triticum aestivum*). J Exp Bot 67:287–299

Zou S, Wang H, Li Y, Kong Z, Tang D (2018) The NB-LRR gene *Pm60* confers powdery mildew resistance in wheat. New Phytol 218:298–309

Chapter 3
Increasing Genetic Gains in Maize in Stress-Prone Environments of the Tropics

B. M. Prasanna, Sudha K. Nair, Raman Babu, Manje Gowda, Xuecai Zhang, Yunbi Xu, Mike Olsen, Vijay Chaikam, Jill E. Cairns, Mainassara Zaman-Allah, Yoseph Beyene, Amsal Tarekegne and Cosmos Magorokosho

Abstract Maize (*Zea mays* L.) provides food security, income, and livelihoods to millions of smallholders in the developing world. However, maize yields in the tropical rainfed environments, especially in sub-Saharan Africa and South Asia, are affected by an array of abiotic and biotic stresses, thereby limiting national maize yields to 1–3 tons per hectare (t/ha), while the global average is around 5 t/ha. Therefore, developing and deploying high-yielding, climate-resilient maize (with tolerance to drought, heat, waterlogging, and biotic stresses), coupled with climate-smart agricultural practices, are critical for improving maize yields, and reducing the high risk and vulnerability of the maize-growing smallholder farmers in the tropics to the climate variability. CIMMYT (International Maize and Wheat Improvement Center) has been intensively engaged since 1970s in breeding elite tropical maize germplasm with tolerance to important abiotic stresses, especially drought, using managed-stress screening and selection for key secondary traits.

B. M. Prasanna (✉) · M. Gowda · M. Olsen · V. Chaikam · Y. Beyene
International Maize and Wheat Improvement Center (CIMMYT), Nairobi, Kenya
e-mail: b.m.prasanna@cgiar.org

S. K. Nair
CIMMYT, ICRISAT Campus, Patancheru Greater Hyderabad, Telangana, India

R. Babu
Corteva Multi-Crop Research Center, Hyderabad, Telangana, India

X. Zhang
CIMMYT, El Batan, Texcoco, Mexico D.F., Mexico

Y. Xu
CIMMYT-China, Institute of Crop Science, Chinese Academy of Agricultural Sciences, Beijing, China

J. E. Cairns · M. Zaman-Allah · A. Tarekegne · C. Magorokosho
CIMMYT, P.O. Box MP163, Harare, Zimbabwe

A. Tarekegne
Zamseed, Lusaka, Zambia

C. Kole (ed.), *Genomic Designing of Climate-Smart Cereal Crops*,
https://doi.org/10.1007/978-3-319-93381-8_3

This formed the base for successful development, testing, and deployment of CIMMYT-derived abiotic stress-tolerant maize varieties in sub-Saharan Africa, Latin America, and Asia, in partnership with an array of public and private sector institutions. Increasing genetic gains and breeding efficiency, especially in developing elite multiple stress-tolerant maize germplasm, requires: (a) carefully undertaken field-based phenotyping at several relevant sites as well as under technically demanding managed-stress screens; (b) better understanding of the genetic architecture of traits; and (c) utilization of modern breeding tools/strategies, including high-throughput and reasonably precise field-based phenotyping, doubled haploid (DH) technology, molecular marker-assisted breeding, genomic selection, transgenics, breeding information management, and decision-support system. At the same time, it has become imperative to defend the genetic gains from devastating transboundary diseases (e.g., maize lethal necrosis or MLN) and insect-pests (e.g., Fall armyworm). Multi-institutional efforts, especially public–private alliances, are key to ensure that the improved varieties effectively reach the climate change-vulnerable farming communities, and to develop and deploy technologies that can protect the maize crops of the smallholders in the tropics from the emerging biotic threats.

Keywords Maize · Climate resilience · Genetic gains · Phenotyping · Molecular markers · Seed systems · Tropics

3.1 Introduction

Maize is cultivated on more than 180 million hectares (M ha) globally, contributing ~50% (1,170 million metric tons or MMT) to the global grain production. About 60–70% of the cultivated area under maize is located in the developing world, with a predominant proportion in the low and lower middle income countries (FAOSTAT 2018). The crop provides over 20% of total calories in human diets in 21 countries, and over 30% in 12 countries that are home to a total of more than 310 million people (Shiferaw et al. 2011).

Maize has been adapted in diverse agro-ecologies worldwide. Temperate maize is grown in cooler climates beyond 34°N and 34°S, while tropical maize is grown in warmer environments located between 30°N and 30°S latitudes. An intermediate type, subtropical maize, is grown between the 30° and 34° latitudes. The tropical maize can be further classified into lowland (sea level to $\leq$ 1000 m.a.s.l.), mid-altitude (1000–1600 m.a.s.l.) and highland ($\geq$ 1600 m.a.s.l.). Tropical maize is grown in over 60 countries, occupying about 60% of the area harvested and representing 40% of the world production.

Abiotic stresses, especially drought, heat, waterlogging, poor soil fertility, acidity, and combinations of various abiotic stresses (e.g., drought and heat) occur quite frequently in the rainfed maize-based cropping systems in the tropics, with substantial negative impact on yields, especially in sub-Saharan Africa (SSA) and

South Asia. In addition, tropical maize is highly exposed to an array of potentially devastating diseases and insect-pests. Therefore, development and deployment of improved maize varieties with tolerance to multiple abiotic stresses, coupled with disease and insect-pest resistance, is critical for building resilience and adaptive capacity of the farming communities in the tropics to climatic variability (Shiferaw et al. 2014). Tesfaye et al. (2017, 2018) highlighted the potential benefits of incorporating drought, heat, and combined drought and heat tolerance into maize varieties in the climate-vulnerable tropical environments. The magnitude of the simulated benefits from drought tolerance, heat tolerance, and combined drought and heat tolerance and potential acceptability of the varieties by farmers could vary across sites and climate scenarios, indicating the need for proper targeting of varieties where they fit best and benefit the most.

While enhanced adoption of multiple stress-tolerant maize varieties is undoubtedly important, one must not ignore the need for complementary adoption of other climate-smart, sustainable intensification practices. Drought-tolerant maize can only tolerate (not resist) spells of drought, especially during the most sensitive crop stage at flowering, but such varieties cannot effectively tolerate prolonged drought spanning the vegetative, reproductive, and grain-filling stages. Therefore, building climate resilience in crops as well as farming communities in the stress-prone tropics requires a multi-disciplinary and multi-institutional strategy. This includes more extensive awareness creation and adoption of climate-smart agronomic management practices, strengthening of local capacities, and focusing on sustainability.

In this article, we focus on some key approaches to increase genetic gains in maize, especially in the stress-prone environments of the tropics, based on CIMMYT's extensive experience of breeding improved maize varieties adapted to SSA, Latin America, and Asia.

3.2 Breeding for Climate Resilience in Tropical Maize

The food security and livelihoods of several million smallholder farmers in the tropics are based on a great extent on effective development and deployment of high-yielding and multiple stress-tolerant maize varieties. Understanding the environmental conditions that contribute to major abiotic and biotic stresses, and effectively unraveling genetic variability for such stresses in appropriate environments are two critical factors for the success of breeding for climate resilience. CIMMYT's work since 1970s on characterization of drought-prone environments in the tropics, identification of suitable secondary traits and trait donors in breeding for drought tolerance, optimizing procedures for undertaking managed-drought stress phenotyping trials, developing drought-tolerant maize germplasm through extensive multi-location and multi-year experiments, and disseminating the stress-tolerant cultivars in partnerships with various public and private organizations, holds considerable significance for improving the livelihoods of the resource-poor farmers in the developing world (Cairns and Prasanna 2018). The success in developing and deploying elite, farmer-preferred, multiple stress-tolerant maize varieties in the tropics has been built through dedicated efforts over the last six decades (Fig. 3.1).

CIMMYT has used two major approaches for developing sources of abiotic stress tolerance that have been widely used in maize breeding programs in SSA, Asia, and Latin America. The first was the constitution of drought-tolerant populations for undertaking recurrent selections and derivation of elite inbred lines. The DTP-Y, DTP-W, and La Posta Sequia are examples of such populations. The second approach was full-sib recurrent selection under managed-drought stress within elite populations to increase the frequency of drought tolerance alleles in germplasm already adapted to the lowland tropics (e.g., Edmeades et al. 1999). Both approaches have generated lines that have become important sources of drought and heat tolerance, such as DTPYC9-F46-1-2-1-2 and La Posta Sequía C7F64-2-6-2-2 (Cairns et al. 2012). Thus, population formation and improvement have resulted in an increase in the frequency of drought-adaptive alleles and the identification of superior sources of drought tolerance (Edmeades et al. 2017).

Besides constitution of appropriate maize populations for implementing recurrent section for improving drought stress tolerance, CIMMYT also identified and used suitable secondary traits (e.g., anthesis-silking interval or ASI), and developed managed-drought stress screening protocol and a breeding scheme that is designed to develop products that perform well under both optimal and stressed environments (Cairns and Prasanna 2018). CIMMYT's maize product advancement process typically includes not only regional on-station trials of promising

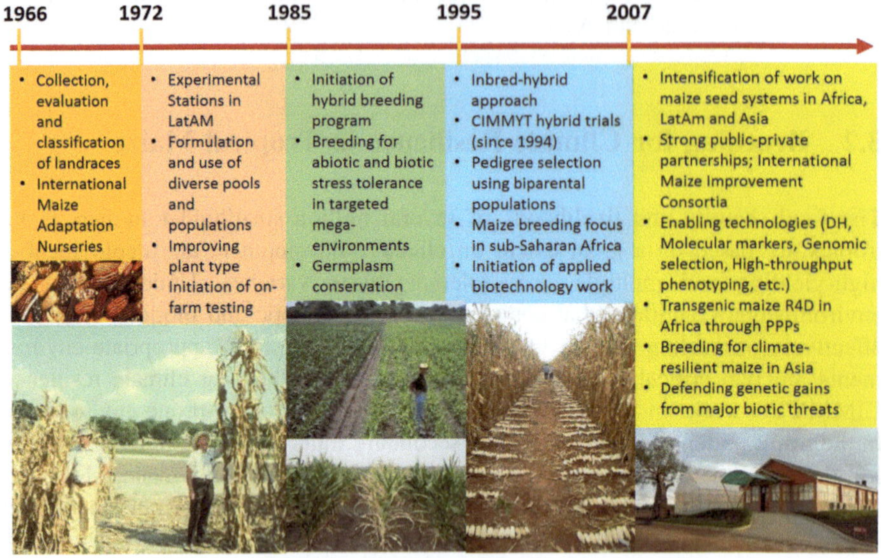

Fig. 3.1 Evolution of maize breeding at CIMMYT

pre-commercial hybrids coming out of the breeding pipeline vis-à-vis internal genetic gain checks and commercial checks but also extensive regional on-farm varietal trials to ascertain the performance of the promising pre-commercial hybrids under farmer-managed conditions. This also provides the opportunity for the socioeconomic team to assess farmers' own product as well as their trait preferences. The best entries coming out of this rigorous process are then allocated to public/private sector partners for varietal registration, scale-up, and delivery in the target geographies.

An array of projects are being implemented by CIMMYT and partners for developing and deploying climate-resilient maize varieties adapted to SSA, Asia, and Latin America Under the Drought-Tolerant Maize for Africa (DTMA) project (now continuing as Stress-Tolerant Maize for Africa or STMA Project), jointly implemented by CIMMYT and IITA, in close collaboration with NARS and private sector institutions in 13 countries in Africa, more than 300 elite drought-tolerant (DT) and disease-resistant maize varieties have been released since 2007, with more than 60% of these being hybrids. These varieties perform as well as or better than the commercial varieties currently available on the market under optimum (no water deficit stress) conditions and out-perform the best commercial checks by at least 25–30% under drought stress and low-input conditions. The DT maize varieties developed by CIMMYT/IITA typically have a combination of traits that confer tolerance to drought conditions; these include reduced barrenness under drought stress, short anthesis-silking interval, reduced leaf senescence (as compared to susceptible germplasm) and longer leaf area duration during grain filling (Edmeades 2008; Bruce et al. 2002). In 2018, more than 100 seed companies in SSA produced an estimated 75,000 tons of certified seed of MAIZE-derived improved DT maize varieties. In South Asia, through the USAID-funded Heat Tolerant Maize for Asia (HTMA) project, a large heat-stress phenotyping network, comprising 23 sites in the four countries (India, Nepal, Bangladesh, and Pakistan), has been established. Several drought-tolerant and heat-tolerant CIMMYT-derived elite maize varieties have been released during 2016–2018 by public and private sector partners in South Asia, and several more are in pipeline. It is possible to further increase genetic gains in maize grain yield in stress-prone environments of the tropics through a clear product development and deployment strategy (Cairns and Prasanna 2018).

Targeted deployment of improved climate-resilient varieties by GIS-based prediction of areas of climate vulnerability, improving varietal turnover (with newer and better genetics), appropriate agronomic management practices for realizing the genetic potential of improved varieties, and creating better linkages for the smallholder maize farmers to output markets are critical for strengthening maize value chains in the developing world. Delivering low-cost improved maize seed to smallholder farmers with limited purchasing capacity and market access requires stronger public–private partnerships, and enhanced support to the committed local seed companies, especially in terms of information on access to new products, adequate and reliable supplies of early-generation (breeder and foundation) seed, and training on hybrid seed production, quality assurance/quality control (QA/QC), seed business management, market segmentation, and territory planning.

Appropriate government policies and adoption of progressive seed laws and regulations are critical for improving smallholder farmers' access to improved climate-resilient seed, and for overcoming key bottlenecks affecting the seed value chains, particularly in the areas of policy, credit availability, seed production, germplasm, and marketing (Cairns and Prasanna 2018).

3.3 Increasing Genetic Gains Through Maize Breeding in the Tropics

"Genetic gain" in simple terms refers to the gain in population mean achieved through each breeding cycle. Breeding being a cyclical process of crossing, evaluating, selecting, and crossing again, the efficiency (both in terms of time and cost) with which breeding programs can make considerable shift in population mean from one cycle to the other determines the extent of genetic gains. Regardless of the trait of interest, or the breeding methods employed, genetic gain represented by "ΔG" serves as a simple universal expression for expected genetic improvement (Falconer and Mackay 1996). The genetic gain equation that represents the factors leading to genetic gain (rather than a formula to calculate the same) has come to be known as the "breeder's equation". The genetic gain equation can be represented as $\Delta G = i.r.\sigma A/t$, where 'i' represents selection intensity, 'r' represents selection accuracy, 'σA' represents genetic variance and 't' represents cycling time ((Falconer and Mackay 1996).

Under optimal conditions, genetic gain for maize grain yield was estimated at 94.7 kg ha^{-1} yr^{-1} in China over a period of 30 years (Ci et al. 2011), 132 kg ha^{-1} yr^{-1} in Argentina over a period of 32 years (Luque et al. 2006), 80 kg ha^{-1} yr^{-1} in Canada over a period of 100 years (Bruulsema et al. 2000), and 65 to 75 kg ha^{-1} yr^{-1} in the United States over a period of 70 years (Duvick 2005). Badu-Apraku et al. (2013, 2016) estimated genetic gain by era studies for a period of 23 years in improved open-pollinated varieties (OPVs) in West and Central Africa, and estimated a genetic gain of 40 kg ha^{-1} yr^{-1} under optimal conditions and 13.5 kg ha^{-1} yr^{-1} under random stress. Over a period of 10 years, an era study conducted by CIMMYT in eastern and southern Africa (ESA) estimated the genetic gains from the CIMMYT hybrid maize breeding program at 109.4, 32.5, 22.7, 20.9, and 141.3 kg ha^{-1} yr^{-1} under optimal, managed drought, random drought, low N, and maize streak virus (MSV) infection, respectively (Masuka et al. 2017a). Within the CIMMYT OPV breeding program for ESA genetic gain in the early maturity group under optimal, random drought, low N and MSV was 109.9, 29.2, 84.8, and 192.9 kg ha^{-1} yr^{-1} (Masuka et al. 2017b). In the intermediate-late maturity group genetic gain under optimal, random drought, low N and MSV was 79.1, 42.3, 53.0, and 108.7 kg ha^{-1} yr^{-1}. There are few reports evaluating genetic gain for key traits achieved by maize breeding programs in the tropics. Beyond the era studies, it is imperative to constantly monitor the gains in the breeding programs using internal

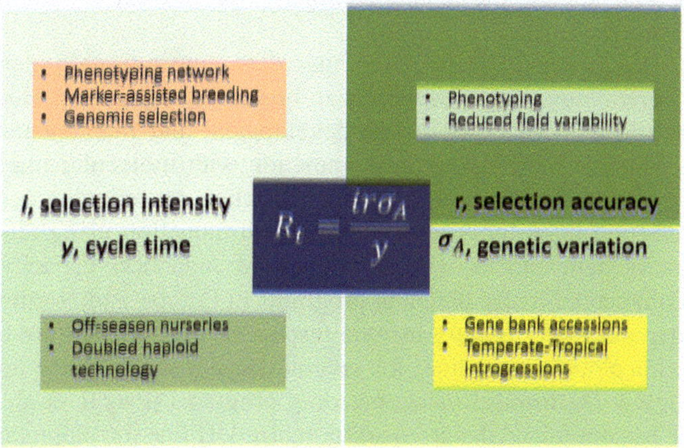

Fig. 3.2 Some of the key components for increasing genetic gains for maize yields in the stress-prone environments of the tropics

genetic checks and/or commercial checks, and take appropriate measures to maintain the gains in the face of emerging threats and constraints.

Increasing genetic gain in maize grain yield in the stress-prone environments of the tropics is indeed a challenge, but possible nevertheless with a clear product development and deployment strategy (Cairns and Prasanna 2018). Conventional maize breeding, although successful, is a relatively slow and resource-intensive process. The increasing demand for high-yielding, multiple stress-tolerant and nutritionally enriched maize varieties warrants accelerated breeding that makes use of modern tools and technologies, including doubled haploids, molecular markers, high-throughput and reliable phenotyping, off-season nurseries and decision-support tools. The "breeder's equation" provides the focus around which new technologies can contribute to increased genetic gains (Fig. 3.2).

3.4 Enabling Tools and Technologies for Increasing Genetic Gains

3.4.1 Doubled Haploid (DH) Technology

One of the conceptually simplest ways to increase genetic gain is to reduce the breeding cycle time—if selection intensity, accuracy, and variability remain constant, halving cycle time will double the genetic gain (Xu et al. 2017a). Breeding cycle times are typically 10 years or more in the tropics, compared to less than five in temperate regions (Challinor et al. 2016). Faster product cycle times are not only important for adaptation to the changing climates, but also for countering emerging pests and diseases.

Derivation and use of DH lines, compared to conventionally-derived inbred lines, offers several advantages to the maize breeding programs, including simplified logistics and reduced costs in line development and maintenance (Prasanna et al. 2012). The use of DH lines in conjunction with molecular markers significantly improves genetic gains and breeding efficiency, by reducing cycle time and enhancing selection intensity. The in vivo haploid induction using temperate haploid inducers (genetic stocks with high haploid induction capacity) has been adapted by commercial maize breeding programs in Europe and North America for over a decade, and more recently in Asia (especially China), but the lack of tropically adapted haploid inducer lines for several decades impeded the application of DH technology in the tropical maize breeding programs (Prigge et al. 2012).

CIMMYT has optimized the protocol for maize DH line development in tropical genetic backgrounds, reducing significantly the time taken to develop parental lines (Prasanna et al. 2012). Tropically adapted first-generation haploid inducers with a haploid induction rate (HIR) of 5–8% were first developed by CIMMYT, in collaboration with the University of Hohenheim (Prigge et al. 2011; Prasanna et al. 2012) by transferring the maternal haploid induction trait from the temperate haploid inducers developed by University of Hohenheim. These tropicalized haploid inducers with better agronomic performance than the temperate haploid inducers in tropical conditions, were released in 2012, enabling the National Agricultural Research Systems (NARS) and small- and medium-enterprise (SME) private sector maize breeding programs in sub-Saharan Africa, Asia, and Latin America to more easily adopt DH technology.

Recognizing the scope to further improve the first-generation tropically adapted inducer lines (TAILs) for various traits, CIMMYT initiated the development of second-generation haploid inducers for the tropics by transferring the haploid induction trait from first-generation TAILs to elite CIMMYT Maize Lines (CMLs), marker-assisted selection for higher haploid induction rate, and phenotypic selection for superior agronomic performance (Chaikam et al. 2018). These second-generation tropically adapted inducer lines (called CIM2GTAILs) have high haploid induction rates ($\sim$10–13%), better agronomic performance in terms of plant vigor, synchrony with tropical populations, better standability, resistance to tropical foliar diseases, and resistance to ear rots compared to first-generation TAILs in trials at different locations in Mexico and Kenya. Inducer hybrids developed using these CIM2GTAILs exhibit greater heterosis for plant vigor and pollen production while maintaining similar haploid induction rates as the parents and are well suited for open pollinations in isolation nurseries. Taller stature inducers with greater pollen production reduce the cost of DH production by enabling the shift from hand pollination to isolations as well as by reducing the land cost since the area required for the inducer line in the induction nursery is lowered. For example, if an inducer male with less pollen requires a 2 male: 4 female planting plan while an inducer with greater pollen production enables a 2 male: 6

female planting plan, the land cost for producing haploid kernels is reduced by 25%.

While the DH technology is the primary mode of deriving new inbred lines by several large private sector breeding programs, NARS and SMEs seed companies in several African and Asian countries are yet to fully derive the benefits out of maize DH technology for various reasons. CIMMYT, in partnership with Kenya Agricultural and Livestock Research Organization (KALRO) established a centralized maize DH facility at Kiboko (Kenya) in 2013. The facility is now producing nearly 80,000 DH lines each year, serving the maize breeding programs of CIMMYT, national partners, and SME seed companies in sub-Saharan Africa. Through dedicated maize DH facilities in Kenya and Mexico, CIMMYT's Global Maize Program produces annually over 100,000 DH lines (up from less than 5000 in 2011) and selects the best out of these lines in breeding pipelines. These DH lines were tested for per se performance and hybrids combinations under drought stress, artificial and hot spot disease conditions and several drought-tolerant and disease-resistant DH line have been developed and released as CMLs (CIMMYT Maize Lines). More than 80 DH-based hybrids were released by CIMMYT through partners for commercialization in eastern and southern Africa (Beyene et al. 2017).

3.4.2 Genomics-Assisted Breeding

For effectively meeting the challenge of developing improved cultivars with combinations of relevant adaptive traits, including biotic and abiotic stress tolerance, and nutritional quality, it is imperative that breeding programs routinely use molecular tools in product development. With the rapid reduction in genotyping costs, new genomic selection technologies have become available that allow the maize breeding cycle to be greatly reduced, facilitating the inclusion of information on genetic effects for multiple stresses in selection decisions.

3.4.2.1 Maize Genome and Its Complexity

The maize genome is approximately 2500 Mb in size, comparable to the size of the human genome. Maize inbred lines have an average nucleotide diversity of around 1% in the genic regions (Tenaillon et al. 2001; Wright 2005), similar to the divergence between humans and chimpanzees (Mikkelsen et al. 2005). read depth variants, representing moderate-sized deletions, insertions and duplications add to the complexity of the maize genome (Chia et al. 2012), along with the structural changes leading to gene non-collinearity among inbred lines caused by activities of transposable elements such as Helitrons (Fu and Dooner 2002). Whereas the genetic diversity shown by this amazing crop has helped to drive its domestication and adaptation to diverse agro-ecologies worldwide, and helped in improvement through breeding and selection, the size and complexity of the genome poses a challenge in terms of discovery and deployment of genomic tools for improving key traits relevant for smallholders, and for continuously increasing genetic gains.

To chart and utilize the genetic diversity in maize, resequencing of a vast number of germplasm accessions has been undertaken, based on which the first-, second- and third-generation maize Haplotype Maps (HapMaps) have been constructed. These efforts involved resequencing of a large number of tropical and temperate maize inbreds, landraces, and wild relatives (Z. *mays* ssp. *parviglumis* and Z. *mays* ssp. *mexicana*), with 3.3, 55, and 83 M variants identified in HapMap 1, HapMap2, and HapMap3, respectively, to capture the whole genome variation (Gore et al. 2009; Chia et al. 2012; Bukowski et al. 2018). Based on the HapMaps, sequencing data from a large number of maize lines, and population relationship, a pangenome could be assembled via sequence alignment. By reduced representation sequencing of 14,129 maize inbred lines, 26 million tags were generated, 4.4 million of which were accurately mapped as sequence anchors (Lu et al. 2015), providing a foundation for pangenome construction. With the availability of high-quality whole genome sequences from two temperate inbreds (B73 and Mo17) and several tropical inbreds, a comprehensive maize pangenome is being constructed.

All pangenomic information including single nucleotide polymorphisms (SNPs), insertion/deletions (indels), non-coding RNAs, and transposable elements, needs to be incorporated into efficient databases. By integrating genomic and gene expression data, the core genome, variable genome, and the expression levels can be linked (Golicz et al. 2016). On the other hand, the exome (the subset of the genome which includes coding regions) represents the genome space where mutations are likely to affect protein structure and function. The maize exome is much smaller than the whole genome, making exome sequencing data more easily manageable and applicable for plant breeding (Warr et al. 2015). These efforts can be integrated with other functional genomics approaches, including insertional mutation, expressed sequence tag (EST) development, gene cloning, transcriptional profiling, transformation, tilling, and could be further used to discover genes and their functions.

Through large-scale resequencing of maize germplasm including over 30 tropical maize lines, a 55 K SNP array has been developed with improved genome coverage, containing over 4000 SNPs that do not exist in the B73 reference genome (Xu et al. 2017). Based on the 55 K SNP array, a set of high-throughput marker panels, containing 20, 10, 5, and 1 K SNP markers, were developed through genotyping-by-target sequencing (GBTS). Through this approach, genotyping cost can be significantly reduced by genotyping using the same 20 K marker panel at different sequencing depths (Z. Guo and Y. Xu, personal communications).

The genetic architecture of maize is more dispersed, as compared to self-pollinated crops like rice; the majority of agronomically important traits in cross-pollinated crops such as maize appear to be controlled by many small-effect genes (Morrell et al. 2012; Wallace et al. 2014). The most important consequence of the mating system is the impact on the levels of heterozygosity and the amount of effective recombination, factors which induce fundamental differences in trait architecture and patterns of linkage disequilibrium or LD (Morrell et al. 2012), and thus, breeding strategies. This could also apply to genomics-assisted breeding strategies, as such strategies differ among various crop species, based on their genome complexities and trait architecture. Although highly diverse maize populations show rapid decline of LD, as in humans, it is possible to develop populations with strong LD (Rafalski and Morgante 2004), such as closed breeding pools, that are more amenable to genomics-assisted breeding.

Despite the complexity of the maize genome, large multi-national seed companies are now routinely using applied genomics information and tools to (i) dissect the genetic structure of their germplasm to understand gene pools and germplasm (heterotic) groups, (ii) provide insights into allelic content of potential germplasm for use in breeding, (iii) screen early generation breeding populations to select segregants with desired combinations of marker alleles associated with beneficial traits (in order to avoid costly phenotypic evaluations), and (iv) establish genetic identity (fingerprinting) of their products.

3.4.2.2 Trait-Marker Discovery and Marker Deployment

Most of the agronomically and economically important traits in maize are quantitatively inherited (Wallace et al. 2014), with several small effect genetic loci with epistatic and environmental interaction. When two genetically diverse parents are crossed producing a progeny with a maximum genetic variation for a particular trait, linkage mapping can be done to elucidate linkage/association between specific markers and the genetic loci controlling the trait; this is called quantitative trait locus (QTL) mapping. Considering its major limitation, that only allelic diversity that segregates between the parents of the particular population can be assayed, genome-wide association study (GWAS) in a panel of breeding-relevant diverse lines came into vogue. After the first genome-wide association study (GWAS) reported in maize a decade ago (Beló et al. 2008), there have been numerous publications on GWAS in maize for traits ranging from nutritional quality to abiotic and biotic stress tolerance, and grain yield (Xao et al. 2017). The plethora of articles on GWAS, especially after the whole genome sequencing in maize, created an unfounded perception that GWAS is the method of choice for genetic mapping, as opposed to QTL mapping. However, it must be emphasized that QTL mapping and GWAS are quite complementary, and when effectively combined, can overcome each other's limitations (Korte and Farlow 2013). Based on this understanding, large genetic resources for joint linkage and association mapping have been created in temperate maize (Yu et al. 2008; Dell'Acqua et al. 2015). To date, such populations based on tropical maize germplasm, which could be genotyped at high density and shared across the tropical maize-based breeding programs, are not developed. Several association mapping panels have been assembled in CIMMYT, including the DTMA (Drought-Tolerant Maize for Africa) panel, IMAS (Improved Maize for African Soils) panel, CAAM (CIMMYT Asia Association Mapping) panel and HTMA (Heat Tolerant Maize for Asia) panel. These panels have been used in GWAS of many traits relevant to the tropics (Suwarno et al. 2015; Nair et al. 2015; Gowda et al. 2015; Zaidi et al. 2016; Rashid et al. 2018; Cao et al. 2017; Hindu et al. 2018). Similarly, hundreds of articles have been published on QTL mapping for various traits relevant to tropical maize, but there are few studies where both methods are judiciously employed to discover and validate trait-associated markers (Bernardo 2016).

The routine deployment of trait-based markers in the forward breeding pipeline requires high-throughput low-density marker system at an economical cost with a fast turn-around time. The cost efficiency will be decided based on the accuracy of the markers being used in the breeding pool and the relative cost and time advantage with respect to existing phenotyping methods. High-throughput platform for genotyping (HTPG) is an initiative funded by the Bill and Melinda Gates Foundation (B&MGF) to broker access to low-cost and fast turn-around genotyping facilities to Consultative Group on International Agricultural Research (CGIAR) institutions, and NARS and SME seed company partners. This initiative envisions that lowering the genotyping cost will enable CGIAR and other NARS/SME breeders to utilize marker-based selection in forward breeding and to take advantage of low-cost genotyping in breeding applications. As part of this initiative, a set of 10 SNPs can be genotyped in a turn-around time of two weeks at a cost of 1.5 USD (http://cegsb.icrisat.org/high-throughput-genotyping-project-htpg/).

Trait-based markers are developed and deployed in tropical maize for a number of quality traits such as *opaque2*, *opaque16*, and provitamin-A in various breeding programs in the tropics (Babu et al. 2005; Gupta et al. 2009; Muthusamy et al. 2014; Zunjare et al. 2018; Sarika et al. 2018). Two of these are causal gene-based markers and are associated with large-effect size. It is noteworthy that some of the Asian countries, especially China and India, have made significant progress in the development of biofortified maize varieties using molecular marker-assisted breeding. For example, at the ICAR-Indian Agricultural Research Institute, marker-assisted introgression of *opaque2* has recently led to the commercial release of three QPM hybrids viz., "Pusa HM4 Improved," "Pusa HM8 Improved," and "Pusa HM9 Improved" (Hossain et al. 2018a). These hybrids possessed 3.49 and 0.84% lysine and tryptophan in protein, respectively. Also, pyramiding of *opaque2* and *opaque16* showed an increase of 64% lysine and 86% tryptophan over *opaque2*-based hybrids. "Pusa Vivek QPM9 Improved," India's first provitamin-A rich maize hybrid was developed through introgression of *crtRB1*. This hybrid showed 8.15 μg/g of provitamin-A compared to 1–2 μg/g in normal maize (Muthusamy et al. 2014; Hossain et al. 2018b). Similarly, other trait-based markers based on causal genes are being used in marker-assisted breeding for waxy and sweet corn (Yang et al. 2013; Feng et al. 2015).

There are many efforts to discover, validate, and deploy trait-based markers for resistance to major diseases in the tropics. *Msv1*, a major effect locus contributing to MSV resistance has been fine-mapped (Nair et al. 2015) and is being deployed widely in a forward breeding pipeline in ESA. Maize lethal necrosis (MLN) is another devastating disease where genomic tools were efficiently discovered and are being deployed in the CIMMYT maize breeding programs in Africa (Gowda et al. 2015, 2018).

Molecular markers have also been deployed at CIMMYT for the development of improved tropical haploid inducers (Chaikam et al. 2018). MAS for *qhir1*, a QTL with significant positive effect on haploid induction rate improved the efficiency of haploid inducer development as the trait haploid induction is difficult to phenotype, besides the fact that segregation distortion occurs in families having *qhir1*. MAS for *qhir1* increased the selection intensity and the frequency of plants/families with *qhir1* and high HIR.

CIMMYT's Global Maize Program in Africa employed marker-assisted recurrent selection (MARS) to improve grain yield under drought stress in 34 biparental tropical maize populations. For example, Beyene et al. (2016) conducted a study to evaluate the performance of C1S2-derived hybrids obtained after three MARS cycles (one cycle of recombination [C1], followed by two generations of selfing (S2), under both drought stress (DS) and well-watered (WW) conditions in 10 bi-parental maize populations. They evaluated hybrids developed by crossing 1) 47–74 C1S2 lines advanced through MARS, 2) the best five S5 lines developed through pedigree selection, and 3) the founder parents with a single-cross tester from a complementary heterotic group. The hybrids and five commercial checks were evaluated in Kenya under 3 managed-drought stress and 3–5 optimum-moisture conditions. Results showed that across the managed-drought stress locations, the top 10 C1S2-derived hybrids from each of the 10 biparental populations produced 0.5–46 and 11–55% higher mean grain yields than mean yield of hybrids developed using pedigree selection and the commercial checks, respectively. Across the optimum-moisture locations, the best 10 hybrids derived from C1S2 of each population produced 3–13 and 8–37% higher grain yields than the yield of hybrids derived using conventional breeding.

3.4.2.3 Genomic Prediction

Trait-linked markers that were discovered and deployed by way of marker-assisted selection (MAS) require prior knowledge of the precise location and effect size of QTLs and germplasm specificity. This also requires that the innate genetic architecture of the species supports such interventions by way of having large-effect genes/QTLs controlling agronomically important traits. As discussed before, having very few of such loci identified in maize, genomic prediction becomes an important tool to improve genetic gains. "Genomic prediction" is a form of MAS where genome-wide markers are used to estimate the breeding value of individuals (Meuwissen et al. 2001). The concept that was originally developed in dairy cattle found its way to crop plants, especially in crops like maize. While the public-sector maize breeding programs in the tropics have been slow to make use of genomic prediction, multi-national companies have been routinely practicing genomic prediction in their breeding pipelines. Genomic prediction brought a paradigm shift in the way plant breeding is done, shifting the unit of selection from individual lines to alleles. Several factors like heritability, trait architecture, marker density, training population size and relationship between the training and prediction populations are critical for the accuracy of the predicted breeding value (Combs and Bernardo 2013).

When medium- to high-density genotyping costs and turn-around times decrease sufficiently, to at least partially replace resource-intensive field-based phenotyping, genomic prediction will be highly beneficial and cost-efficient in driving genetic gains in breeding programs. Recently, two medium-density genotyping options have been proposed and deployed with reduced cost; rAmpSeq (Buckler et al. 2016), and rhAMPSeq from Integrated DNA technologies, the costs of which are expected to stabilize at USD 5 per sample. These advancements have substantial scope for deploying genomic prediction in public-sector breeding programs in the tropics. Practical haplotype graph (PHG), which represents a simplified pangenome graph of maize is currently under development. The maize PHG enables the integration of genotypic data derived from multiple genotyping platforms to a common marker set, facilitating the combination of legacy data sets with data coming from newer genotyping methods. By combining data sets over time, more robust training models can be developed to improve prediction accuracy. Low-cost sequencing technologies, coupled with the PHG, facilitate the genotyping of large numbers of samples to increase the size of training populations for genomic prediction models (https://bitbucket.org/bucklerlab/practicalhaplotypegraph/wiki/Home). In addition to low-cost marker systems, implementing whole genome prediction models in routine breeding pipelines require careful development of relevant training sets and their phenotyping at high precision, to have an impact on continued and enhanced genetic gains (Cooper et al. 2014).

Genomic prediction could be applied to source population improvement by way of rapid cycling and could lead to an improvement in genetic gains primarily due to changing allele frequencies through the use of markers in a time-efficient manner. In a study of rapid cycle genomic selection (RCGS) in eight biparental populations in eastern Africa, the average gain from genomic selection per cycle across eight populations was 0.086 t ha^{-1}. The average grain yield of Cycle 3-derived hybrids was significantly higher than that of hybrids derived from Cycle 0. Hybrids derived from C3 produced 7.3% higher grain yield under drought than those developed through the conventional pedigree breeding method (Beyene et al. 2015). In two biparental RCGS for deriving improved stress-tolerant lines in the CIMMYT-Asia breeding program, a gain of 10–20% in grain yield under drought was observed after two cycles of genomic selection, compared to phenotypic selection (Vivek et al. 2017). RCGS is also applied in multi-parent synthetic populations in CIMMYT breeding programs to increase the efficiency of line derivation. In a multi-parent population, a 7.74% increase in genetic gains was observed under optimal conditions for grain yield (Zhang et al. 2017). The CIMMYT maize breeding program in Asia is currently working with RCGS of six biparental populations for deriving doubled haploid (DH) lines from improved cycles (Sudha Nair, unpublished).

CIMMYT's Global Maize Program in Kenya evaluated a total of 734 Stage II hybrids (produced by crossing the top elite lines developed through MARS/GS with three elite single-cross testers from opposite heterotic group) and evaluated these under 4–6 optimum-moisture, two random random-drought stress (rainfed) locations, and two managed-drought stress sites in eastern Africa. Results showed that when averaged across optimum-moisture locations, the top 10 hybrids gave yield advantage ranging from 21–56% compared with the mean yield of the checks. When averaged across two random-drought stress locations, the top 10-performing hybrids had yield advantage of 56–111% compared with the mean yield of the commercial hybrids. Under managed-drought sites, the top 10 hybrids out-yielded the checks by 27–107% (Yoseph Beyene, unpublished).

Apart from population improvement, genomic prediction based on early-stage yield testing (Stage 1) is an important tool in the modern maize breeding pipeline, enabling increased selection intensity and reduced cost and time. In a proof of concept study in a set of 22 biparental populations evaluated for grain yield and other agronomic traits, moderate to high prediction accuracies were obtained with higher heritability and with a training population size that was at least 50% of the total population (Zhang et al. 2017). CIMMYT has started routine breeding program-wide genomic predictions in biparental maize populations in 2017, represented by 15,000 breeding lines entering Stage 1 testing; at least 50% of the CIMMYT breeding pipelines are expected to be based on genomic prediction by 2021–2022 (Xuecai Zhang, unpublished). Additional applications of genome-wide marker strategies, including hybrid prediction and population selection can be implemented, but the methodologies are at a nascent stage in the tropical breeding programs.

3.4.2.4 Seed Chipping and Genotyping

Maize is highly amenable to nondestructive seed-based genotyping, owing to its large seed size and clear separation of the endosperm from the embryo. Seed (endosperm) chipping, DNA extraction, and genotyping of maize seeds by manual methods were reported (Gao et al. 2008) and have been used routinely in CIMMYT breeding programs. Manual seed chipping is meticulous, relatively labor-intensive, and not highly scalable. Monsanto Technology LLC holds the patent for the automatic seed chipping and extraction process (US Patent No. US007591101B2), which enables efficient handling of large volumes of seeds. CIMMYT is currently pilot testing this application in collaboration with Monsanto, which when optimized could facilitate expanded application of genomic technologies in maize breeding programs in the tropics. The advent of automated seed DNA genotyping has the potential to further augment genetic gains through increased selection intensity and reduction of land and labor costs required for tissue sampling. The ability to enrich source populations for favorable alleles prior to the DH process holds promise for improving the genetic merit of DHs delivered, further enhancing genetic gains.

3.4.3 High-Throughput Field-Based Phenotyping

Accurate and high-quality phenotyping is one of the most critical pieces of a successful breeding program. Methods for the measurement of most required traits have largely remained unchanged over the past few decades and are manual, laborious, and time-consuming; with some being prone to human error or lacking repeatability. The use of data collection methods that allow reliable assessment of crop traits at a reasonable cost and faster than the methods currently in use, can significantly improve resource use efficiency and contribute to increased genetic gain through improved selection efficiency (Araus et al. 2018). The selection cost reduction will allow resources to be allocated to the generation and management of larger populations, enabling an increase in selection intensity within a fixed budget (Araus et al. 2018). Sensor technology coupled with progress in image processing offers radically new perspectives for field-based high-throughput phenotyping (HTP) and is anticipated to enable a better integration of phenotyping approaches into breeding programs by helping to (a) extract more value from every research plot, and (b) improve phenotypic data quality.

Breeding programs of the majority of the NARS and SME seed companies in the tropics have limited capacity for undertaking high-throughput and reasonably precise phenotyping, particularly under repeatable and representative levels of abiotic and biotic stresses in the field. This is indeed a major constraint for increasing genetic gains, and the capacity to breed better cultivars with higher grain yield and stress resilience (Prasanna et al. 2013). Appropriate trial management and spatial variability handling, definition of key constraining conditions prevalent in the target population of environments, and the development of more comprehensive data management, including crop modeling, are all integral components of phenotyping (Araus et al. 2018).

The development of low-cost, high-throughput phenotyping tools has the potential to play an important role in reducing breeding costs, thus allowing resources to be allocated to generation and management of larger populations, enabling an increase in selection intensity within a fixed budget (Araus et al. 2018). In partnership with advanced research institutions, the CIMMYT's Global Maize Program has made significant progress in validating and deploying proximal and remote sensing tools for measuring key traits relevant for maize breeding programs, including plant height, ear height, and ear-related phenotypes using proximal sensing; and stand count, leaf senescence, and canopy cover through remote sensing. Some examples of current developments on low-cost and high-throughput field phenotyping tools for important traits in the CIMMYT's Global Maize Program are presented here.

3.4.3.1 Plant and Ear Height Measurements Using Hand-Held Laser Distance Meter

Plant and ear height have routinely been measured in the field using a telescopic stick with decimeter marks or a ruler and the data captured manually; making the process prone to errors while consuming time and resources. In the case of maize, the process is even more difficult because most hybrids are from 2.5 to 3.5 m in height. Various sensors are currently available for measuring plant height but not all are applicable for measuring ear height. These include light detection and ranging (LiDAR), ultrasonic sensors, and RGB camera (Crommelinck and Höfle 2016; Friedli et al. 2016; Hämmerle and Höfle 2016). Assessment of plant height from images is relatively complex and the level of accuracy of the data still needs improvement before routine implementation in tropical maize breeding. This can be done through the use of real-time kinematic (RTK) GPS (Xiong et al. 2017) but the associated cost is still very high. Recently at CIMMYT, the use of sensors like the laser distance meter (Hand-held Leica Disto D110, Leica Geosystems AG, Heerbrugg, Switzerland) has provided new possibilities for plant/ear height data collection. The sensor which can be used directly or mounted on a phenopole, connects to mobile phones or tablets through Bluetooth, providing a simple, very low-cost estimation of plant and ear height (<200 USD per sensor). Estimated plant and ear height are highly correlated with measurements collected using a ruler. Only one person is required for measuring and capturing data with a laser distance sensor compared to two people when a ruler is used, thus reducing the cost of data collection by 50%. The data are also automatically captured electronically, which significantly reduces the time required for measuring plant or ear height (Fig. 3.3).

Fig. 3.3 Plant and ear height data collection using the hand-held Leica Disto D110 (Leica Geosystems AG, Heerbrugg, Switzerland) at the CIMMYT Maize Research Station in Harare, Zimbabwe

3.4.3.2 Aerial Sensing for Early Vigor and Canopy Senescence Assessment

Crop early vigor and canopy senescence are often assessed based on visual scores that are qualitative, and often subjective. Imaging methods can provide a standardized, rapid, cost-effective and more objective way of collecting these data (Fig. 3.4). The common methods include the use of canopy reflectance (for example normalized difference vegetation index (NDVI)) to monitor crop cover or leaf senescence but the cost of sensors is often higher than that of an RGB camera (Zaman-Allah et al. 2015). Recently at CIMMYT, RGB images taken with consumer-grade digital cameras onboard low-cost unmanned aerial vehicle (UAV) were used to derive a senescence index based on the ratio of senesced canopy to total canopy cover under low nitrogen conditions. The senescence index was highly correlated with grain yield compared to visual measurements of canopy senescence, while broad-sense heritability was equal to or higher than visual measurements (Makanza et al. 2018a). The time required for phenotyping using a UAV was reduced by 95% relative to visual measurements. With advances in image analysis methods, the rapid cost reduction of sensors, and effective image processing software, there is still potential for wider applications of field-based phenotyping by UAVs.

3.4.3.3 Digital Ear Phenotyping

Recent open-source image analysis protocols have been developed to measure maize yield components using both a line scanner and conveyor belt and flatbed scanner (Liang et al. 2016; Miller et al. 2017). These methodologies are generally slow (only a couple of ears per photo, i.e., 1–5) and are not easy to use in the field. Additionally, they do not provide a comprehensive data set of the relevant kernel and ear traits from a single image of unshelled ears. Ear digital imaging (EDI) is a simple, low-cost, high-throughput and robust method for extracting yield

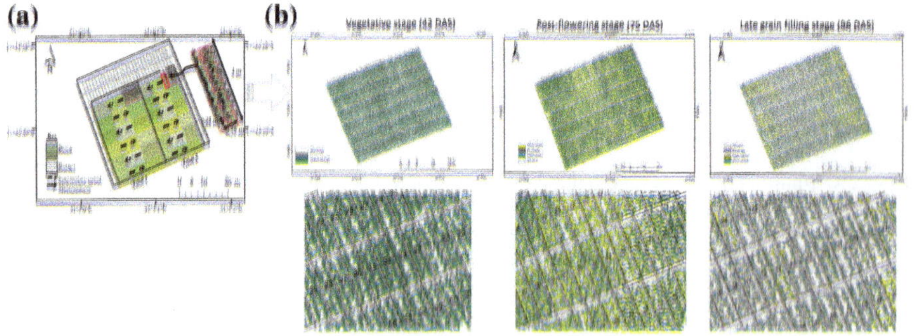

Fig. 3.4 **a** Pre-processed details of a portion of a maize field with plot details and, **b** time sequence processed aerial images of maize hybrids at three different developmental stages grown at the CIMMYT Maize Research Station in Harare, Zimbabwe

components (ear and kernel attributes) from harvested maize ears developed at CIMMYT (Makanza et al. 2018b). The method provides estimates of ear and kernel attributes, i.e., ear number and size, kernel number and size as well as kernel weight from photos of ears harvested from field trial plots (Fig. 3.5). The image processing method uses a script that runs in a batch mode on ImageJ; an open-source software. Kernel weight was estimated using the total kernel number and the average kernel size. Estimated yield components (including kernel weight and number) were significantly correlated with manual measurements of yield components (r > 0.80). Current investment in combine harvesters at key breeding locations may supersede this technology. However, it will continue to provide an important quality control feedback in on-farm trials where yields are often measured by non-researchers. Furthermore, the cost and maintenance of harvest equipment do not make them accessible for small breeding programs, especially within national programs, and the use of EDI as a surrogate measurement for grain yield will continue to have potential application in many parts of Africa for the foreseeable future.

3.4.4 *Breeding Informatics and Artificial Intelligence*

The increasing availability of breeding related information, including pedigree, phenotypic and genotypic, coupled with environmental data, brings both opportunities and challenges in effectively managing and utilizing such information in breeding programs (Xu 2010). This necessitates the development of integrated platforms or one-stop data portals that can effectively bring together high-density genotyping, high-throughput and precision phenotyping, and multi-dimensional environment profiling along with a suite of decision-support tools to drive modern breeding programs. Data integration from multiple sources is one of the key components in developing breeding informatics systems. Efficient breeding

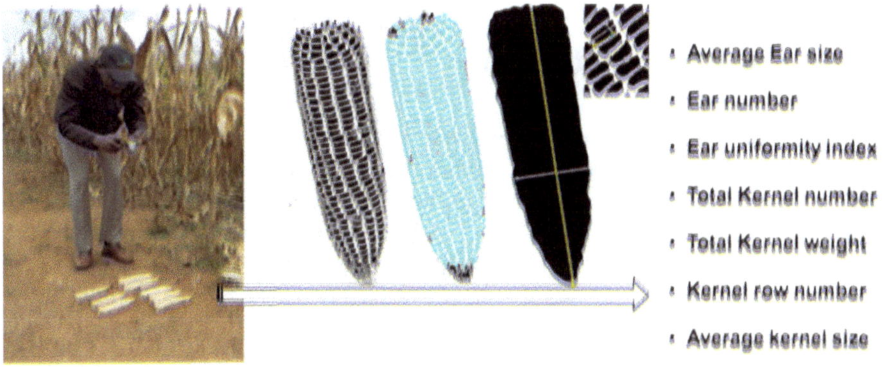

Fig. 3.5 Photo acquisition, simplified processing steps, and ear and kernel attributes can be generated using the ear digital imaging (EDI) method

informatics systems will need to include data curation tools, automated quality control workflows, data processing pipelines, visualization tools, and simple and user-friendly data analytical and mining tool kits. This is one area where the tropical maize breeding programs are clearly lagging behind, since most of the data accumulated as part of the breeding cycles are maintained as flat files which are not in queryable databases, hence, breeding programs are not able to make maximum use of the data developed through multiple breeding cycles. This is especially true in cases of using genomic and enabling tools, as the quantity of data produced is high and converting them into selection units requires breeding informatics support.

Vast amounts of data have been generated, and will be generated, by institutions worldwide through several high-throughput platforms of genotyping, phenotyping, and envirotyping. Breeding informatics has been revolutionized with significant changes in data generation, storage, scale, dimension, throughput, and precision, distinctly different from other big data in data properties, collection, treatment, analysis, mining, and utilization. Modern breeding is now becoming increasingly integrated with a programmed breeding pipeline, agricultural engineering, facilities with artificial or controlled environments and biological modeling/simulation, to meet human demands for high yielding, improved quality, resource-use efficient and environment-friendly crop varieties. These trends will rely on artificial intelligence (AI)-guided agriculture to complete the conversion of breeding from big data-driven to AI-driven. Among the four major factors that affect AI, data and knowledge are the two that shape the distinct properties of AI-assisted breeding. On one hand, AI will have significant influence on breeding information system because AI-equipped robots will interact with all the processes relevant to data collection, storage, analysis, sharing and utilization. On the other hand, historical experience and relevant knowledge achieved and accumulated in breeding programs need to be incorporated into AI system to make breeding more impactful. Thus, future breeding programs, including those in maize, could become AI system-based, with strong influence of both breeding informatics and breeders' knowledge of germplasm.

3.4.5 Transgenic Technology

Transgenic technologies have received tremendous attention from the commercial seed sector since the first transgenic hybrid became commercial in 1996. Globally in 2017, out of a total area of 189.8 million hectares under genetically modified (GM) crops, GM maize occupied 59.7 million hectares, which was 32% of the global maize production in 2017 (ISAAA 2017). Of this, the majority area was under stacked (herbicide-resistant + insect-resistant) maize cultivars, followed by herbicide-resistant or insect-resistant (*Bt*) cultivars. Transgenic technology is now a major part of maize improvement, especially in countries such as USA, Brazil, and Argentina. Effective transgenic technologies for weed management and stem borer control in maize are already available and are being used by maize farmers in several other countries, including developing countries like Philippines and Vietnam.

South Africa is the only country in Africa that is presently commercializing insect-resistant (*Bt*) maize. One example of public breeders' access to proprietary transgenes is the Water Efficient Maize for Africa (WEMA) project, which is presently being implemented as the TELA Project, coordinated by the African Agricultural Technology Foundation (AATF). The drought-tolerance transgenic event DT MON87460 from Monsanto (presently Bayer) and the widely used *Bt* events MON810 and MON89034 are available for royalty-free use under a humanitarian license for selected countries in Africa (Ethiopia, Kenya, Mozambique, Tanzania, Uganda, and South Africa) once these transgenes are deregulated in those countries. DT and *Bt* traits have been introgressed into elite, Africa-adapted maize inbreds. Confined field trials (CFTs) of transgenic drought-tolerant event (DT MON87460), transgenic insect-resistant events (*Bt* MON810 and *Bt* MON89034) and stacked DT + *Bt*) have been conducted in most of the project target countries in eastern and southern Africa. Existing regulatory systems in these countries supported the testing of these transgenes under CFTs in South Africa, Kenya, and Uganda and the efficacy of single trait events has been demonstrated. Combined analyses of CFTs of DT MON87460 maize data using 34 hybrids with same base genetics was done in three WEMA countries for three or more years. The results showed that five traited hybrids gave 8–14% greater yield than the non-traited versions, indicating a strong positive gene effect with ample scope for selection and breeding in germplasm of similar genetic background as these hybrids. Efficacy trials were also carried out on *Bt* MON810 in controlling the spotted stem borer (*Chilo partellus*) in the CFT, and the African stem borer (*Busseola fusca*) in the lab, in Kenya (CFT I–CFT III) and in Uganda (CFT II). Results of CFT II in Kenya showed that 75% of the hybrids evaluated had significantly greater yield, ranging from 26 to 113% with the *Bt* trait than without the trait. Results of CFT II in Uganda showed that all the seven *Bt* maize hybrids had significantly greater yield (49–201%) due to the *Bt* trait compared with the non-traited hybrids. Five registered TELA® (*Bt* MON89034) hybrids are already being cultivated by smallholder farmers in six provinces of the Republic of South Africa (WEMA, 2018). The process of deregulation is in progress in other project target countries.

3.5 Utilizing Genomics and Other Enabling Technologies to Enhance Genetic Gains

Molecular markers in the public-sector NARS maize breeding programs have so far been largely limited to the use of trait-specific markers, majority of which are for quality traits deployed through marker-assisted backcross programs (Babu et al. 2005; Gupta et al. 2009; Muthusamy et al. 2014; Zunjare et al. 2018; Sarika et al.

2018). Though this strategy is helpful in improving the elite breeding materials for specific traits, they do not lead to an improvement of the genetic gains in the overall breeding program. This is possible only by targeted use of available trait-specific markers, wherever possible, in the forward breeding pipeline and a breeding program-wide streamlining of genomic prediction schemes. In this section, we will highlight how molecular marker-assisted selection, especially through forward breeding, can help improve genetic gains. Considering the genetic gain equation (discussed before), molecular marker-based breeding can have significant impact on several parameters in the famous breeder's equation (Lynch and Walsh 1998).

3.5.1 Selection Intensity

Selection intensity is described as the proportion of the total population selected for advancement or for further recombination. In conventional breeding process, the number of plants/families handled during early generations is severely limited due to the paucity of land, labor, and money. On the other hand, using early generation screen with trait-specific markers, breeders could select for quality and adaptive traits (disease resistance, secondary traits for abiotic stress tolerance etc.), for which trait-linked markers are selected on in a large population to improve their favorable allele frequency before various stages of yield testing. Larger population sizes allow greater selection intensity and increase the probability of identifying superior progenies (Moose and Mumm 2008). Similarly, genomic prediction in early yield testing stage (Stage 1) at a reduced cost relative to phenotypic selection schemes enables evaluation of a larger number of progenies and a higher selection intensity, considering the resources required for test crossing and field evaluations. Doubled haploids and seed DNA genotyping are some of the most potent accompanying tools that can be effectively integrated with genomic tools for increasing genetic gains through selection intensity.

CIMMYT Maize Program routinely deploys markers in the forward breeding pipeline for resistance to diseases like MSV and MLN in Africa (Nair et al. 2015; Gowda et al. 2018), and for quality traits like provitamin-A (Babu et al. 2013). In Asia, markers for forward breeding shall soon be deployed by CIMMYT for resistance to TLB, after ongoing pilot testing. This has been made possible by high-throughput, low-cost, low-density genotyping platforms like HTPG. Genomic prediction within biparental populations on lines entering Stage 1 testing has also been deployed in a large scale from 2017, through coordinated efforts among maize breeders, biometricians, software developers, and innovative high-throughput medium-density genotyping efforts (Buckler et al. 2016).

3.5.2 Selection Accuracy

Selection accuracy in the genetic gains equation could entail multiple facets of accuracy, such as (1) heritability (repeatability) of the trial; (2) accuracy of MAS depending on linkage of the marker with the gene/QTL; (3) accuracy of genomic prediction; and (4) accuracy of correlation between the tested locations to the target populations of environments (TPEs). Among these, molecular marker-based interventions are directly related to the second and third aspects, related to marker-assisted forward breeding and genomic prediction, respectively. The accuracy with which trait-specific markers can be used in forward breeding depends on the diagnostic nature of the marker, where factors like linkage of the marker with the causal polymorphism and the effect size associated with the marker under selection are important.

Diagnostic markers for traits in the strict sense would mean the functional polymorphism itself, which is responsible for a large-effect size and are identified after fine mapping and cloning (Bortiri et al. 2006). There are relatively few cloned genes for agronomically important traits in the public domain in maize, including for quality traits like the maize starch pathway genes (reviewed in Whitt et al. 2002), *opaque 2* (Schmidt et al. 1990), fatty acid composition (Zheng et al. 2008) and provitamin-A (Harjes et al. 2008; Yan et al. 2010), resistance to leaf blight (Johal and Briggs 1992), rust (Collins et al. 1999), TLB (Hurni et al. 2015), head smut (Zuo et al. 2015), SCMV (Tao et al. 2013; Liu et al. 2017a), Anthracnose stalk rot (Frey 2006), and haploid induction (Kelliher et al. 2017; Gilles et al. 2017; Liu et al. 2017b). Due to the genetic architecture of the crop, some of the major genes like *htn1* for TLB resistance act as a QTL with low to moderate effect in certain genetic backgrounds and environments (e.g., race constitution of pathogens) (Hurni et al. 2015). Hence there are not many large-effect haplotypes that could be used as diagnostic markers in the strict sense. Still, marker-assisted forward breeding can be successful, depending on the informativeness of the haplotype under selection for the trait in question in the specific breeding pool. The relative efficiency of marker-only selection compared to phenotypic selection depends on the proportion of the total additive genetic variance due to the known loci relative to the heritability of the trait (Smith 1967).

The selection accuracy factor in the genetic gains equation is also applicable in the context of genomic prediction, where the prediction accuracy shows the relationship of the genomic estimated breeding value to the true breeding value of the breeding lines. One of the most important factors that influences this is the relationship between the training population and the breeding populations. If breeding programs create relevant training sets for the breeding populations and recalibrate training models by dynamically updating the training populations to ensure high prediction accuracy, improved genetic gains can be achieved by genomic predictions. Prediction accuracy was found to be the highest within biparental families, followed by half-sib families and is very low among unrelated families (Riedelsheimer et al. 2013). For achieving high selection accuracy, it is of utmost importance to have high-precision phenotyping technologies, experimental designs, and biometrical analyses that can effectively handle field experimentation errors, if any. For the creation of dynamically calibrated training sets which are predictable for breeding populations, it is also important to have a very strong breeding informatics management system to connect breeding pedigrees, their genotypes and phenotypes; this is currently a sizable gap in the public-sector maize breeding programs in the tropics.

3.5.3 Genetic Variance

Molecular markers guide the introduction of regulated and useful variation into the breeding programs. They offer an excellent opportunity to accelerate breeding by selecting either for known genes or QTL using genetic markers, rather than effects on phenotypes. Many gene bank accessions or exotic germplasm carry rare alleles that could potentially improve various traits breeders work with. Marker-based approaches like marker-assisted backcrossing (MABC) have been proven successful in carefully introgressing such favorable alleles without disturbing the overall constitution of the breeding pool. In tropical maize, this has been proved in the case of provitamin-A (Menkir et al. 2017) and MLN resistance (Olsen et al. 2018). Also, marker-assisted pyramiding assists in accurately transferring many favorable loci for a trait, without altering the genomic background of elite lines in the breeding pool. Gene bank accessions could be used more effectively by a combination of speed breeding and genomic selection (Li et al. 2018). Apart from specific donor germplasm for improving desired traits, genetic diversity analyses within breeding pools guide decisions regarding the diverse founder lines within the genetic pools that are crossed to develop progeny having high additive genetic variance, from which progenies could be selected in such a way that they move the population mean in a significant positive direction in ensuing generations.

3.5.4 Cycle Time

When genetic gain is measured by unit time, accelerating the breeding process becomes critical as it will shorten the cycle time and thus increase the total genetic gain per unit time. When markers are available for traits of interest, MAS for target loci through forward breeding or backcross introgression has particular advantages, especially in terms of saving time and resources in introgression of recessive traits like *opaque2* (Babu et al. 2015), *qMLN06.157* for MLN resistance (Olsen et al. 2018), or quality traits like provitamin-A which are phenotyped after harvesting (Muthusamy et al. 2014). RCGS for population improvement (as against recurrent selection) helps to save every intervening cycle between intermating cycles, where testcross progenies are evaluated (Beyene et al. 2015; Vivek et al. 2017) and enhances genetic gains per unit time. When used in combination with enabling tools like seed DNA genotyping and doubled haploidy, they will have an enormous impact on reducing the breeding cycle time, thereby increasing genetic gain.

3.6 Strengthening Maize Seed Systems in the Tropics

Targeted deployment of improved climate-resilient varieties by GIS-based prediction of areas of climate vulnerability, emphasis on QA/QC throughout the seed value chain (Gowda et al. 2017), improving varietal turnover (with newer and better genetics), recommendations on appropriate agronomic management practices for realizing the genetic potential of improved varieties (especially in stress-prone environments), and creating better linkages of the smallholder maize farmers to output markets (for providing greater incomes to the farmers) are all absolutely critical for strengthening maize value chains in the tropics.

For new climate-resilient maize varieties to contribute toward smallholders' adaptation to climate variability in the tropics, it is important to further strengthen the seed systems. Delivering affordable improved maize seed to smallholder farmers with limited purchasing capacity and market access requires stronger public–private partnerships, and enhanced support to the committed local seed companies, especially in terms of information on access to new products, adequate and reliable supplies of early-generation (breeder and foundation) seed, and training on hybrid seed production, QA/QC, seed business management, market segmentation, and territory planning.

While an array of climate-resilient maize varieties have been released in recent years, especially through the CIMMYT's product pipeline, much remains to be done in the multiple stress-prone rainfed tropics in sub-Saharan Africa and Asia in terms of market-oriented adoption of these varieties, and replacing the obsolete climate-vulnerable varieties that are presently grown by resource-poor farmers. Appropriate government policies and adoption of progressive seed laws and regulations are critical for improving smallholder farmers' access to improved climate-resilient seed, and for overcoming key bottlenecks affecting the seed value chains, particularly in the areas of policy, credit availability, seed production, germplasm, and marketing (Cairns and Prasanna 2018).

3.7 Protecting Genetic Gains from Devastating Diseases and Insect-Pests

There is increasing evidence that the changing climates and global trade are two of the major factors responsible for increased incidence of transboundary plant and animal pathogens and insect-pests. Deutsch et al. (2018) highlighted that global yield losses of three of the most important cereal staples—rice, maize, and wheat—are projected to increase by 10–25% per degree of global mean surface warming, and consequent changes in the spectrum of insect-pests. This will be most acute in areas where increase in temperatures may lead to increases in both population growth and metabolic rates of insects.

3.7.1 Maize Lethal Necrosis (MLN)

The MLN disease, caused by a combination of maize chlorotic mottle virus (MCMV) and sugarcane mosaic virus (SCMV), first appeared in Kenya in 2011, and was subsequently reported in various eastern African countries, including Tanzania, Uganda, Rwanda, and Ethiopia. MLN has caused extensive crop losses in eastern Africa (Mahuku et al. 2015; Marenya et al. 2018); for example, these were estimated at US$261 million in Ethiopia (Fentahun et al. 2017), and US$198 million in Kenya (De Groote et al. 2016). Most of the elite Africa-adapted CIMMYT lines used in commercial products were highly susceptible to MLN, and commercial hybrids in Kenya showed 70–100% yield loss under heavy MLN pressure.

CIMMYT, in partnership with KALRO, established a centralized MLN screening facility in Naivasha, Kenya in 2013. This enabled intensive screening of over 150,000 maize germplasm entries over the last five years, resulting in the identification of sources of resistance to MLN (Beyene et al. 2018), and release of 15 CIMMYT-derived MLN-tolerant/resistant maize hybrid varieties so far in Kenya, Tanzania, and Uganda.

CIMMYT team in Africa also discovered and validated genomic regions in maize conferring MLN resistance (Gowda et al. 2015, 2018). Since the elite germplasm pool was highly susceptible to MLN and the important donor lines were non-elite and/or non-adapted, CIMMYT undertook a large-scale introgression effort using MABC to simultaneously introduce and validate genomic regions conferring MLN tolerance/resistance. Twenty-five RP lines were prioritized based on the importance of their hybrids in the market and in the advanced breeding pipeline. RP and donor parent (DP) lines were organized into heterotic groups (A and B) so that DP alleles would be deployed uniquely within each heterotic group (HG). Using markers linked to putative QTL from six donor lines (Gowda et al. 2015, 2018), from 2 to 4 loci per project (RP × DP combination) were targeted for introgression. In parallel, four of the best yellow-grained MLN tolerant lines were converted to white grain versions while simultaneously introgressing the favorable allele for MSV resistance at *msv1*. One to six versions were advanced per RP with yield advantage ranging from 1.0 to 3.1 Mg ha^{-1} over the respective control (original RP testcrossed to a common MLN-tolerant tester). Mean grain yield of advanced versions was 5.3 Mg ha^{-1} compared to 3.1 Mg ha^{-1} mean grain yield of the original RP control hybrids. Mean grain yield of three elite commercial hybrids which were widely grown at the time of the MLN epidemic was 0.9 Mg ha^{-1} and grain yield of a recently released CIMMYT MLN tolerant three-way cross hybrid was 4.6 Mg ha^{-1} (Olsen et al. 2018).

Protecting farmers' maize crops from deadly transboundary diseases like MLN (with the causal viruses transmitted by multiple means, including insect-vectors and contaminated seed) requires a multi-pronged strategy and integrated solutions, not just breeding and deployment of improved varieties. Proactive preparedness and prevention are far better than cure!

3.7.2 Fall Armyworm

A new and complex challenge called fall armyworm (*Spodoptera frugiperda*; FAW), a highly aggressive and invasive insect-pest with devastating effect, has been officially reported in the beginning of 2016 in Nigeria, and since then, rapidly spread across the African continent. Presently, more than 40 countries in Africa have officially reported the incidence of FAW. Since 2018, FAW incidence has been also been reported in India, Bangladesh, Nepal, Sri Lanka, Myanmar, Thailand, and southern China.

FAW moth populations migrate very fast (almost 100 km per night, and nearly 500 km before laying eggs), and thus, can invade new areas quickly. The pest can complete its life cycle within 1–2 months (depending on weather conditions), with each female moth capable of laying on average 1500 eggs. Moreover, the pest has a wide spectrum of host range, including maize, sorghum, sugarcane, soybean, vegetables, etc. Based on a recent study (Early et al. 2018), the strongest climatic limits on FAW's year-round distribution are the coldest annual temperature and the amount of rain in the wet season. Much of sub-Saharan Africa can host year-round FAW populations. South and Southeast Asia and Australia have climates that would permit FAW to invade. Current trade and transportation routes reveal that Australia, China, India, Indonesia, Malaysia, Philippines, and Thailand face high threat of FAW invasions originating from Africa. FAW has a strong appetite for maize; therefore, the implications of the incidence of this pest in maize-growing countries in Africa and Asia is indeed a major concern. CIMMYT is at the forefront in the fight against FAW in Africa, in collaboration with several national, regional and global partners, focusing on an integrated pest management (IPM) strategy (Prasanna et al. 2018).

An effective IPM strategy for control of FAW will employ host plant resistance, biological control, cultural control, and environmentally safer synthetic and biopesticides to protect the crops from economic injury while minimizing negative impacts on people, animals, and the environment. Many organizations, including both public and private sector actors, have been intensively working on identifying/validating/developing technologies/management practices that can help manage the pest in Africa, as well as creating awareness among the stakeholders on monitoring, surveillance, and IPM-based FAW control in Africa and Asia. Sources of native genetic resistance (partial, polygenic resistance) to FAW have been developed through intensive work at CIMMYT-Mexico from 1970s to 1990s (Mihm 1997), and through research work conducted by USDA-ARS, University of Florida, and Embrapa-Brazil. Some of these sources of insect-pest resistance were specifically tested against FAW, while others were tested for resistance to other insect-pests but have the potential to confer resistance to FAW. CIMMYT is presently undertaking intensive screening of tropical/subtropical maize germplasm against FAW under artificial infestation (in screenhouses) in KALRO-Kiboko, Kenya. Some of the CMLs (e.g., CML71), as well as insect-resistant maize inbred lines and hybrids derived through the insect resistant maize for Africa (IRMA) project show significant promise; validation experiments are in progress. IITA team in Nigeria is also undertaking similar efforts on germplasm screening for native genetic resistance and has initiated the selection of maize lines under FAW infestation.

Deploying transgenic or genetically modified (GM) crop varieties that express lepidopteran resistance genes is another strategy to effectively control FAW damage in maize. Several different *cry* genes are available—e.g., *cry1A, cry1Ab, and cry1F* —and have been exploited in commercial *Bt* maize varieties globally for over 20 years. In addition, *Bt* produces another class of lepidopteran-specific proteins termed vegetative insecticidal proteins (VIP). These VIPs are encoded by *vip* genes, the most notable of which is the *vip3A* gene used to confer FAW resistance. Numerous GM maize hybrids, including various combinations of *cry* and *vip* genes, are commercially available in Brazil and North America, where over 80% of the total maize production area is cultivated with *Bt* maize (Horikoshi et al. 2016).

In Africa, *Bt* maize is currently commercially available only in South Africa, where regulatory authorities have overseen multiple approvals, with more than 15 years of deployment of such products. Two GM products are available that provide protection against FAW: (a) The MON810 event, which is intended to control stem borer but also confers partial resistance to FAW, has been cultivated in South Africa since 1997; and (b) the MON89034 event, which has demonstrated efficacy for control of both FAW and stem borer, has been cultivated in South Africa since 2010. MON89034 is particularly recommended for FAW control due to its high efficacy against the pest, as well as anticipated durability of control over time due to its incorporation of "stacked" or "pyramided" insect resistance traits.

Under the TELA Project, NARS institutions in South Africa, Kenya, Tanzania, Uganda, Mozambique, and Ethiopia have been testing the performance of MON810 *Bt* and stacked *Bt* + drought tolerance (DT) transgenes introgressed into locally adapted African maize varieties. The emerging results are consistent with the performance of *Bt* maize in other countries: When introduced into locally preferred African maize varieties, the MON810 event is demonstrating strong control of stem borers and partial control of FAW in Kenya, Mozambique, Uganda, and Ethiopia. An application for approval of MON810 in Kenya is pending finalization, and applications for approvals in other TELA partner countries are expected to be ready for submission in 2018.

3.8 Conclusions

Intensive multi-institutional efforts are required to identify and utilize climate-resilient tropical/subtropical maize germplasm in product development pipelines. There is an increasing body of evidence confirming the benefits of climate-resilient maize varieties to increase yields, reduce yield variability, and ultimately, increase food security. To increase genetic gains through maize breeding in the stress-prone tropics, and for enhancing the pace, precision, and efficiency of breeding progress, judicious and effective integration of modern tools/strategies, especially high-density genotyping, high-throughput and precision phenotyping, DH technology, molecular marker-assisted and genomic selection-based breeding, and knowledge-led decision-support systems, are vital.

Intensive efforts are required to build the capacity of the institutions on methods to characterize and control field site variation (for improving repeatability), adopting appropriate experimental designs, selection of "right" traits for phenotyping, proper integration, analysis and application of heterogeneous data sets, and increasing the genetic signal-to-noise ratio to detect real differences between genotypes (Prasanna et al. 2013). There is also a distinct need for the public and private institutions to come together and establish stronger "phenotyping networks" for comprehensive and efficient characterization of breeding materials for important target traits.

Emerging seed enterprises in sub-Saharan Africa, Asia, and Latin America need to be strengthened to become more market-oriented and dynamic, and for providing smallholders with greater access to affordable climate-resilient improved seed. Understanding the smallholder farmers' constraints for adoption of modern maize varieties, enhancing affordability and access to quality seed, and improving linkages of resource-poor farming communities to the input and output markets should be accorded top priority.

Genetics and breeding alone cannot completely solve the complex challenge of enhancing maize productivity at the smallholder farm level, especially in the face of depleting/degrading natural resources and increasing climatic variabilities. There is a distinct need for effective complementation of improved maize cultivars with suitable precision-conservation agriculture practices, integrated nutrient management, scale-appropriate mechanization, as well as institutional and policy innovations for strengthening maize value chains in the developing world.

Acknowledgements The work presented in this article was supported by the CGIAR Research Program on Maize (MAIZE), and several multi-institutional projects implemented by CIMMYT with partners in sub-Saharan Africa, Asia, and Latin America. The CGIAR Research Program MAIZE receives W1&W2 support from the Governments of Australia, Belgium, Canada, China, France, India, Japan, Korea, Mexico, Netherlands, New Zealand, Norway, Sweden, Switzerland, U.K., U.S., and the World Bank.

References

Araus JL, Kefauver SC, Zaman-Allah M, Olsen MS, Cairns JE (2018) Translating high throughput phenotyping into genetic gain. Trends Plant Sci 23:451–466

Babu R, Nair SK, Kumar A, Venkatesh S, Sekhar JC, Singh NN, Srinivasan G, Gupta HS (2005) Two-generation marker-aided backcrossing for rapid conversion of normal maize lines to quality protein maize (QPM). Theor Appl Genet 111:888–897

Babu R, Rojas NP, Gao S, Yan J, Pixley K (2013) Validation of the effects of molecular marker polymorphisms in LcyE and CrtRB1 on provitamin a concentrations for 26 tropical maize populations. Theor Appl Genet 126:389–399

Badu-Apraku B, Fakorede MAB, Talabi AO, Oyekunle M, Akaogu IC, Akinwale RO, Annor B, Melaku G, Fasanmade Y, Aderounmu M (2016) Gene action and heterotic groups of early white quality protein maize inbreds under multiple stress environments. Crop Sci 56:183–199

Badu-Apraku B, Yallou CG, Oyekunle M (2013) Genetic gains from selection for high grain yield and *Striga* resistance in early maturing maize cultivars of three breeding periods under *Striga*-infested and *Striga*-free environments. Field Crops Res 147:54–67

Beló A, Zheng P, Luck S, Shen B, Meyer DJ, Li B, Tingey S, Rafalski A (2008) Whole genome scan detects an allelic variant of fad2 associated with increased oleic acid levels in maize. Mol Genet Genom 279:1–10

Bernardo R (2016) Bandwagons I, too, have known. Theor Appl Genet 129:2323–2332

Beyene Y, Mugo S, Oikeh SO, Juma C, Olsen M, Prasanna BM (2017) Hybrids performance of doubled haploid lines derived from 10 tropical bi-parental maize populations evaluated in contrasting environments in Kenya. Afr J Biotechnol 16:371–379. https://doi.org/10.5897/AJB2016.15697

Beyene Y, Semagn K, Crossa J, Mugo S, Tarekegne A, Babu R, Meisel B, Sehabiague P, Olsen M, Prasanna BM, Atlin GN, Vargas M, Banziger M (2015) Genetic gains in grain yield through genomic selection in eight bi-parental maize populations under drought stress. Crop Sci 55:154–163

Beyene Y, Semagn K, Mugo S, Prasanna BM, Tarekegne A, Gakunga J, Sehabiague P, Meisel B, Oikeh SO, Olsen M, Crossa J (2016) Performance and grain yield stability of maize populations developed using marker-assisted recurrent selection and pedigree selection procedures. Euphytica 208:285–297

Bortiri E, Jackson D, Hake S (2006) Advances in maize genomics: the emergence of positional cloning. Curr Opin Plant Biol 9:164–171

Bruulsema T, Tollenaar M, Heckman J (2000) Boosting crop yields in the next century. Better Crop 84:9–13

Buckler E, Ilut DC, Wang XY, Kretzschmar T, Gore MA, Mitchell SE (2016) rAmpSeq: using repetitive sequences for robust genotyping. bioRxiv, https://doi.org/10.1101/096628. https://doi.org/10.1016/j.molp.2017.06.008

Bukowski R, Guo X, Lu Y, Zou C, He B, Rong Z, Wang B, Xu D, Yang B, Xie C, Fan L, Gao S, Xu X, Zhang G, Li Y, Jiao Y, Doebley JF, Ross-Ibarra J, Lorant A, Buffalo V, Romay MC, Buckler ES, Ware D, Lai J, Sun Q, Xu Y (2018) Construction of the third-generation *Zea mays* haplotype map. Gigascience 7:1–12. https://doi.org/10.1093/gigascience/gix134

Cairns JE, Prasanna BM (2018) Developing and deploying climate-resilient maize varieties in the developing world. Curr Opin Plant Biol 45. https://doi.org/10.1016/j.pbi.2018.05.004

Cairns JE, Sonder K, Zaidi PH, Verhulst N, Mahuku G, Babu R, Nair SK, Das B, Govaerts B, Vinayan MT, Rashid Z, Noor JJ, Devi P, San Vicente F, Prasanna BM (2012) Maize production in a changing climate: impacts, adaptation and mitigation strategies. Adv Agron 114:1–58

Cao S, Loladze A, Yuan Y, Wu Y, Zhang A, Chen J, Huestis G, Cao J, Chaikam V, Olsen M, Prasanna BM, San Vicente F, Zhang X (2017) Genome-wide analysis of tar spot complex resistance in maize using genotyping-by-sequencing SNPs and whole-genome prediction. Plant Genome 10:1–14. https://doi.org/10.3835/plantgenome2016.10.0099

Chaikam V, Nair SK, Martinez L, Lopez LA, Utz HF, Melchinger AE, Prasanna BM (2018) Marker-assisted breeding of improved maternal haploid inducers in maize for the tropical/subtropical regions. Front Plant Sci 9:1527. https://doi.org/10.3389/fpls.2018.01527

Challinor AJ, Koehler A-K, Ramirez-Villegas J, Whitfield S, Das B (2016) Current warming will reduce yields unless maize breeding and seed systems adapt immediately. Nat Clim Change 6:954–958

Chia J-M, Song C, Bradbury PJ, Costich D, de Leon N, Doebley J, Elshire RJ, Gaut B, Geller L, Glaubitz JC, Gore M, Guill KE, Holland J, Hufford MB, Lai J, Li M, Liu X, Lu Y, McCombie R, Nelson R, Poland J, Prasanna BM, Pyhäjärvi T, Rong T, Sekhon RS, Sun Q, Tenaillon MI, Tian F, Wang J, Xu X, Zhang Z, Kaeppler SM, Ross-Ibarra J, McMullen MD, Buckler ES, Zhang G, Xu Y, Ware D (2012) Maize HapMap2 identifies extant variation from a genome in flux. Nat Genet 44:803

Ci X, Li M, Liang X, Xie Z, Zhang D, Li X (2011) Genetic contribution to advanced yield for maize hybrids released from 1970 to 2000 in China. Crop Sci 51:13–20

Collins N, Drake J, Ayliffe M, Sun Q, Ellis J, Hulbert S, Pryor T (1999) Molecular characterization of the maize Rp1-D rust resistance haplotype and its mutants. Plant Cell 11:1365–1376

Combs E, Bernardo R (2013) Accuracy of genome-wide selection for different traits with constant population size, heritability, and number of markers. Plant Genome 6:1–7

Cooper M, Messina CD, Podlich D et al (2014) Predicting the future of plant breeding: complementing empirical evaluation with genetic prediction. Crop Pasture Sci 65:311–336

Crommelinck S, Höfle B (2016) Simulating an autonomously operating low-cost static terrestrial LiDAR for multitemporal maize crop height measurements. Remote Sens 8:205. https://doi.org/10.3390/rs8030205

De Groote H, Oloo F, Tongruksawattana S, Das B (2016) Community-survey based assessment of the geographic distribution and impact of maize lethal necrosis (MLN) disease in Kenya. Crop Protec 82:30–35

Dell'Acqua M, Gatti DM, Pea G, Cattonaro F, Coppens F, Magris G, Hlaing AL, Aung HH, Nelissen H, Baute J, Frascaroli E, Churchill GA, Inzé D, Morgante M, Pè ME (2015) Genetic properties of the MAGIC maize population: a new platform for high definition QTL mapping in *Zea mays*. Genome Biol 16:1–23. https://doi.org/10.1186/s13059-015-0716-z

Deutsch CA, Tewksbury JJ, Tigchelaar M, Battisti DS, Merrill SC, Huey RB, Naylor RL (2018) Increase in crop losses to insect pests in a warming climate. Science 361:916–919

Duvick DN (2005) The contribution of breeding to yield advances in maize (*Zea mays* L.). Adv Agron 86:83–145

Early R, Gonzalez-Moreno P, Murphy ST, Day R (2018) Forecasting the global extent of invasion of the cereal pest *Spodoptera frugiperda*, the fall armyworm. bioRxiv http://dx.doi.org/10.1101/391847

Edmeades GO, Bolaños J, Chapman SC, Lafitte HR, Banziger M (1999) Selection improves drought tolerance in tropical maize populations: I. Gains in biomass, grain yield, and harvest index. Crop Sci 39:1306–1315

Edmeades GO, Trevisan W, Prasanna BM, Campos H (2017) Tropical maize (*Zea mays* L.). In: Genetic improvement of tropical crops, Springer, Cham, pp 57–109

Falconer DS, Mackay TFC (1996) Introduction to quantitative genetics, 4th edn. Addison Wesley Longman, Harlow

FAOSTAT (2018) FAO statistical database. FAO, Rome, Italy

Feng F, Wang Q, Liang C, Yang R, Li X (2015) Enhancement of tocopherols in sweet corn by marker-assisted backcrossing of ZmVTE4. Euphytica 206:513–521

Fentahun M, Feyissa T, Abraham A, Kwak HR (2017) Detection and characterization of maize chlorotic mottle virus and sugarcane mosaic virus associated with maize lethal necrosis disease in Ethiopia: an emerging threat to maize production in the region. Eur J Plant Pathol 149:1011–1017

Frey TJ (2006) Fine-mapping, cloning, verification, and fitness evaluation of a QTL, Rcg1, which confers resistance to *Colletotrichum graminicola* in maize. PhD Dissertation, University of Delaware, USA

Friedli M, Kirchgessner N, Grieder C, Liebisch F, Mannale M, Walter A (2016) Terrestrial 3D laser scanning to track the increase in canopy height of both monocot and dicot crop species under field conditions. Plant Meth 12:9. https://doi.org/10.1186/s13007-016-0109-7

Fu H, Dooner HK (2002) Intraspecific violation of genetic colinearity and its implications in maize. Proc Natl Acad Sci USA 99:9573–9578

Gao S, Martinez C, Skinner DJ, Krivanek AF, Crouch JH, Xu Y (2008) Development of a seed DNA-based genotyping system for marker-assisted selection in maize. Mol Breed 22:477–494

Gilles LM, Khaled A, Laffaire J, Chaignon S, Gendrot G, Laplaige J, Bergès H, Beydon G, Bayle V, Barret P, Comadran J, Martinant JP, Rogowsky PM, Widiez T (2017) Loss of pollen-specific phospholipase NOT LIKE DAD triggers gynogenesis in maize. EMBO J 36:707–717

Golicz AA, Batley J, Edwards D (2016) Towards plant pangenomics. Plant Biotechnol J 14:1099–1105

Gore MA, Chia JM, Elshire RJ, Sun Q, Ersoz ES et al (2009) A first-generation haplotype map of maize. Science 326:1115–1117

Gowda M, Das B, Makumbi D, Babu R, Semagn K, Mahuku G, Olsen MS, Bright JM, Beyene Y, Prasanna BM (2015) Genome-wide association and genomic prediction of resistance to maize lethal necrosis disease in tropical maize germplasm. Theor Appl Genet 128:1957–1968

Gowda M, Beyene Y, Makumbi D, Semagn K, Olsen MS, Bright JM, Das B, Mugo S, Suresh LM, Prasanna BM (2018) Discovery and validation of genomic regions associated with resistance to maize lethal necrosis in four biparental populations. Mol Breed 38:66. https://doi.org/10.1007/s11032-018-0829-7

Gowda M, Worku M, Nair SK, Palacios-Rojas N, Prasanna BM (2017) Quality Assurance/Quality Control (QA/QC) in maize breeding and seed production: theory and practice. Nairobi, CIMMYT, p 28

Gupta HS, Agrawal PK, Mahajan V, Bisht GS, Kumar A, Verma P, Srivastava A, Saha S, Babu R, Pant MC, Mani VP (2009) Quality protein maize for nutritional security: rapid development of short duration hybrids through molecular marker assisted breeding. Curr Sci 96:230–237

Hämmerle H, Höfle B (2016) Direct derivation of maize plant and crop height from low-cost time-of-flight camera measurements. Plant Meth 12:50. https://doi.org/10.1186/s13007-016-0150-6

Harjes CE, Rocheford TR, Bai L, Brutnell TP, Kandianis CB, Sowinski SG, Stapleton AE, Vallabhaneni R, Williams M, Wurtzel ET, Yan J, Buckler ES (2008) Natural genetic variation in lycopene epsilon cyclase tapped for maize biofortification. Science 319:330–333

Hindu V, Palacios N, Babu R, Suwarno WB, Rashid Z, Usha R, Saykhedkar GR, Nair SK (2018) Identification and validation of genomic regions influencing kernel zinc and iron in maize. Theor Appl Genet. https://doi.org/10.1007/s00122-018-3089-3

Horikoshi RJ, Bernardi D, Bernardi O, Malaquias JB, Okuma DM, Miraldo LL, Amara FSD, Omoto C (2016) Effective dominance of resistance of *Spodoptera frugiperda* to Bt maize and cotton varieties: implications for resistance management. Sci Rep 6:34864

Hossain F, Muthusamy V, Pandey N, Vishwakarma AK, Baveja A et al (2018a) Marker-assisted introgression of opaque2 allele for rapid conversion of elite maize hybrids into quality protein maize. J Genet 97:287–298

Hossain F, Muthusamy V, Zunjare RU (2018b) Molecular breeding for development of biofortified maize hybrids in India. In: Prasanna BM, Das A, Kaimenyi KK (eds) Book of extended summaries, 13th Asian maize conference and expert consultation on maize for food, feed, nutrition and environmental security. Ludhiana, India, 8–10 Oct 2018. CIMMYT, Mexico D.F., pp 210–220

Hurni S, Scheuermann D, Krattinger SG et al (2015) The maize disease resistance gene Htn1 against northern corn leaf blight encodes a wall-associated receptor-like kinase. Proc Natl Acad Sci USA 112:8780–8785. https://doi.org/10.1073/pnas.1502522112

ISAAA (2017) Global status of commercialized Biotech/GM crops: 2017. ISAAA Brief No. 53. ISAAA: Ithaca, NY

Johal GS, Briggs SP (1992) Reductase activity encoded by the HM1 disease resistance gene in maize. Science 258:985–987

Kelliher T, Starr D, Richbourg L et al (2017) MATRILINEAL, a sperm-specific phospholipase, triggers maize haploid induction. Nature 542:105–109

Korte A, Farlow A (2013) The advantages and limitations of trait analysis with GWAS: a review. Plant Meth 9:1–9

Li H, Rasheed A, Hickey LT, He Z (2018) Fast-forwarding genetic gain. Trends Plant Sci 23:184–186

Liang X, Wang K, Huang C, Zhang X, Yan J, Yang W (2016) A high-throughput maize kernel traits scorer based on line-scan imaging. J Intl Measure Confed 90:453–460

Liu C, Li X, Meng D, Chen C, Dong X, Xu X, Chen B, Li W, Li L, Tian X, Zhao H, Song W, Luo H, Zhang Q, Lai J, Jin W, Yan J, Chen S (2017a) A 4-bp insertion at ZmPLA1 encoding a putative phospholipase a generates haploid induction in maize. Mol Plant 10:520–522

Liu Q, Liu H, Gong Y, Tao Y, Jiang L, Zuo W, Yang Q, Ye J, Lai J, Wu J, Lübberstedt T, Xu M (2017b) An atypical thioredoxin imparts early resistance to sugarcane mosaic virus in maize. Mol Plant 10:483–497

Lu F, Romay MC, Glaubitz JC, Bradbury PJ, Elshire RJ et al (2015) High-resolution genetic mapping of maize pan-genome sequence anchors. Nat Commun 6:6914

Luque SF, Cirilo AG, Otegui ME (2006) Genetic gains in grain yield and related physiological attributes in argentine maize hybrids. Field Crops Res 95:383–397

Lynch M, Walsh B (1998) Genetics and analysis of quantitative traits. Sinauer, Sunderland, Mass, USA, 980 p. https://doi.org/10.1086/318209

Mahuku G, Lockhart BE, Wanjala B, Jones MW, Kimunye JN, Stewart LR, Cassone BJ, Sevgan S, Johnson N, Kusia E, Lava Kumar P, Niblett CL, Wangai A, Kiggundu A, Asea G, Pappu H, Prasanna BM, Redinbaugh MG (2015) Maize lethal necrosis (MLN), an emerging threat to maize-based food security in sub-Saharan Africa. Phytopathology 105:956–965

Makanza R, Zaman-Allah M, Cairns JE, Magorokosho C, Tarekegne A, Olsen M, Prasanna BM (2018a) High-throughput phenotyping of canopy cover and senescence in maize field trials using aerial digital canopy imaging. Remote Sens 10(2):330. https://doi.org/10.3390/rs10020330

Makanza R, Zaman-Allah M, Cairns JE, Eyre J, Burgueño J, Pacheco A, Diepenbrock C, Magorokosho, Tarekegne A, Olsen M, Prasanna BM (2018b) High-throughput method for ear phenotyping and kernel weight estimation in maize using ear digital imaging. Plant Meth 14:49. https://doi.org/10.1186/s13007-018-0317-4

Marenya P, Erenstein O, Prasanna BM, Makumbi D, Bright J, Beyene Y (2018) Maize lethal necrosis disease: evaluating agronomic and genetic control strategies for Ethiopia and Kenya. Agri Syst 162:220–228

Masuka B, Atlin GN, Olsen M, Magorokosho C, Labuschagne M, Crossa J, Bänziger M, Pixley K, Vivek B, Biljon AV, Macrobert J, Prasanna BM, Makumbi D, Tarekegne A, Das B, Zaman-Allah M, Cairns JE (2017a) Gains in maize genetic improvement in eastern and southern Africa: I CIMMYT hybrid breeding pipeline. Crop Sci 57:168–179

Masuka B, Magorokosho C, Olsen M, Atlin GN, Bänziger M, Pixley K, Vivek B, Labuschagne M, Matemba-Mutasa R, Burgenõ J, Macrobert J, Prasanna BM, Das B, Makumbi D, Tarekegne A, Crossa J, Zaman-Allah M, Cairns JE (2017b) Gains in maize genetic improvement in Eastern and Southern Africa (ii) CIMMYT open pollinated varieties (OPVs) breeding pipeline. Crop Sci 57:180–191

Menkir A, Maziya-Dixon B, Mengesha W, Rocheford T, Alamuet EO (2017) Accruing genetic gain in pro-vitamin a enrichment from harnessing diverse maize germplasm. Euphytica 213. https://doi.org/10.1007/s10681-017-1890-8

Meuwissen THE, Hayes BJ, Goddard ME (2001) Prediction of total genetic value using genome-wide dense marker maps. Genetics 157:1819–1829

Mihm JA (ed) (1997) Insect resistant maize: recent advances and utilization. In: Proceedings of an international symposium held at the international maize and wheat improvement center (CIMMYT), Nov 27–Dec 3 1994. CIMMYT, Mexico, D.F

Mikkelsen TS, Hillier LW, Eichler EE, Zody MC, Jaffe DB et al (2005) Initial sequence of the chimpanzee genome and comparison with the human genome. Nature 437:69–87

Miller ND, Haase NJ, Lee J, Kaeppler SM, de Leon N, Spalding EP (2017) A robust, high-throughput method for computing maize ear, cob, and kernel attributes automatically from images. Plant J 89:169–178

Moose SP, Mumm RH (2008) Molecular plant breeding as the foundation for 21st century crop improvement. Plant Physiol 147:969–977

Morrell PL, Buckler ES, Ross-Ibarra J (2012) Crop genomics: advances and applications. Nat Rev Genet 13:85

Muthusamy V, Hossain F, Nepolean T, Choudhary M, Saha S, Bhat JS, Prasanna BM, Gupta HS (2014) Development of β-carotene rich maize hybrids through marker-assisted introgression of β-carotene hydroxylase allele. PLoS ONE 9:1–22. https://doi.org/10.1371/journal.pone.011358310.1007/BF00020113

Nair S, Babu R, Magorokosho C, Mahuku G, Semagn K, Beyene Y, Das B, Makumbi D, Kumar L, Olsen M, Prasanna BM (2015) Fine mapping of *Msv1*, a major QTL for resistance to maize streak virus leads to development of production markers for breeding pipelines. Theor Appl Genet 128:1839–1854

Olsen MS, Cairns JE, Gowda M, Chaikam V, Zaman-Allah M, Beyene Y, Nair SK, Gao SY, Bright JM, Burgueño J, Crossa J, Gedil MA, Ndhlela T, N'gan'ga M, Ogugo V, Ramu P, Robbins K, Suresh LM, Zhang X, Prasanna BM (2018) Accelerating maize breeding in the (sub)tropics - Innovative tools to reduce time and costs. Theor Appl Genet (Under Review)

Prasanna BM, Araus JL, Crossa J, Cairns JE, Palacios N, Das B, Magorokosho C (2013) High throughput and precision phenotyping in cereal breeding programs. In: Gupta, PK Varshney RK (eds) Cereal Genomics, 2nd edn. Springer, Berlin

Prasanna BM, Chaikam V, Mahuku G (2012) Doubled haploid technology in maize breeding: theory and practice. CIMMYT, Mexico D.F

Prasanna BM, Huesing JE, Eddy R, Peschke VM (eds) (2018) Fall Armyworm in Africa: A Guide for Integrated Pest Management, 1st edn. CIMMYT, Mexico, CDMX

Prigge V, Sánchez C, Dhillon BS, Schipprack W, Araus JL et al (2011) Doubled haploids in tropical maize. I. Effects of inducers and source germplasm on *in vivo* haploid induction rates. Crop Sci 51:1498–1506

Prigge V, Xu X, Li L, Babu R, Chen S, Atlin GN, Melchinger AE (2012) New insights into the genetics of *in vivo* induction of maternal haploids, the backbone of doubled haploid technology in maize. Genetics 190:781–793

Rafalski A, Morgante M (2004) Corn and humans: recombination and linkage disequilibrium in two genomes of similar size. Trends Genet 20:103–111. https://doi.org/10.1016/j.tig.2003.12.002

Rashid Z, Singh PK, Hindu V, Zaidi PH (2018) Genome-wide association study in Asia-adapted tropical maize reveals novel and explored genomic regions for sorghum downy mildew resistance. Sci Rep 1–12. https://doi.org/10.1038/s41598-017-18690-3

Riedelsheimer C, Endelman JB, Stange M, Sorrells ME, Jannink JL, Melchinger AE (2013) Genomic predictability of interconnected biparental maize populations. Genetics 194:493–503

Sarika K, Hossain F, Muthusamy V, Zunjare RU, Baveja A, Goswami R, Thirunavukkarasu N, Jha SK, Gupta HS (2018) Opaque16, a high lysine and tryptophan mutant, does not influence the key physico-biochemical characteristics in maize kernel. PLoS ONE 13:1–14. https://doi.org/10.1371/journal.pone.0190945

Schmidt RJ, Burrt FA, Aukerman MJ, Burr B (1990) Maize regulatory gene opaque-2 encodes a protein with a "leucine-zipper" motif that binds to zein DNA. Proc Natl Acad Sci USA 87:46–50

Shiferaw B, Prasanna B, Hellin J, Banziger M (2011) Crops that feed the world. 6. Past successes and future challenges to the role played by maize in global food security. Food Secur 3:307–327

Shiferaw B, Tesfaye K, Kassie M, Abate T, Prasanna BM, Menkir A (2014) Managing vulnerability to drought and enhancing livelihood resilience in sub-Saharan Africa: technological, institutional and policy options. Weather Clim Extrem 3:67–79

Smith C (1967) Improvement of metric traits through specific genetic loci. Anim Prod 9:349–358

Suwarno WB, Pixley KV, Palacios-Rojas N, Kaeppler SM, Babu R (2015) Genome-wide association analysis reveals new targets for carotenoid biofortification in maize. Theor Appl Genet 128:851–864

Tao Y, Jiang L, Liu Q, Zhang Y, Zhang R, Ingvardsen CR, Xu M (2013) Combined linkage and association mapping reveals candidates for Scmv1, a major locus involved in resistance to sugarcane mosaic virus (SCMV) in maize. BMC Plant Biol 13:162

Tenaillon MI, Sawkins MC, Long AD, Gaut RL, Doebley JF, Gaut BS (2001) Patterns of DNA sequence polymorphism along chromosome 1 of maize (*Zea mays* ssp. *mays* L.). Proc Natl Acad Sci USA 98:9161–9866

Tesfaye K, Zaidi PH, Gbegbelegbe S, Boeber C, Rahut DB, Getaneh F, Seetharam K, Erenstein O, Stirling C (2017) Climate change impacts and potential benefits of heat-tolerant maize in South Asia. Theor Appl Clim 130:959–970

Tesfayea K, Kruseman G, Cairns JE, Zaman-Allah M, Wegary D, Zaidi PH, Bootee KJ, Rahut D, Erenstein O (2018) Potential benefits of drought and heat tolerance for adapting maize to climate change in tropical environments. Clim Risk Manag 19:106–119

Vivek BS, Krishna GK, Vengadessan V, Babu R, Zaidi PH, Kha LQ, Mandal SS, Grudloyma P, Takalkar S, Krothapalli K, Singh IS, Ocampo ETM, Xingming F, Burgueño J, Azrai M, Singh RP, Crossa J (2017) Use of genomic estimated breeding values results in rapid genetic gains for drought tolerance in maize. Plant Genome 10:0. https://doi.org/10.3835/plantgenome2016.07.0070

Warr A, Robert C, Hume D, Archibald A, Deeb N, Watson M (2015) Exome sequencing: current and future perspectives. Genes Genomes Genet 5:1543–1550

Whitt SR, Wilson LM, Tenaillon MI, Gaut BS, Buckler ES (2002) Genetic diversity and selection in the maize starch pathway. Proc Natl Acad Sci USA 99:12959–12962

Wright SI (2005) The effects of artificial selection on the maize genome. Science 308:1310–1314

Xiong X, Yu L, Yang WN, Liu M, Jiang N, Wu D, Chen G, Xiong L, Liu K, Liu Q (2017) A high throughput stereo-imaging system for quantifying rape leaf traits during the seedling stage. Plant Meth 13:17. https://doi.org/10.1186/s13007-017-0157-7

Xu C, Ren Y, Jian Y, Guo Z, Zhang Y, Xie C, Fu J, Wang H, Wang G, Xu Y, Li P, Zou C (2017a) Development of a maize 55 K SNP array with improved genome coverage for molecular breeding. Mol Breed 37:20. https://doi.org/10.1007/s11032-017-0622-z

Xu Y (2010) Molecular plant breeding. CAB International, Wallingford, p 734

Xu Y, Li P, Zou C, Lu Y, Xie C, Zhang X, Prasanna BM, Olsen MS (2017) Enhancing genetic gain in the era of molecular breeding. J Exp Bot 68:2641–2666

Yan J, Kandianis CB, Harjes CE, Bai L, Kim EH, Yang X, Skinner DJ, Fu Z, Mitchell S, Li Q, Fernandez MG, Zaharieva M, Babu R, Fu Y, Palacios N, Li J, Dellapenna D, Brutnell T, Buckler ES, Warburton ML, Rocheford T (2010) Rare genetic variation at *Zea mays crtRB1* increases β-carotene in maize grain. Nat Genet 42:322–327

Yang L, Wang W, Yang W, Wang M (2013) Marker-assisted selection for pyramiding the waxy and opaque-16 genes in maize using cross and backcross schemes. Mol Breed 31:767–775

Yu J, Holland JB, McMullen MD, Buckler ES (2008) Genetic design and statistical power of nested association mapping in maize. Genetics 178:539–551

Zaidi PH, Seetharam K, Krishna G, Krishnamurthy L, Gajanan S, Babu R, Zerka M, Vinayan MT, Vivek BS (2016) Genomic regions associated with root traits under drought stress in tropical maize (*Zea mays* L.). PLoS One. 11: e0164340. https://doi.org/10.1371/journal.pone.0164340

Zaman-Allah M, Vergara O, Araus JL, Tarekegne A, Magorokosho C, Zarco-Tejeda PJ, Hornero A, Hernandez Alba A, Das B, Craufurd P, Olsen M, Prasanna BM, Cairns JE (2015) Unmanned aerial platform-based multi-spectral imaging for field crop phenotyping in maize. Plant Meth 11:35

Zhang X, Pérez-Rodríguez P, Burgueño J, Olsen M, Buckler E, Atlin G, Prasanna BM, Vargas M, San Vicente F, Crossa J (2017) Rapid cycling genomic selection in a multiparental tropical maize population. Genes Genomes Genet 7:2315–2326

Zheng P, Allen WB, Roesler K, Williams ME, Zhang S, Li J, Glassman K, Ranch J, Nubel D, Solawetz W, Bhattramakki D, Llaca V, Deschamps S, Zhong GY, Tarczynski MC, Shen B (2008) A phenylalanine in DGAT is a key determinant of oil content and composition in maize. Nat Genet 40:367–372

Zunjare RU, Hossain F, Muthusamy V, Baveja A, Chauhan HS, Bhat JS, Thirunavukkarasu N, Saha S, Gupta HS (2018) Development of biofortified maize hybrids through marker-assisted stacking of β-carotene hydroxylase, lycopene-ε-cyclase and *opaque2* genes. Front Plant Sci. 9:1–12. https://doi.org/10.3389/fpls.2018.00178

Zuo W, Chao Q, Zhang N, Ye J, Tan G, Li B, Xing Y, Zhang B, Liu H, Fengler KA, Zhao J, Zhao X, Chen Y, Lai J, Yan J, Xu M (2015) A maize wall-associated kinase confers quantitative resistance to head smut. Nat Genet 47:151–157

Chapter 4
Genomic Approaches for Climate Resilience Breeding in Oats

Julio Isidro-Sánchez, Elena Prats, Catherine Howarth, Tim Langdon and Gracia Montilla-Bascón

Abstract Oat (*Avena sativa* L.), ranking sixth in world cereal production, is primarily produced as a multipurpose crop for grain, pasture, and forage or as a rotation crop in many parts of the world. Recent research has elevated its potential dietary value for human nutrition and health care. Oats are well adapted to a wide range of soil types and can perform on acid soils. World oat production is concentrated between latitudes 35–65° N, and 20–46° S. *Avena* genomes are large and complex, in the range of 4.12–12.6 Gb. Oat productivity is affected by many diseases, although crown rust (*Puccinia coronate* f. sp. *avenae*) and stem rust (*Puccinia graminis* f. sp. *avenae*) are the key diseases worldwide. The focus of this chapter is to review the major developments and their impacts on oat breeding, especially on the challenges posed by climate or environmental changes (biotic and abiotic stresses mainly) for oat cultivation. Next-generation breeding tools will help to develop approaches to genetically improve and manipulate oat which would aid significantly in oat enhancement efforts. Although, oat biotechnology has been advanced at a similar pace as the rest of cereals, it lags still behind. More genomic tools, from genomic assisted breeding to genome editing tools are needed to improve the resources to improve oats under climate change in the next few decades.

Keywords Oat · Resilience · B-glucans · Genomic assisted breeding · Disease resistance

J. Isidro-Sánchez (✉)
School of Agriculture and Food Science, University College Dublin, Dublin, Ireland
e-mail: j.isidro@ucd.ie

E. Prats · G. Montilla-Bascón
CSIC, Institute for Sustainable Agriculture, Córdoba, Spain

C. Howarth · T. Langdon
Institute of Biological, Environmental and Rural Sciences, Aberystwyth University Wales, Aberystwyth, UK

C. Kole (ed.), *Genomic Designing of Climate-Smart Cereal Crops*,
https://doi.org/10.1007/978-3-319-93381-8_4

4.1 Introduction

Oat is a self-pollinated cereal grain crop that belongs to the family Gramineae (Poaceae) with a haploid chromosome number of 7. The genus *Avena* has diploid, tetraploid, and hexaploid ploidy levels and comprises about 70 species. The world's sequencing databases hold about 80,000 accessions of cultivated oat species and preserve about 20,000 accessions of wild *Avena* species (Badaeva et al. 2010).

Hexaploid oat, *Avena sativa* L (common oat) and *Avena byzantina* L are the main oats grown for grain and fodder. It is believed that hexaploid oats appeared from a spontaneous polyploidization event involving the fusion of an ancestral diploid and a tetraploid (Ladizinsky 2012). The genome designations of A, C, and D were giving to hexaploid oat (AACCDD) in 1974 (Rajhathy and Thomas 1974). The A and D genomes are very similar, while the C genome is quite different. This difference is easily observable at the chromosome level in C-banded karyotypes. The C genome chromosomes stain much darker than the A and D genomes (Yan et al. 2016a). It is hypothesized that the polyploidization events (tetra andhexa) occurred in the northwest of Africa (Chew et al. 2016). Recent studies (Lippi et al. 2015) have shown that the hunter-gatherer population (32.600 BP) preferred genus *Avena* to process food. This finding is currently the most ancient evidence of the processing cereals.

Currently, oat remains an important crop in many countries as it serves as a dual-purpose crop. i.e., for forage and fodder and for animal feed. As a forage/fodder oat is a great source of minerals, protein, and fiber. It can be fed green and the surplus can be converted into silage or hay for use during the scarcity period. In agriculture, oat is highly suitable for extensive and sustainable production systems. Oats are well suited for use as cover or break crops in winter rotations since they are not susceptible to the major root diseases of wheat and barley. Due to their high biomass production, they also reduce weed growth. Cultivation of oats also assures production of maximum quantities of early and nutritious forage supplies in the fodder-deficit periods during freezing temperatures in the winter season. It is also one of the major sources of green forage available in winter for the starving cattle. As animal feed consumption oat is a key particularly for calves, horses, sheep, and poultry.

Oats are well adapted to a wide range of soil types and can perform well on acid soils. World oat production is concentrated between latitudes 35–65° N, and 20–46° S. Most of the world's production comes from spring-sown cultivars, although autumn sowing is practiced in Mediterranean climates and in regions with mild winters as in Ireland, where spring oats are planted as a winter habit growth. Where winters are long and severe, such as in Canada, northern states of USA and Russian federation, as well as in Scandinavia, Finland, and Norway, short season to mild maturing oat are preferably sown. In regions with temperate climates, oats are variously spring, winter, and/or autumn sown depending on regional climatic conditions, crop rotation requirements, end use, and other farming practices. In warmer regions, spring type oats are sown in autumn to avoid summer heat and drought.

Grain oats are classified as high in soluble fiber, β-glucans, lipids, protein, and antioxidants. Dietary guidelines and nutritional policies have advised the consumers at least three servings of whole grains daily. This recommendation is based on evidence gathered from observational studies on the effects of consumption of whole grains on chronic disease (Jones and Engleson 2010).

Although oats have many health and nutritional benefits, the recognition when comparing to other cereals like wheat, corn, and rice, is limited. As a result, oat had received far less research funding and had fewer genomic resources available for genetic and genomic research. The oat genome is approximately 11.3 Gbp and harbors many chromosomal rearrangements including major translocations (Rajhathy and Thomas 1974; Jellen and Leggett 2006). These rearrangements are often population specific, making it difficult to transfer data between populations and to develop genomic tools that can be used for any given population. Nevertheless, new tools and genomics resources have become available. For oats, several marker systems had been developed including, the Oat 6 K single nucleotide polymorphism (SNP) chip and diversity array (DArT) markers (Oliver et al. 2011; Tinker et al. 2014). The DArT marker system is based on a microarray platform where reduced representation genomic libraries are created and hybridized to microarrays. The current DArT platform has 2,349 polymorphic markers. The Oat 6 K custom Infinium iSelect BeadChip (Illumina, SanDiego, CA) has 4,975 SNP markers. The SNP chip provides a quick method of getting high-density marker data for genome-scale projects. Several oat populations have also been genotyped using genotyping by sequencing (GBS), providing an additional source of genetic markers. The combination of these technologies and marker systems provides a large volume of genetic markers making it possible to do genome-scale projects (Smith 2017).

4.1.1 Economic Importance of Oats

Although, the oat crop is primarily produced for animal feed, recent research has elevated its potential dietary value for human consumption due to its economic value in human nutrition and health care.

The global oat crop has retained its place as an important cereal, despite a decline in hectare. Oat was produced for animal feed and for human food in over 70 countries during the past decade (FAO 2017a). Recent recognition of health components within the grain, such as β-glucan and antioxidants, has bolstered consumer interest in oat. Cereal β-glucan has been shown to be beneficial in both helpings reduce cholesterol and stabilize blood serum in type II diabetes. Research has also shown that oat produces unique antioxidants, which when fully characterized and capitalized on through breeding and biotechnology, may further increase the dietary value of the crop.

4.1.1.1 Oat Production and Value

Oats rank sixth in the world cereal statistics following maize, rice, wheat, barley, soybean, and sorghum. Although oat grain has always been an important livestock feed, the world oat grain declines as farm mechanization increased from 1930. In the developing world, oats remain an important crop in marginal ecologies and for specialist uses in developed countries. In many parts of the world, oats are grown for grain as well as for forage and fodder, hay, silage, straw for bedding and chaff.

The global average oat crop (Fig. 4.1) area harvested has decreased from 1961 with 38 millions of hectares to 23 million in 2016 while the world production decreased from 50 to 22 million tons in 2017 (FAO 2017b). In turn, oat production is centralized in Europe with 62.3% of the total global production followed by America (29.6%), Asia and Oceanía (3.8%), and Africa (0.3%). Currently, the Russian Federation is the highest producer worldwide with an average of more than 10 million tons, followed by United States, Canada, and Germany.

Within the top 10 producers, the oat price per ton has generally increased except in Spain, Finland, and the Russian Federation. It is worth to note that there has been a significant increase (32%) in price per ton in Australia from 74 to 227$ US/ton.

4.1.1.2 Health Grain Quality Oat

Oats are known for their cholesterol-lowering properties, which has led to health claims both by the US Food and Drug Administration (FDA) and by the European

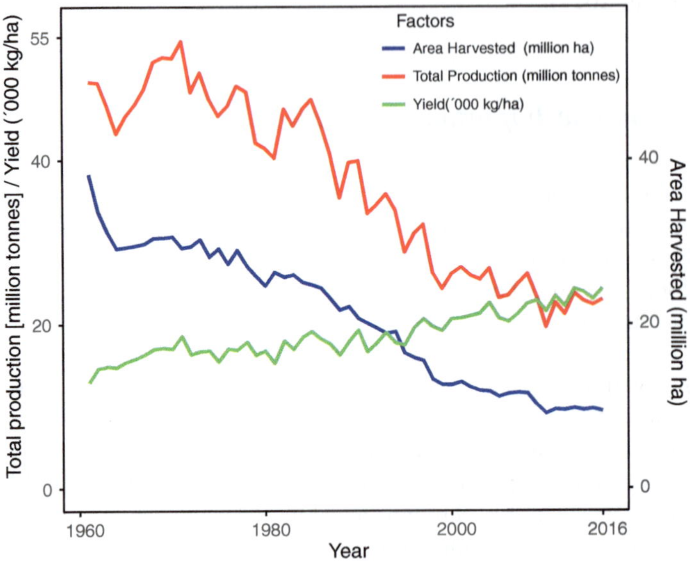

Fig. 4.1 Area Harvested, total production, and yield of oats from 1960 to 2016 (FAO 2017a)

Food Safety Authority (EFSA) (FDA 2012; Thies et al. 2014a). Usually, the reduction of plasma cholesterol is attributed to the soluble mixed linked (1,3), (1,4)-β-D-glucan fibers of oats, often referred to only as β-glucans. Oats do, however, also contain other potential bioactive components, such as lipids, phenolic acids, sterols, flavonoids, and a group of polyphenols unique for oats, avenanthramides (Tripathi et al. 2018). Results in recent animal studies have revealed that both oat lipids and oat proteins have cholesterol-lowering properties (Thies et al. 2014b). The avenanthramides have not been reported to reduce plasma cholesterol, but to have other beneficial properties in the prevention of cardiovascular diseases, such as antioxidative, anti-inflammatory, antiproliferative, and vasodilatory effects (Meydani 2009). Consumption of oats is associated with reduced LDL levels in blood and with increased cholesterol and bile acid output in feces, as shown in both human and animal studies (Slavin et al. 1999; Sengupta et al. 2006). This suggests that the lowered blood cholesterol may be due to reduced intestinal bile acid or cholesterol uptake, to increased hepatic bile acid or reduced LDL synthesis, or to combinations of these effects.

4.1.2 Oat Composition and Structure

4.1.2.1 Oat Grain Structure

Whole grains were defined as "the intact, ground, cracked or flaked caryopsis, whose principal anatomical components, the starchy endosperm, germ and bran are present in substantially the same relative proportion as they exist in the intact caryopsis" by the American Association of Cereals Chemists International (AACC 1999).

Oat grain has several and well-differentiated parts, the most external are the hulls, a rich source of dietary fiber composed of cellulose, hemicelluloses, and lignin and contributing up to 10% of the total grain (Dhingra et al. 2012). These structures provide protection to oat kernel except for naked oats since they lack them. Oat groat is separated into the trichomes, bran, germ o embryo, and starchy endosperm. Trichomes are removed to the commercial fractions.

Aleurone layer is the most important part of bran, it´s a single layer around the endosperm. Oat aleurone cells are characterized by a prominent nucleus, a gross cell wall and a cytoplasm containing aleurone grains. These grains are composed by protein bodies (storage proteins), starch grains and lipid bodies which change during grain development and maturation (Peterson et al. 1985). Aleurone layer supports the embryo's early growth due to its capacity to synthesize and secrete the digestive enzymes necessaries to solubilize storage components accumulated in the starchy endosperm.

The embryo is the reproductive part of the kernel and it is composed of rudimentary leaves and roots, coleoptile and coleorhizae. The embryo is the smallest part of the oat grain.

The starchy endosperm comprises around 80% of oat groat and acts as a food store for the developing embryo during germination and beyond, with normally 1% of free sugars available in grain (MacArthur-Grant 1986). Cereals differ in the shapes of the starch granules of the endosperm and in the nature of the endosperm proteins (Peterson et al. 1985).

About 55–60% of the endosperm component in oat is starch (Table 4.1). This compound is known as a storage polysaccharide, composed of amylose and amylopectin. Oat starchy endosperm also contains an important part of dietary fiber (β-glucan), and protein and lipid bodies different from bran and ambry layers.

4.1.2.2 Nutritional Composition

Traditionally, oat has been used as livestock or as a winter crop in no-till rotations around the world. Moreover, oats have several functionalities for animal health, the most important is related to the positive effects on the immune system of farm animals. Nevertheless, after its recognition as a healthy food in the mid-1980s the use of oat as human food has increased progressively. Oat grain highlights among cereals due to its multifunctional characteristics and nutritional profile and is considered an important source of soluble dietary fiber, well-balanced proteins, unsaturated fatty acids, and antioxidant essentials for human health (Singh et al. 2013; Clemens and van Klinken 2014; Rasane et al. 2015; Montilla-Bascón et al. 2017). Oat is a good source of natural antioxidants and a unique source of avenanthramides and avenalumic acid (Bryngelsson et al. 2002; Mattila et al. 2005; Meydani 2009). Consumption of oats possesses various benefits for human health such as hypocholesterolaemic (Anderson 2003; Okarter and Liu 2010), anti-cancerous (Kasum et al. 2002; Haas et al. 2009) and prevention of type 2 diabetes and cardiovascular diseases (Meyer et al. 2000; Liu et al. 2000; Tapola et al. 2005). Besides recent reports point out to this cereal as suitable in the most diet of celiac patients (Janatuinen et al. 2002; Peräaho et al. 2004). The principal healthy properties of oat can be attributed to different biocompounds:

Table 4.1 Composition of oat grain

Component	Average value (%)	Range (%)
Starch[a]	51.1	44–61
Protein[d]	15.2	11–20
Moisture	10.0	9–14
Fiber (cellulose, hemicellulose, lignin)	8.9	7–11
Fat[b]	7.6	5–10
B-glucans[a]	4.2	2.2–6.6
Free sugars[c]	1.1	0.9–1.3

[a]Welch (1995); [b]Youngs (1986); [c]MacArthur-Grant (1986); [d]Singh et al (2013)

Fiber (β-Glucans)

Oat has a high content of unique soluble fibers (1,3;1,4)-β-D-glucans (β-glucans), which are the principal components that involve in healthy properties. β-glucans are non-starch polysaccharides formed by glucoses connected by β (1–3)–(1–4) binds located in the endosperm and aleurone cell walls (Butt et al. 2008; Clemens and van Klinken 2014). β-glucans improve food degradation by retarding stomach emptying with a dampening effect on glucose oscillation in the small intestine and blood serum causing a cholesterol-lowering effect and enhance the balance of the intestinal microflora (Guo et al. 2014; Frid et al. 2017; Khan et al. 2018). Several studies have also shown that oat β-glucans moderate the glycemic and insulin response, due to its viscosity glucose and sterol absorption are retarding contributing to attenuated postprandial plasma glucose and insulin levels (Butt et al. 2008). Moreover, this compound improves the immune system against foreign invaders by enhancing the ability of macrophages, neutrophils, and natural killer cells (Rondanelli et al. 2009). The starch of oats degrades completely but slowly, which also contributes to stabilizing the blood glucose levels and prolonging satiety. The germ especially contains a variety of vitamins and minerals (Webster and Wood 2011).

On the other hand, the wall polysaccharides of grasses have a recognized interest and potential for large-scale biomass production of bioethanol industries have been recognized and have generated renewed interest in enzymes and corresponding genes that direct the biosynthesis and degradation of (1,3;1,4)-b-D-glucans (Fincher 2009).

Oil

Oat is characterized by the highest oil quality and content among all cereals containing more than 10% as compared to about 2–3% in wheat (Peterson and Wood 1997).

Oat lipids are stored majorly in the endosperm and are characterized by high content (>80%) of unsaturated fatty acids. These lipids confer stability, longevity, and are not only implicated in the flavor and flavor-off and pasting oat grain properties but also determine its energy content and nutritional qualities (Zhou et al. 1999a). Oats are particularly rich in unsaturated fatty acids such as oleic (C 18:1) and linoleic acid (C 18:2), counting for approximately 40 and 36% of oil total content (Halima et al. 2015). Both fatty acids are involved in the biosynthesis of long-chain polyunsaturated fatty acids such as arachidonic acid (AA), eicosapentaenoic acid (EPA) and docosahexaenoic acid (DHA) very important for human health (Kalbasiashtari and Hammond 1977). The highest levels of unsaturated acid together with the low amounts of saturated fat may reduce the risk of heart and vascular diseases, the high oil and fiber content induce a long-lasting feeling of satiety that helps to weight loss (Guo et al. 2014; Rasane et al. 2015; Surampudi et al. 2016). Recent studies have demonstrated the beneficial dermatological effects of oat oil due to the role in the maintenance of the epidermal water barrier (Nebus et al. 2009; Halima et al. 2015). Moreover, oat is essential to the establishment of body cells through the antioxidant capacity exerted by its phospholipids fraction (Koprivnjak et al. 2010; Szydlowska-Czerniak et al. 2010).

Proteins

The quality of oats protein is higher than that of other cereals due to the balance of the amino acids, particularly Lysine, and protein composition (Youngs 1986; Welch 1995; Shewry and Casey 1999). Groat oat protein concentration is higher in the aleurone layer but the total contribution is equal in the bran and endosperm (Butt et al. 2008). The four principal proteins of oat kernel are albumin, globulin, prolamine, and glutenin, differing from other cereals in their concentration. Albumin and globulin are the main storage protein in oat which have the highest balance of essential amino acids for human body and specially rich in Lysine, while prolamin is lowest (Singh et al. 2013). Recent studies have demonstrated that oats can be tolerated by most people suffering from celiac disease (Janatuinen et al. 2002; Peräaho et al. 2004). Twenty percent of oat proteins are similar to gluten in wheat but have a different composition, these are the avenins. The immunogenicity of avenins depends on specific oat varieties and oat-sensitivity patients (Londono et al. 2013).

Antioxidants Components

Oats also contain a varied range of phenolic compounds including ester linked glycerol conjugates, ester linked alkyl conjugates, ether and ester linked glycerides, anthranilic acids and avenanthramides (AVAs)(Mattila et al. 2005; Liu 2007; Martinez-Villaluenga and Penas 2017; Gramza-Michalowska et al. 2018).

The bran fraction also contains polyphenols, of which the avenanthramides are especially well known for their antioxidant activity and retarding effects on arteriosclerosis and inflammation. On the other hand, oats antioxidants may also contribute to the stability and the taste of food products (Peterson 2001).

Oats are an important source of antioxidant components such as tocols, phenolics acids, and avenanthramides (Bryngelsson et al. 2002; Bratt et al. 2003; Chen et al. 2008). Tocols are lipid-soluble compounds especially important for their antioxidant capacity from their capacity to scavenge free radicals (Suarna et al. 1993). Tocopherols, tocotrienols, and vitamin E are the principal tocols found in oat and their concentration depends on location and genotype (Peterson and Qureshi 1993). These antioxidant components have the capacity to reduce the serum cholesterol concentration and also have the ability to inhibit the growth of cancer cells (Halima et al. 2015).

Oat is not only a good source of natural antioxidants, it is also a unique source of avenanthramides and avenalumic acids (Bryngelsson et al. 2002; Mattila et al. 2005; Meydani 2009). Avenanthramides are polyphenolic, conjugated of hydroxycinnamic acids with anthranilic or 5-hydroxy anthranilic acids, and more than 23 different phenolic conjugates have been described and identified by Collins (1989), and (Ishihara et al. 2014). The major avenanthramides' forms found in oat are Avn-A (*N*-p-coumaroyl-5-hydroxyanthranilic acid), Avn-B (*N*-p-feruloyl-5-hydroxyanthranilic acid) and Avn-C (*N*-caffeoyl-5-hydroxyanthranilic acid) (Sur et al. 2008).

The antioxidant activity of Avns in oat is higher than the antioxidant activity of the other phenolic antioxidants such as ferulic and caffeic (Dimberg et al. 1992), but also these antioxidant activity differences exist among Avns.

Avenanthramides play a biological role as phytoalexins in oat leaves display an antifungal activity with a pathogen-infection induced synthesis. The avenanthramides may also protect against rancidity and other sensory properties such as color, flavor, and texture, contributing to the stability and the taste of food products (Peterson 2001; Halima et al. 2015). In addition these compounds can exert anti-inflammatory, antiproliferative, and antiatherosclerotic effect, anti-itching, and anti-aging properties and could reduce the risk of colorectal cancer (Vitaglione et al. 2008; Meydani 2009; Feng et al. 2013).

4.2 Genetic Diversity Within Oats and Evolution

4.2.1 Oat Genome Structure and Evolution

Avena genomes are large and complex, in the range of 4.12 Gb (*A. damascena*, the smallest diploid) to 12.6 Gb (*A. sterilis*, the largest hexaploid) (Yan et al. 2016b). These sizes are comparable to the equivalent genomes of wheat, barley, and their wild relatives, which diverged from Avena some 25–50 myr, with the most recent analyses supporting the older of these limits (Schubert et al. 2019). At the time of writing, there are no publically available genome references available for *Avena* species, although a number of diploid genomes have been completed and hexaploid reference assemblies are underway (https://avenagenome.org/). An initial survey of repetitive sequences in diploid and hexaploid genomes has been made (Liu et al. 2019) which confirms the expected high proportion of repeat elements (72%) in *Avena* genomes, and identifies families which show sub-genome specificity in the hexaploid. The hexaploid C sub-genome is larger than those of the A and D lineages, and contains a larger proportion of heterochromatic regions and of repetitive families in common with the most closely related diploids (Fominaya et al. 1995; Yan et al. 2016b) allowing their direct use as probes for C sub-genome identification by in situ hybridization (Jellen et al. 1994). The A and D sub-genomes are very similar to each other and have been difficult to distinguish cytogenetically. Liu et al. (2019) report a retrotransposon which has been preferentially amplified in the D sub-genome, which provides a complement to a retrotransposon fragment found earlier to have been preferentially amplified in the A sub-genome (Linares et al. 1998). The small number of identified repeats which show differential amplification between the A and D sub-genomes is consistent with molecular phylogeny and high-throughput marker analyses which indicate that the hexaploid D is a relatively recent variant of the A (Peng et al. 2008, 2010, 2018; Yan et al. 2016a) which can be expected to have a similar sequence content and organization. A number of karyotype variants are known, however, with the most significant hexaploid polymorphism being associated with growth habit in progenitor *A. sterilis* lineages and their likely domesticates (Zhou et al. 1999b). The ancestral karyotype is typically winter type while the derived A/C genome translocation is of the spring type. Other rearrangements between sub-genomes are seen in the hexaploid (Chaffin et al. 2016). The D sub-genome appears to have formed the hexaploid progenitor tetraploid before subsequent addition of the A (below), so maybe expected to be further along the process of diploidization (Mandakova and Lysak 2018).

There are a considerable number of differences in inflorescence architecture and grain composition between oat and Triticeae species (Kellogg et al. 2013; Stewart and McDougall 2014) that may be expected to be reflected in divergent gene content and expression. For example, in wheat some 105 gluten genes contribute to 80% of the seed storage protein while in oat a single related family, the avenins has only 10–11 members and provides 10–15% of seed storage protein (Londono et al. 2013; Clavijo et al. 2017). Gluten genes and avenins are found in clusters so large changes in copy number may require only local amplification of genes rather than genome-wide changes. Production of other oat-specific metabolites may also require only relatively minor changes in specificity of common structural genes. For example, production of the anti-inflammatory avenanthramides may have required only two new enzyme activities (Li et al. 2019). Critical processes such as flowering time regulation also appear to be frequently conserved between *Avena* and *Triticeae* species (Nava et al. 2012). However, a quite unexpected genome reorganization was found for the avenacin biosynthetic cluster (Qi et al. 2004). Avenacins are saponins with antifungal properties, produced in roots of *Avena* species but not found in other cereals. Multiple structural genes had been recruited by duplication and diversification from other pathways, and physically relocated to a single chromosomal region. Other examples of functionally related gene clusters for metabolic pathways in plants have been found subsequently (Nutzmann et al. 2016). While there are obvious adaptive advantages to the functional clustering of genes (Nutzmann et al. 2018), it is still not clear how such significant reorganization is carried out in plant genomes. A comparison of the forthcoming *Avena* genomes with their Triticeae relatives may shed light on these processes.

4.2.2 Genetic Diversity Within Oats

The *Avena* genus contains several dozen diploid, tetraploid, and hexaploid species, with a center of diversity in the Western Mediterranean (Baum 1977; Loskutov and Rines 2011; Ladizinsky 2012). All are annuals with the exception of a single species, *A. macrostachya*, which has also been considered as an "intermediate" taxon between *Avena* and Helictotrichon (Winterfeld et al. 2009). Two distinct lineages may be easily defined (Fu and Williams 2008; Loskutov 2008) and extant diploids may be assigned unambiguously to either the A or the C branch. The A and C lineage divergence may have occurred from 4 to 20 myr ago (Liu and Chen 2017;

Fu 2018). There is less certainty about the origin of sub-genomes within extant polyploids, although there is now a consensus that a variant lineage of the A genomes, designated as the D, is found with C genome lineages in most extant tetraploids, with one of these DC species subsequently giving rise to today's ADC hexaploids (Yan et al. 2016a). These hexaploids include the domesticated *A. sativa* (white oat) and *A. byzantina* (red oat), as well as relatives such as the widespread wild species *A. sterilis* and the weed species *A. fatua*. A number of DC tetraploids are found in the western Mediterranean but the closest relative to the hexaploid progenitor, *A. insularis*, occurs at the eastern end of their distribution, and was found first in Sicily (Ladizinsky 1998). No extant D genome diploid has been found.

Another variant of the A genome, designated the B, is found in the common and widespread tetraploid *A. barbata* which has proved to be so successful in North America that it has been used as a model for invasive species (Crosby et al. 2014). Related tetraploids (*A. abyssinica and A. vaviloviana*) are found in Ethiopia, while a tetraploid with a different A genome configuration (*A. agadiriana*) is found in Morocco. Both *A. agadiriana* genomes appear to be A lineage variants (Badaeva et al. 2010). One of the probable components of *A. agadiriana, A. damascena*, appears to have been more widespread than its descendant at some stage, being found at either end of the Mediterranean, although now rare in the west and possibly extinct in the east. Yet another A genome diploid, *A. canariensis*, is an endemic island species, displaying high levels of intraspecific variation with some evidence of incipient speciation based on karyotype differences (Morikawa and Leggett 1996, 2008). The exact nature of the B, D, and other A variants, such as the components of *A.agadiriana*, have been the subject of some discussion and uncertainty, which is likely to be resolved only by detailed genomic analyses.

4.3 Challenges Posed by Climate Changes for Oat Cultivation

Biotic and abiotic stresses are crucial players determining oat yield and are mainly responsible for yield gaps in rainfed crops (Anderson et al. 2016). They are in a certain way interrelated since environmental conditions are crucial for fungal development and spread, so both, biotic and abiotic factors affect the differential distribution of the oat crop across different types of environments (Chaves et al. 2003). Furthermore, climate change is exacerbating abiotic stress on a global scale, increasing irregularity and unpredictability of the stress episodes. Thus, there is an urgent need to develop adaptation strategies to target oat crop to specific environments (Marshall and Ohm 1987; Beebe et al. 2011).

4.3.1 Abiotic Factors

The most important abiotic factors constraining oat yield are related to water availability and temperature. Temperature stress is the major abiotic factor affecting growth and development of cultivated oat (Marshall 1992). Indeed, this factor delineates the area of adaptation of the crop. Thus, areas with relatively mild winters and warm summers are used for growing winter oat whereas in areas with relatively colder winters and cooler summers spring oat is cultivated. The limit between the winter and the spring zone is especially risky because low-temperature stress in the winter may cause injury of winter-sown oats whereas heat stress in the summer may severely reduce the grain yield and quality of spring-sown oats (Marshall 1992).

High temperatures favor drought episodes and indeed, the effects of drought and heat are usually confounded. Both drought and high temperatures usually cause primary crop losses worldwide, with an average yield loss >50% for most major crop plants and so in oats (Qin et al. 2011; Bailey-Serres et al. 2012). Although oats have a vigorous root system that exploit the soil well, their transpiration rates, and hence, water requirements, are higher than that of other small grain cereals (Ehlers 1989). Thus, oats are especially susceptible to grain abortion caused by drought and high temperatures showed as empty white spikelets (Sánchez-Martín et al. 2017). Although few data on practical yield losses are available, severe levels of heat or drought or both obviously cause large reductions in the quantity and quality of oat grain. Both high temperature and drought are increasingly affecting oats in the last years, not only due to climate change. During the last 20 years and due to the good adaptation of oat to a wide range of soil types including marginal soils, where oats can perform better than other small-grain cereals (Stevens et al. 2004; Buerstmayr et al. 2007; Løes et al. 2007; Ren et al. 2007), cultivated oat area is spreading to southern regions. Thus, a steady-state increase of the oat cultivated area has been observed within the Mediterranean area, with approximately a rate of 7500 Ha increase per year (FAO 2017b). Therefore, there is a need of oat germplasm adapted to the Mediterranean like agroclimatic conditions, characterized by mild and moderately rainy winters, and warm and dry springs; and of implementing specific breeding programs based on the particular requirements of the southern areas (Sánchez-Martín et al. 2014).

In the other extreme, violent wind and rainstorms also may cause severe damage to an oat crop by causing lodging (flattening) of the plants. Lodging may inhibit grain filling if it occurs before physiological maturity and there will be additional grain losses during harvesting operations. Lodging losses may reach 50% of yield (Bailey-Serres et al. 2012). In addition, the risk of lodging increases with soil fertility level, and growers experience intangible losses because they are compelled to practice suboptimal fertilization of oat (Marshall 1992).

4.3.2 Biotic Factors

Oat fungal diseases are major constraints for this crop. Among the pathogenic fungi infecting oats, biotrophic pathogens such as the powdery mildew or rust are the most difficult to control by mean of crop management in part due to their very efficient mechanisms of spread (i.e., under ideal conditions, spores can infect oat plants even after being carried in the wind for hundreds of miles and one spore may generate up to 200,000 new spores). The use of resistant varieties is one of the best alternatives to control these pathogens, since it avoids the use of expensive chemicals that threat consumers and environment (Stevens et al. 2004). However, the resistance obtained is often overcome by emerging pathogenic races. This is mainly due to inappropriate use of resistance sources, which are of monogenic nature and for which the underlying resistance mechanisms are unknown. Thus, it is necessary to identify novel sources of resistance, and furthermore, characterize the specific resistance responses/mechanisms at cellular and molecular level. This will allow the combination of responses acting during different fungal developmental stages, the use of mechanisms with a polygenic base and it would also ease the selection process (Prats et al. 2007; Sánchez-Martín et al. 2012).

4.3.2.1 Crown Rust

Crown rust disease, caused by *Puccinia coronate* f. sp. *avenae*, is the most widespread and damaging disease of oat. There are severe epidemics in virtually every oat-growing region of the world reducing overall yield from 10 to 40% (Behnken et al. 2009), albeit individual oat fields may suffer total crop failures (Simons 1985; Chong 2006). Crown rust epidemics are particularly important in the Mediterranean rim where rust populations are more virulent than in the center and north of Europe (Herrmann and Roderick 1996), probably favored by the optimum climatic conditions for its development with mild to warm (20–25 °C) sunny days and mild nights (15–20 °C). Damage to leaves, particularly the flag leaf, reduces photosynthesis and interferes with the transport of photosynthesized sugars from leaves to the developing grain. This not only limits growth and reduce forage quality but also causes thin kernels with low test weight - factors which greatly reduce milling quality. Badly rusted plants have stunted root systems and have a higher predisposition to suffer from drought stress.

The use of race-specific (Pc) genes for resistance has been the primary means of control. Up to date, more than 90 genes for crown rust resistance have been assigned with permanent designations (Chong et al. 2000). Unfortunately, these genes are rapidly defeated by the emergence of new pathogenic populations, some of them in a short time as two years (Nazareno et al. 2018). The gene *Pc94* transferred from *A. strigosa* is currently regarded as the most effective gene for resistance to *P. coronate* (Chen et al. 2007a). However, virulence against this gene, albeit at low frequency, has already been detected in Canada and in the

Mediterranean area (Chong et al. 2011; Sánchez-Martín et al. 2012). Thus, there is an urgent need to find additional sources of effective resistance. In addition to race-specific resistance, non-specific quantitative resistance or partial resistance may hamper rust development. This kind of resistance often manifest at higher level in adult plants, being named adult plant resistance. One of their main advantages is that is usually effective against a wide spectrum of isolated, characteristic that is highly attractive in cases such as crown rust where so many pathotypes exist. In contrast with race-specific resistance, non-specific resistance may remain effective in the field for extended time period. Such, it is the case of the variety "Red Rustproof" and MN841801 which has remained partially resistance to crown rust for more than 30 years (Leonard 2002). Thus, screenings for resistance considering mechanisms other than hypersensitive-based race-specific resistance and their use in breeding are necessary despite that usual complexity of inheritance disfavoured the use of this kind of resistance in breeding programs in the past.

4.3.2.2 Powdery Mildew

Oat powdery mildew caused by *Blumeriagraminis* DC. f. sp. *Avenae* Em. Marchal is one of the most important foliar diseases of common oat in the cooler and humid regions of Europe. Cool, moist weather condition with high relative humidity and temperatures ranging between 15 and 21 °C is optimum for its development. Temperatures climbing above 25 °C, limit powdery mildew disease development.

Damage caused by the oat powdery mildew fungus increases the number of nonproductive tillers, reduces kernel size, number of seeds per unit area and test weight, resulting in yield loss even at low levels of infection. Severe infections can result in oat plant stunting. Heads on the later tillers may show more heavy infection because they reside lower in the wheat canopy, where humidity remains high. The annual crop losses from mildew infections are estimated to range from 5 to 10%, but greater yield loss occurring when the oat flag leaf becomes severely diseased by powdery mildew by the time of heading may raise up to 40% (Jones 1977; Clifford 1995).

As happened with most plant diseases the use of resistance varieties is one of the most environmentally friendly and effective means for controlling and limiting the consequences of oat powdery mildew disease. Race-specific resistance has been described in oat against powdery mildew. The resistance to oat powdery mildew encoded by major genes has been determined based on the reaction of differential oat cultivars and lines to various pathotypes of *B. graminis* f. sp. avenae and has been termed the oat mildew resistance (OMR) group (Jones and Jones 1979). To date, five OMR groups (OMR1 e OMR5) and seven resistance genes have been characterized (Roderick et al. 2000; Yu and Herrmann 2006; Hsam et al. 2014; Okon 2015). In addition, partial and adult plant resistance has also been described in oat against powdery mildew. Up to seven additive genetic factors have been described to contribute to this resistance in the oat cv. Maldwyn (Jones 1986). However, quantitative assessment of powdery mildew in detached leaves has not demonstrated sufficiently precise to be used alone in selecting segregants for high levels of APR. Identification of the components of resistance that account for this enhancement would be highly desirable to ease breeding programs. A particular form of non-specific resistance to powdery mildew is the *mlo* resistance. Plant genotypes carrying the recessive *mlo* gene display highly effective papilla-based penetration resistance to powdery mildew which, in the case of barley, has proved durable for over 25 years. Due to this effectiveness and durability ca. 60% of spring barley cultivars in central and Western Europe carry *mlo*. This kind of resistance against powdery mildew has also been described recently in pea (Pavan et al. 2011) and tomato (Bai et al. 2008). However up to date no *mlo*genotypes has been reported in oat.

4.3.2.3 Barley Yellow Dwarf Virus

Barley yellow dwarf virus (BYDV) is a phloem-limited virus transmitted by aphids that has become increasingly widespread in USA and Europe and it is now a serious concern for oat producers. Actually, BYDV includes several related viruses, grouped in five strains that can be transmitted by 23 species of aphids and affect almost 100 species of grasses (Watkins and Lane 2004). Damaging outbreaks are most likely at high light intensities during cool and moist periods (15 °C).

Early infection results in severe stunting. Almost all part of the plant is affected. Leaves become shorter and bronzed reducing the photo-assimilation capacity. Tillering is often reduced and ripening delayed. BYDV infection also reduces the number of florets or seeds per head (tiller), and seed weight (Jensen and D'Arcy 1995). In addition, root system is reduced increasing the possibility of lodging and drought susceptibility (Watkins and Lane 2004). These symptoms may lead to losses of up to 50%, and yield gaps ranging from 1700 to 2700 kg/ha as reported by (McKirdy et al. 2002).

Two types of resistance to BYDV have been distinguished: virus resistance and field resistance. Virus resistance refers to low virus titer in infected plants whereas field resistance (tolerance) refers to the reduction of symptoms of infection independent of the virus titer (Foresman et al. 2016). Up to date and based on QTL analysis two to four genomic regions comprising two to four genes are thought to be responsible for BYDV tolerance in oats, but phenotyping for BYDV is still the most reliable method for screening breeding material for tolerance. This highlights the urgent need of genetic studies that allow marker-assisted breeding for a more efficient selection of resistant lines.

4.3.2.4 Fusarium Head Blight

Concerns about Fusarium head blight (FHB) or scab is increasing rapidly over the last years. This disease is caused by several *Fusarium* spp., including *F. graminearum, F. culmorum, F. poae, and F. avenaceum* (Langseth and Elen 1996; Tekauz et al. 2004). The disease is rarely apparent or recognizable in a standing oat crop, in contrast to barley or wheat in which damage to spikes, i.e., "blighting" is normally distinct and can be quantified. This probably lead to overlook the incidence of the disease in the past. However, identification of the *Fusarium* spp. and mycotoxins recovered from samples of oat seed have increased the concern on this disease. There are several mycotoxins that can be detected on oats, but the ones considered most important are deoxynivalenol (DON, vomitoxin), zearalenone (ZON), HT-2 toxin (HT2), and T-2 toxin (T2). DON and ZON are produced by several *Fusarium* species but the dominant species is *F. graminearum*. HT2 and T2 are produced by several other *Fusarium* species and the dominant one is *F. langsethiae*. In addition, secondary damage of this pathogen can manifest as a reduction in the protein content of the grain (Mauler-Machnik and Zahn 1994; Mesterházy and Bartók 1996) and decreased germination and seed vigor (Tekle et al. 2013).

The disease mainly develops in areas of humid climate. FHB appears especially where high rainfall occurs for more than 48 h consecutively during, or soon after, anthesis, and when the temperature lies between 20 and 25 °C and the relative humidity is above 90%. Overall, losses expressed as reduction in the oat grain weight may reach 5% far lower than that found in rye (43.9%) or wheat (17.5%) (Panisson et al. 2003). This is probably due to the spike morphology, with widely spaced pedicels, which make difficult the spread of the disease between spikelets. However, European Commission has recently set legislative limits for the mycotoxins DON and ZON for unprocessed cereals, intermediate products (e.g., *flour)* and finished products for human consumption and guideline limits for these mycotoxins in feed. This current and future legislation on mycotoxins in cereals and cereal products has the potential for considerable impact on the viability of oat production, and particularly its share of the area of cereals grown.

Control measures for FHB depend not only on the adoption of more resistant cultivars but also the use of crop rotation with or without the application of fungicides (Pirgozliev et al. 2003). Studies regarding the possible sources of FHB resistance in oats have yet to be developed and no oat cultivars are presently immune to scab, most oat cultivars being classified as FHB susceptible or, at most, moderately resistant (Martinelli et al. 2014).

4.4 Genomics in Breeding Oat for Resilience

4.4.1 Molecular Marker-Assisted Breeding

Traditional plant breeding programs based on hybridization of selected parents followed by selection of the resulting genetic recombination over a number of generations have a long track record of delivering economic benefits to agriculture, typically by enhancing crop yields and grain quality. Oats are a self-pollinating cereal and as oat varieties are pure inbred lines, conventional breeding programs require at least eight generations to ensure that the resulting plants and their progeny are genetically identical. Production of new varieties is limited by the generation time between each round of selection and with selection being based predominantly on phenotypes evaluated initially in only one environment with only later generations assessed in replicated multi-site trials. Plant breeders aim to develop improved crop varieties that are adapted to produce high yields of quality grain over a wide range of environments with the adaptability of a variety usually tested by the degree of interaction with different environments under which it is planted (Ahmad et al. 2013). Analysis of the genotype x environment interaction on grain yield and quality is therefore essential in a variety evaluation (Becker and Leon 1988) and to understand the adaptability and stability of varieties (Hongyu et al. 2014) for different environments.

The primary goal of all oat breeding programs is to maximize yield. Irrespective of whether the end use is grain or forage, high grain yield is necessary for oats to be economically competitive against other cereals, such as wheat and the main crop competitors such as soybean and maize (in America) and oilseed rape in the UK. In oats, plant breeding has been reported to improve yields by up to 40% (Valentine et al. 2011). Although there are different end uses for oats, plant breeding programs aim to develop "well rounded" varieties with high yield, good resistance to biotic and abiotic stresses (see Sect. 4.3) and with high end use characteristics. Understanding the genetic control of these traits is key to their improvement by plant breeding. A major challenge is to efficiently combine large numbers of traits that are polygenic in nature. Improved lodging resistance is another breeding target. This is a complex trait, as described earlier, dependent upon straw strength and rooting characteristics. A lodged oat crop has stems lying horizontally on the ground which results in reduced yield with poor grain quality, together with considerably increased harvest time. Lodging resistance is an important trait and selection is generally for shorter stiff strawed lines which stand better against wind and rain. Crop genetic improvement is now being used to address wider environmental targets such as stress tolerance and nutrient-use efficiency. By allowing reduced use of inputs such as fertilizers and pesticides, these traits lead to reduced environmental impacts and support the long-term economic sustainability of the oat crop and other sectors.

Given the length of time that it takes to breed a new variety, considerable effort to develop new breeding methods that accelerate progress are underway. This includes speed breeding methods (Valentine 1984; Watson et al. 2018), shuttle breeding the development of doubled haploids (Tanhuanpää et al. 2008) and the incorporation of molecular marker technologies. Opportunities exist to enhance genetic gain in crop breeding by combining phenotypic selection with faster molecular breeding approaches. The ability to directly establish the presence of a trait such as a disease resistance gene or grain chemical composition at a seedling stage without the need to undertake artificial inoculation of disease or to harvest, mill and undertake chemical analyses of grain would increase the speed and precision of the breeding process. Marker-assisted backcrossing (MABC) represents one route to achieve this. It has been successful for the introgression of major traits controlled by one or a few genes of large effect but is difficult with more complex traits governed by many genes, each with a small effect. One main challenge of MABC is to remove unfavorable alleles of closely linked genes, that is, to eliminate linkage drag, particularly in the case of exotic introgression from crop wild relatives.

Molecular markers have been an essential tool in the discovery of genes that affect the quality and adaptation of many crop species and for the characterization and use of genetic diversity. However, until recently, the lack of large numbers of polymorphic DNA markers along with the size and complexity of the oat genome (Rines et al. 2006) has limited their application in oat breeding. Progress in genomic research in oats has been revolutionized by technological advances in DNA sequencing. Single nucleotide polymorphisms (SNPs) have become the marker of choice due to their ease of use and scoring and their ability to be automated with relative ease. Combined with recent development of technologies such as next-generation sequencing (NGS), this has radically changed approaches to marker development in complex plant genomes such as oats. The development of an Illumina 6 K oat chip containing highly informative SNP markers and data from twelve different bi-parental populations resulted in the first physically-anchored consensus map of oats (Oliver et al. 2013). NGS combined with the reduction in genome complexity enables genotype-by-sequencing (GBS) approaches. This combines marker discovery and genotyping to produce high-density markers at a relatively low sample cost (Elshire et al. 2011; Poland et al. 2012) and has been used to expand the oat consensus map (Chaffin et al. 2016). Recently, the marker density of the oat consensus map was increased by the use of additional populations and the addition of more than 70000 loci (Bekele et al. 2018) generated by the GBS program "Haplotag" (Tinker et al. 2016). The development of cheap, high throughput of markers overcomes many previously insurmountable difficulties involved in the breeding of complex traits; the inability to identify individual genes, the masking effects of environment and the presence of linked undesirable genes.

It is not simply the development of high-throughput multiplex markers that is required; associations have to be made between markers and the phenotypic traits of interest. This is no trivial task. Suitable mapping populations are required and accurate phenotyping methods are needed in an appropriate environment(s) particularly for the assessment of difficult or complex characters. Quantitative trait loci (QTLs) are often identified in experimental mapping populations, and subsequently have to be verified in the breeding germplasm. This limits their usefulness in marker-assisted selection (MAS) (Snowdon and Friedt 2004). Involving at least one parent currently or recently in use in the breeding programs increases the direct applicability of marker associations for use in selection. Bi-parental populations have a further disadvantage in that recombination tends to be limited resulting in QTL being mapped at low resolution. The development of novel populations incorporating a far larger portion of genetic and phenotypic variation available than current bi-parental mapping populations include association mapping, multi-parent advanced generation inter-cross (MAGIC) and nested association mapping

(NAM) populations (Cavanagh et al. 2008; Huang and George 2011; Huang et al. 2012).

Bi-parental populations have allowed the identification of a wide range of QTL associated with important agronomic traits such as height, heading date, yield, and grain quality. For example, De Koeyer et al. (2004) identified three QTLs affecting grain yield in the Terra x Marion population including one on Mrg02 that was also associated with plant height. A number of other studies have identified QTLs associated with plant height on Mrg01, Mrg02, Mrg12, Mrg20, Mrg21, and Mrg28 (Siripoonwiwat et al. 1996; Beer et al. 1997; Holland et al. 1997; Zhu and Kaeppler 2003; De Koeyer et al. 2004; Wooten et al. 2009). QTLs for lodging susceptibility have been also found associated with some of these height QTLs, for example, on Mrg01 and Mrg21 (De Koeyer et al. 2004). However, these authors also identified height QTLs that were independent of lodging susceptibility QTLs. The oat dwarfing gene *dw6* has been mapped to Mrg04 (Tanhuanpää et al. 2008; Molnar et al. 2012) and recently, closely linked PCR markers have been developed (Yao et al. 2018).

Heading date (or flowering time) has a key role in the adaptation of oat varieties to different environments and consequently it, along with associated traits such as vernalization and the effect of photoperiod, has been subject to substantial QTL analysis. Comparative mapping has revealed a small number of genomic regions where multiple heading date QTLs have been mapped in a wide range of bi-parental populations. These include Mrg02, Mrg12, Mrg20, and Mrg21 (Wight et al. 1994; Siripoonwiwat et al. 1996; Beer et al. 1997; Holland et al. 1997, 2002; Zhu and Kaeppler 2003; De Koeyer et al. 2004; Locatelli et al. 2006; Wooten et al. 2009; Nava et al. 2012). Esvelt Klos et al. (2016) using association analysis of phenotypes from 10 locations over two years confirmed these QTLs. QTL for heading date on Mrg02 and Mrg12 were found to overlap with homeologues of the vernalization response gene As-vrn3 gene (Nava et al. 2012). A major day-length response gene, *Di1*, has also been found to be associated in this region of Mrg02 (Wight et al. 1994). Mrg20 and Mrg21 are thought to be homeologues chromosomes (Chaffin et al. 2016) and QTLs for vernalization response have also been mapped to these linkage groups (Holland et al. 1997, 2002) as well as the vernalization gene *vrn1* (Nava et al. 2012).

There exist many single-gene resistances to the major diseases of oat and these have been a focus of considerable genetic analysis. For example, race-specific crown rust (*Puccinia coronate* f. *spavenea*) resistance genes mapped include Pc38, 39 and 48 (Wight et al. 2004), Pc54 and 59 (Bush et al. 1994), Pc58 (Jackson et al. 2007), Pc68 (Penner et al. 1993; Kulcheski et al. 2010; Satheeskumar et al. 2011), Pc71 (Bush et al. 1994), Pc91 and 92 (Rooney et al. 1994; McCartney et al. 2011), Pc94 (Chong et al. 2004) and Pcq2 (Zhu and Kaeppler 2003). A number of these have been confirmed using in a wider range of germplasm using association analysis (Esvelt Klos et al. 2017) and converted into PCR-based markers for deployment in high-throughput MAS in oat breeding programs (Chen et al. 2007b; Gnanesh et al. 2013). Some of these race specific genes are found clustered in the oat genome and markers developed for one gene in a cluster may be useful for the selection of linked genes although it may be that some of these resistance genes are in fact not independent. There is also evidence that some resistance genes interfere with the action of others. An example of this is the suppression of Pc62 and Pc94 by Pc38 (Chong and Seaman 1994; Wilson and McMullen 1997) and markers could be used to precisely select desired allelic combinations. In addition, a major QTL on Mrg06 associated with adult plant partial resistance (APR) to crown rust has been identified using three recombinant inbred line populations. PCR assays were designed for selected SNPs and used to verify the QTL (Lin et al. 2014). Synteny with wheat suggests that this QTL is orthologous with the stripe rust APR gene *Yr16* in wheat. Recently a new unique crown rust resistance gen has been identified in the diploid oat *A. strigosa* and transferred into hexaploid cultivated oat germplasm (Rines et al. 2018). Closely linked markers to the resistance were found on linkage group Mrg20 and successfully converted to PCR assays.

Oat stem rust caused by *Puccinia graminis* Pers. f. sp. *Avenae* Eriks. and E. Henn., is a major disease in some environments and although there are 17 numbered oat stem rust resistance (*Pg*) genes and the *Pg-a* complex that have been described (Martens 1985), relatively few are currently used in oat breeding (Fetch and Fetch 2011). A number of these have been mapped including *Pg3* (Penner et al. 1993), *Pg9* and *Pg11* (O'Donoughue et al. 1997) and *Pg13* (Kebede et al. 2018). *Pg13* was found to be tightly linked to the crown rust resistance gene Pc91.

Currently a number of major genes for powdery mildew (*Blumeriagraminis* f sp. *Avenae*) resistance have been identified using monosomic analysis. These include *Pm1* on chromosome 1D, Pm3 on chromosome 17A, Pm6 on chromosome 10D, *Pm7* on chromosome 13A and Pm8 on chromosome 4C (Hsam and Zeller 1998; Sanz et al. 2010; Hsam et al. 2014). Pm5, which was introgressed into hexaploid oat from *Avena macrostachya*, has been mapped onto a region corresponding to Mrg20 (Yu and Herrmann 2006). Pm4, derived from the diploid oat species *A. barbata* (Thomas and Aung 1978) has recently been mapped using DArTseq and PCR-specific markers developed suitable for MAS (Okoń and Ociepa 2018). QTLs associated with barley yellow dwarf virus resistance have also been mapped (Jin et al. 1998; Barbosa-Neto et al. 2000; Zhu and Kaeppler 2003) as well as for Fusarium head blight resistance (He et al. 2013).

Another requirement for MAS is the cost-effectiveness and ease of use of the identified markers in applied breeding programs. Many early studies in oats reported QTLs associated with markers that are not readily useful in MAS such as RAPD (random amplified polymorphic DNA), AFLP (amplified fragment length polymorphism) and DArT (Diversity Array Technology) markers. These markers are not suitable for high-throughput genotyping and cannot readily detect heterozygous individuals as would be required for MABC. As reported above, however, a number of the markers identified have been converted into easy to use PCR-based assays. Moreover, advances in sequencing technology enable the development of custom multiplex SNP assays suitable for the high throughput required for a breeding program.

Sustainable future crop productivity will ultimately be achieved through an increased use of plant genetic resources, including wild relatives and exotic materials, to supplement the genetic diversity that conventional plant breeding has at its disposal. Domesticated crops represent only a fraction of the genetic variation contained within the original wild germplasm and there are many cases where selective introgression of wild material has led to improved agronomic quality. For example, wild *Avena* species have successfully been used as a rich source of disease resistance genes (see reviews byJellen and Leggett 2006; Loskutov and Rines 2011) that have been incorporated into breeding programs. The knowledge of, access to, and use of genetic diversity in cultivated oats and its wild relatives are essential for broadening the genetic base of cultivars to sustain improvement.

4.4.2 Genome-Wide Association Studies in Oats

The goal of genome-wide association studies (GWAS) is to understand the variation in complex traits and diseases by relating genotypes of large numbers of markers to observed phenotypes. In a nutshell, to identify casual mutations that have an effect on a phenotype (any aspect of an organism that can be measured). Association mapping relies on high-density SNP genotyping of unrelated individuals, giving you point associations in the genome. Failure in implementation can lead to the false detection of associations between unlinked markers and traits. An excellent review of GWAS studies that highlight the widely used of this tool in humans, animals, and plants can be found in Visscher et al. (2017). The identified regions, linked to a causal genetic variant, can be selected in a breeding program with the goal of improving genetic gain per unit time (Lande and Thompson 1990). The complexity of the oat genome and the fact that most useful traits in oats are quantitatively controlled by multiple genes is a limitation in oat genomics, especially because the number of tools developed in oat is few in comparison with other cereals as wheat or rice. From mixed genetic background, the genome association analysis which involves the detection of a random set of genotypes, genes, and QTLs is an effective tool for gene discovery. Currently, with large accessibility of molecular markers and the statistical tools refinement, GWAS has regenerated further interest in oats. High statistical power, low probability of Type I error, and the study of confounding effects due to population structure and cryptic relatedness that can cause spurious associations are the keys for success in GWAS. Population structure can cause LD between unlinked loci and consequently generate spurious marker-phenotype associations. In turn, cryptic relatedness, which refers to the presence of recent common ancestry in a sample of unrelated individuals, can have a confounding effect by unmatched association studies. Most GWAS oat studies focused on quality and resistant traits, such as β-glucans, crown rust, barley yellow Dwarf Virus, and powdery mildew.

The first study on GWAS in oats was performed by Newell et al. (2012). They evaluated β-glucan concentration in a very diverse germplasm (431 genotypes) and found three-independent markers to be associated with the trait (Table 4.2). These markers had sequence homology to rice and one marker opt.0133 was located on

Table 4.2 Summary of studies on genome-wide association analysis in oats

Study	Trait	Sample size	Genotyping
Newell et al. (2012)	β-glucan	431	DArT
Asoro et al. (2013)	β-glucan	470	DArT
Montilla-Bascón et al. (2015)	CrownRust and Powdery mildew	177	DArT and SSRs
Bjørnstad et al. (2017)	FHB, DON	424	Infinium 6 K SNP array
Foresman et al. (2016)	BYDV	428	Infinium 6 K SNP array
Winkler et al. (2016)	BYDV, crown rust, DTA, awn frequency, lemma color	759	Infinium 6 K SNP array
Tumino et al. (2017)	Frost tolerance	138	Infinium 6 K SNP array
Esvelt Klos et al. (2017)	Crown rust	631	Infinium 6 K SNP array
Tumino et al. (2017)	Lodging, plant height	138	Infinium 6 K SNP array

FHB fusarium head blight, *DON* deoxynivalenol, *BYDV* barley yellow dwarf virus, *DTA* days to anthesis

rice chromosome seven adjacent to the Cs1F gene family. The *CslF* gene family was previously shown in rice (Burton et al. 2006) and barley (to have β-glucan (Burton et al. 2008) synthase function. In 2013, Asoro et al. (2013) reported a GWAS study also for β-glucan in Elite North American Oats. Even though they also used DArT markers in this study, none of the significant markers were directly homologous to *CslF* or *CslH* gene families in rice as in Newell et al. (2012). Nevertheless, several regions of four DArT markers corresponded to the same regions of three QTL (cdo346A, cdo82, cdo1340) identified by (Kianian et al. 2000). In addition, five associated DArT markers (oPt.14317, oPt.12704, oPt.5064, oPt.16618, and oPt.16436) were close to a QTL from the Terra × Marion population (De Koeyer et al. 2004). In this study, two significant markers (oPt.12704 and oPt.8758) were adjacent to cellulose synthase A catalytic subunit 2 (*CesA2*), a gene that was identified to be co-expressed with *CslF6* in transcriptional studies for barley (*Hordeum vulgare* L.) (Burton and Fincher 2009).

Montilla-Bascón et al. 2015 (Table 4.2) investigated the association between crown rust and powdery mildew resistant using DArT and SSRs markers on a Mediterranean diverse population. They found five markers, two of them very significantly associated with rust resistance, that showed similarity with autophagy and acyltransferase proteins that have been related to plant immune defence reactions. They also found the marker oPt-5014 to be strongly associated with powdery mildew resistance in adult plants, but not significant markers on the seedling stage. Marker oPt-5014 was associated with hypothetical proteins of sorghum, wheat, and rice containing a Zinc knuckle domain (pfam14392) which has been detected in several plants transcription factors and might, therefore, be involved in the regulation of gene expression. The main QTLs were found on linkage groups Mrg01, Mrg03, Mrg08, Mrg20, Mrg23, and Mrg28. These findings were also found by Winkler et al. 2016 which found five SNPs associated with oat seedling resistance in the same linkage groups. Candidate genes for SNPs on Mrg08 and Mrg28 are PcKM and Pc91 (Gnanesh et al. 2013). The marker on Mrg20 (GMI_GBS_92025) may identify *Pc48*, first mapped by Wight et al. (2004). Winkler et al. (2016) also showed that some previous markers linkages for crown resistance gene *Pc38* (Wight et al. 2004) were situated close to Mrg02 which is the same region they identified crown rust reaction and severity as well as days to anthesis (GMI_ES03_c7453_413). The latest study on crown rust resistant was performed by Esvelt Klos et al. 2017. They found 29 SNPs on 12 linkage group sat a genome-wide level of statistical significance. This study reinforce previous funding such as Montilla-Bascón et al. (2015) and Winkler et al. (2016), since the QTL identified included regions previously shown to be linked with seedling resistance genes such as, *Pc48*, *Pc58a*, *Pc68*, *Pc71*, *Pc91*, and *PcKM*, and also with adult plant resistance and adaptation-related QTL. The novelty in this study was that they found regions not previously associated with crown rust resistance on linkage groups Mrg03, Mrg08, and Mrg23. These studies indicated that the genetic architecture of crown rust resistance in oat is complicated by (i) the high number of contributing loci (>100) already identified, (ii) the existence of both adult plant partial resistance and race-specific seedling resistance, and (iii) the crown rust population diversity around the globe. Interpreting these associations in terms of the type of resistance and identity with previously mapped QTLs or *Pc* genes is complicated by the same factors that create complexity in the genetic architecture. In general, given the overlap in map locations and the variation in methods of assessing resistance, some QTLs may be either the same as, or allelic to the *Pc* genes.

Barley yellow dwarf viruses (BYDVs) are responsible for the disease barley yellow dwarf and affect many cereals including oat. There are two main studies (Foresman et al. 2016; Winkler et al. 2016) trying to identify markers associated with BYDV tolerance. In the most recent study (Winkler et al. 2016) they found three markers linked with BYDV tolerance. They map close together on Mrg17 linkage group, which appears to be homologous with BYDW loci identified in other populations (Jin et al. 1998; Barbosa-Neto et al. 2000). The finding on the same Mrg17 by Foresman et al. (2016) suggests a single QTL in this region. In addition, oat Mrg17 shares a large region of synteny with a segment of rice chromosome Os02 (Lüpken et al. 2013; Chaffin et al. 2016), which is in turn mapped to a generic interval for tolerance in barley (*Ryd3*). This raises the possibility that the BYDV tolerance gene in Mrg17 is a orthologous gene of *Ryd3*.

Winkler et al. 2016 (Table 4.1) also studied panicle and floret morphology. They found a strong association on Mrg20 for lemma color, although they did not demonstrate a linkage between awn frequency and lemma color loci. They also found a significant marker associated (GMI_ES03_c7453_413) with days to anthesis. This region contains heading date QTL regions mapped on other populations (Siripoonwiwat et al. 1996; Esvelt Klos et al. 2016). No significant associations were found on other traits in this study such as, beta glucan, groat lipid, groat protein, plant height, test weight, lodging, yield, panicle length, and density, smut reaction, straw breakage, shattering, greenbug reaction, and bundle weight. Plant height and lodging was also studied in Tumino et al. (2017). Is well known that plant height correlate well to lodging severity explaining about 30% of lodging variation (REF). Tuminos analysis showed eight significant associations for lodging (Mrg01, Mrg02, Mrg13, Mrg19, Mrg21, Mrg24, Mrg28) and 5 for plant height (Mrg01, Mrg08, Mrg09. Mrg11, Mrg13). Tumino et al. 2017 also studied frost tolerance finding significant associations in linkage groups Mrg01, Mrg11, Mrg12 and Mrg21.

The only study on FHB severity and DON level was performed by Bjørnstad et al. 2017. The most important QTLs were detected on Mrg11 and Mrg12. These QTL corresponds in terms of common SNP markers with previous QTL studies for DON in chromosome 17 and 10 (He et al. 2013).

4.4.3 Genomic Prediction in Oats

Genomic selection is a relatively new tool that opened new essential modes of using molecular markers in breeding for multi-genic quantitative traits (Meuwissen et al. 2001). Genomic selection estimates breeding values that provide a direct estimation of the superiority of each individual by using dense markers covering the whole genome to estimate genomic breeding values (GEBVs) for quantitative traits of selection candidates in breeding populations. As the markers are closely spaced, a QTL located anywhere in the genome will be in linkage disequilibrium (LD) with at least one marker (Hayes et al. 2010; Crossa et al. 2017; Isidro-Sánchez et al. 2017). Hence, GS is just marker-assisted selection on a genome-wide scale.

The main difference of these methods with regard to GS is that a limited fraction of the genetic variation is explained by the identification of QTL. In MAS genotyping is limited to a selected set of markers that tag putative genes to predict BVs, and therefore, have a poor predictive ability due to that only a limited proportion of genetic variance can be captured by the markers (Bernardo and Yu 2007; Goddard and Hayes 2009; Bernardo 2013). In GS, all markers are included in the model regardless of effect size and therefore allowed to track and select for complex traits with both small and large effect genes. The concept of genomic selection (GS) was born from the requirement to improve the applicability of MAS to quantitative traits.

Previously, the high cost and low throughput of molecular marker systems were bottlenecks for the application of GS. This has changed due to rapid advancements in DNA sequencing technologies. The opportunities presented by next-generation sequencing technologies include the ability to sequence whole genomes, reduced-complexity genomes, or targeted genomic regions in large numbers of genetic samples.

Genomic selection studies have been focused mainly on wheat, rice, corn, and barley but not in oats. There is a paucity of primary literature on GS in oat, which is currently limited to a single report on accuracy and experimental design based on a relatively low-density DArT marker platform (Asoro et al. 2011, 2013) and a retrospective analysis of prior data from a GWAS study (Bekele et al. 2018). Although, there are researchers who are also developing practical GS-based programs in oat, none of this work has been published in the primary literature.

The first and unique study on GS in oats was published by Asoro et al. (2011). They evaluated the accuracies of GS using data from five traits on 446 oat lines genotyped with DArT markers and two GS methods under various training designs. Their results showed that accuracy increased as the number of markers and training size becomes larger. The inclusion of older lines in the training population increased or maintained accuracy, indicating that older generations retained information useful for predicting validation populations. They also analyzed the effect of populations' structure and observed that when training and validation subpopulations were closely related accuracy was greater than when they were distantly related, implying that linkage disequilibrium relationships changed across subpopulations. In addition, they did not find significant accuracy using differences in statistical models. This study provided evidence that using inexpensive and abundant molecular markers can provide great sources for GS analysis. In fact, there is a great effort to use large-scale discovery and/or assay of genome-wide genetic differences using method such as genotyping by-sequencing with an affordable cost and high marker density. The advantages of GBS include the ability to sample genotypes from a full genome, the ability to genotype crops with or without reference genomes, and simultaneous variant discovery and genotype calling at a relatively low cost (Poland and Rife 2013). The application of GBS employs bioinformatics algorithms to align short DNA sequence tags to reference genomes, or to cluster these tags in the case of non-reference-based analysis. This is followed by the identification of SNPs that follow Mendelian allele segregation (Elshire et al. 2011; Lu et al. 2013; Glaubitz et al. 2014).

4.5 Oats for the Future

The utilization of oats for human consumption has increased progressively, due to its dietary and health benefits, which mainly relies on dietary fiber and β-glucan content. A large body of clinical evidence (Thies et al. 2014b) suggests that the consumption of 3 g or more per d of β-glucan from oats, as part of a diet low in saturated fat and cholesterol, may reduce the risk of coronary heart diseases. Increased consumer recognition of the heart-health benefits associated with the cholesterol-lowering properties of oat soluble fiber and the overall benefits of whole grains has expanded the market for whole-grain oat products and oat bran in human foods. In addition, preliminary research on minor oat constituents, such as the antioxidants avenanthramides, is establishing a link between specific oat components and allergenic responses, asthma, and proliferation of cancer cells.

The high morphological and genetic oat diversity makes oats a valuable resource for plant breeding and agriculture. Plant breeding capabilities and the new biological techniques are making significant progress in developing agronomically between oats with enhanced levels of health components. Resistance to main diseases as crown rust, BYDV, powdery mildew or resistance to mycotoxin-producing organism deserves more attention, especially the latter since the risks associated with mycotoxins are not well defined. As we improve our understanding of the agronomical, resistant, and health benefits associated with oat, we will be presented with the opportunity to enhance oats' attributes, especially on health and functional traits. Molecular marker-based linkage maps, marker-trait associations, QTL, marker-assisted selection and genomic selection have been developed/identified for several oat populations especially for some economic traits. Nevertheless, the advance in oat research has been limited due to the large size of the oat hexaploid genome and the decline in global oat hectarage which decreases oats competitiveness with other cereals for research funding. Advances in oat biotechnology will continue as oat researchers take advantage of advances in technology toward more rapid, less expensive, and higher- throughput marker development, DNA sequencing, and bioinformatic analyses. The development of databases of gene and genome sequences, as in several major crops and model plant species, is key for oat development, as well as, the identification of genes underlying economically important traits will enable the identification of perfect markers within the key genes enabling high-throughput selection and genomic prediction. The most important breakthrough needed in oat research is to have the complete sequencing of an oat genome. Together, molecular and breeding laboratories are promising to continue to expand in oat biotechnology research to address the future challenges that face the crop and to accelerate oat genome investigations.

References

AACC I (1999) American Association of Cereals Chemists International. http://www.aaccnet.org/

Ahmad M, Zaffar G, Razvi SM, Mir SD, Rather MA (2013) Stability properties of certain oats genotypes for major grain yielding characteristics. Int J Plant Breed Genet:1–6

Anderson JW (2003) Whole grains protect against atherosclerotic cardiovascular disease. Proc Nutr Soc 62:135–142. https://doi.org/10.1079/pns2002222

Anderson W, Johansen C, Siddique KHM (2016) Addressing the yield gap in rainfed crops: a review. Agron Sustain Dev 36:18. https://doi.org/10.1007/s13593-015-0341-y

Asoro FG, Newell MA, Beavis WD, Scott M, Jannink JL (2011) Accuracy and training population design for genomic selection on quantitative traits in elite North American oats. Plant Genome J 4:132. https://doi.org/10.3835/plantgenome2011.02.0007

Asoro FG, Newell MA, Beavis WD, Scott M, Jannink JL (2013) Genomic, marker-assisted, and pedigree-BLUP selection methods for β-glucan concentration in elite oat. Crop Sci 53:1894. https://doi.org/10.2135/cropsci2012.09.0526

Badaeva ED, Shelukhina OY, Goryunova SV, Loskutov IG, Pukhalskiy VA (2010) Phylogenetic relationships of tetraploid AB genome *Avena* species evaluated by means of cytogenetic (C-banding and FISH) and RAPD analyses. J Bot. https://doi.org/10.1155/2010/742307

Bai Y, Pavan S, Zheng Z, Zappel N, Reinstaedler A (2008) Naturally occurring broad-spectrum powdery mildew resistance in a central American tomato accession is caused by loss of *Mlo* function. Mol Plant-Microbe Interact 21:30–39. https://doi.org/10.1094/mpmi-21-1-0030

Bailey-Serres J, Lee SC, Brinton E (2012) Waterproofing crops: effective flooding survival strategies. Plant Physiol 160:1698–1709

Barbosa-Neto JF, Siripoonwiwat W, O'donoughue LS, Gray SM, Smith DM, Kolb FL, Gourmet C, Brown CM, Sorrells ME (2000) Chromosomal regions associated with barley yellow dwarf virus resistance in oat. Euphytica 114(1):67–76

Baum BR (1977) Oats: wild and cultivated, a monograph of the genus *Avena* L. (Poaceae). Biosyst Res Institute, Ottawa, Canada

Becker HC, Leon J (1988) Stability analysis in plant breeding. Plant Breed 101:1–23

Beebe S, Ramirez J, Jarvis A, Idupulapati R, Mosquera G et al (2011) Genetic improvement of common beans and the challenges of climate change. Crop Adapt Clim Change 26:356–369

Beer SC, Siripoonwiwat W, O'donoughue LS et al (1997) Associations between molecular markers and quantitative traits in an oat germplasm pool: can we infer linkages. J Agri Genomics 3:1–16

Behnken LM, Breitenbach FR, Miller RP (2009) Impact of foliar fungicide to control crown rust in oats in 2009

Bekele WA, Wight CP, Chao S, Howarth CJ, Tinker NA (2018) Haplotype-based genotyping-by-sequencing in oat genome research. Plant Biotechnol J 16:1452–1463

Bernardo R (2013) Genomewide markers as cofactors for precision mapping of quantitative trait loci. Theor Appl Genet 126:999–1009. https://doi.org/10.1007/s00122-012-2032-2

Bernardo R, Yu J (2007) Prospects for genomewide selection for quantitative traits in maize. Crop Sci 47:1082–1090. https://doi.org/10.2135/cropsci2006.11.0690

Bjørnstad Å, He X, Tekle S, Klos K, Huang YF, Tinker NA, Dong Y, Skinnes H (2017) Genetic variation and associations involving Fusarium head blight and deoxynivalenol accumulation in cultivated oat (*Avena sativa* L.). Plant Breed 136:620–636. https://doi.org/10.1111/pbr.12502

Bratt K, Sunnerheim K, Bryngelsson S, Fagerlund A, Engman L, Andersson RE, Dimberg LH (2003) Avenanthramides in Oats (*Avena sativa* L.) and structure−antioxidant activity relationships. J Agri Food Chem 51:594–600. https://doi.org/10.1021/JF020544F

Bryngelsson S, Mannerstedt-Fogelfors B, Kamal-Eldin A, Andersson R, Dimberg LH (2002) Lipids and antioxidants in groats and hulls of Swedish oats (*Avena sativa* L). J Sci Food Agri 82:606–614. https://doi.org/10.1002/jsfa.10.84

Buerstmayr H, Krenn N, Stephan U, Grausgruber H, Zechner E (2007) Agronomic performance and quality of oat (*Avena sativa* L.) genotypes of worldwide origin produced under Central European growing conditions. Field Crop Res 101:343–351. https://doi.org/10.1016/j.fcr.2006.12.011

Burton RA, Fincher GB (2009) (1,3;1,4)-β-*D*-Glucans in cell walls of the poaceae, lower plants, and fungi: a tale of two linkages. Mol Plant 2:873–882. https://doi.org/10.1093/MP/SSP063

Burton RA, Jobling SA, Harvey AJ, Shirley NJ, Mather DE, Bacic A, Fincher GB (2008) The genetics and transcriptional profiles of the cellulose synthase-like HvCslF gene family in barley. Plant Physiol 146:1821–1833

Burton RA, Wilson SM, Hrmova M, Harvey AJ, Shirley NJ, Medhurst A, Stone BA, Newbigin EJ, Bacic A, Fincher GB (2006) Cellulose synthase-like CslF genes mediate the synthesis of cell wall (1, 3; 1, 4)-β-*D*-glucans. Science 311(80):1940–1942

Bush AL, Wise RP, Rayapati PJ, Lee M (1994) Restriction fragment length polymorphisms linked to genes for resistance to crown rust (*Puccinia coronata*) in near-isogenic lines of hexaploid oat (*Avena sativa*). Genome 37:823–831

Butt MS, Tahir-Nadeem M, Khan MK, Shabir R, Butt MS (2008) Oat: unique among the cereals. Eur J Nutr 47:68–79. https://doi.org/10.1007/s00394-008-0698-7

Cavanagh C, Morell M, Mackay I, Powell W (2008) From mutations to MAGIC: resources for gene discovery, validation and delivery in crop plants. Curr Opin Plant Biol 11:215–221

Chaffin AS, Huang YF, Smith S, Bekele WA, Babiker E, Gnanesh BN, Foresman BJ, Blanchard SG, Jay JJ, Reid RW, Wight CP (2016) A consensus map in cultivated hexaploid oat reveals conserved grass synteny with substantial subgenome rearrangement. Plant Genome 9(2):10–25

Chaves MM, Maroco JP, Pereira JS (2003) Understanding plant responses to drought: from genes to the whole plant. Funct Plant Biol 30:239–264. https://doi.org/10.1071/fp02076

Chen C-YO, Milbury PE, Collins FW, Blumberg JB (2007a) Avenanthramides are bioavailable and have antioxidant activity in humans after acute consumption of an enriched mixture from oats. J Nutr 137:1375–1382

Chen C, Neill K, Wichman D, Westcott M (2008) Hard red spring wheat response to row spacing, seeding rate, and nitrogen. Agron J 100:1296–1302. https://doi.org/10.2134/agronj2007.0198

Chen G, Chong J, Prashar S, Procunier JD (2007b) Discovery and genotyping of high-throughput SNP markers for crown rust resistance gene Pc94 in cultivated oat. Plant Breed 126:379–384. https://doi.org/10.1111/j.1439-0523.2007.01364.x

Chew P, Meade K, Hayes A, Harjes C, Bao Y, Beattie AD, Puddephat I, Gusmini G, Tanksley SD (2016) A study on the genetic relationships of *Avena* taxa and the origins of hexaploid oat. Theor Appl Genet 129:1405–1415. https://doi.org/10.1007/s00122-016-2712-4

Chong J (2006) Crown rust in oat in western Canada in 2005. Can Plant Dis Surv 86:66–67

Chong J, Gruenke J, Dueck R, Mayert W, Fetch JM, Mccartney C (2011) Virulence of *Puccinia coronata* f. sp. *avenae* in the eastern Prairie region of Canada during 2007–2009. Can J Plant Pathol Can Phytopathol 33:77–87. https://doi.org/10.1080/07060661.2010.546957

Chong J, Leonard KJ, Salmeron JJ (2000) A North American system of nomenclature for *Puccinia coronata* f. sp *avenae*. Plant Dis 84:580–585

Chong J, Reimer E, Somers D, Aung T, Penner GA (2004) Development of sequence-characterized amplified region (SCAR) markers for resistance gene *Pc94* to crown rust in oat. Can J Plant Pathol 26:89–96. https://doi.org/10.1080/07060660409507118

Chong J, Seaman WL (1994) Incidence and virulence of *Puccinia coronata* f. sp. *avenae* in Canada in 1993. Can J Plant Pathol 16:335–340. https://doi.org/10.1080/07060669409500740

Clavijo BJ, Venturini L, Schudoma C, Accinelli GG, Kaithakottil G, Wright J, Borrill P, Kettleborough G, Heavens D, Chapman H, Lipscombe J (2017) An improved assembly and annotation of the allohexaploid wheat genome identifies complete families of agronomic genes and provides genomic evidence for chromosomal translocations. Genome Res 27:885–896. https://doi.org/10.1101/gr.217117.116

Clemens R, van Klinken BJW (2014) Oats, more than just a whole grain: an introduction. Br J Nutr 112:S1–S3. https://doi.org/10.1017/s0007114514002712

Clifford BC (1995) Diseases, pest and disorders of oat. In: Welch R (ed) The oat crop. Chapman & Hall, London, pp 252–278

Collins FW (1989) Oat phenolics: avenanthramides, novel substituted n-cinnamoylanthranilate alkaloids from oat groats and hulls. J Agri Food Chem 37:60–66. https://doi.org/10.1021/jf00085a015

Crosby K, Stokes TO, Latta RG (2014) Evolving California genotypes of *Avena barbata* are derived from multiple introductions but still maintain substantial population structure. PeerJ 2: e633. https://doi.org/10.7717/peerj.633

Crossa J, Pérez-Rodríguez P, Cuevas J, Montesinos-López O, Jarquín D, de los Campos G, Burgueño J, González-Camacho JM, Pérez-Elizalde S, Beyene Y, Dreisigacker S (2017) Genomic selection in plant breeding: methods, models, and perspectives. Trends Plant Sci 22:961–975

De Koeyer DL, Tinker NA, Wight CP, Deyl J, Burrows VD, O'Donoughue LS, Lybaert A, Molnar SJ, Armstrong KC, Fedak G, Wesenberg DM (2004) A molecular linkage map with associated QTLs from a hulless × covered spring oat population. Theor Appl Genet 108:1285–1298

Dhingra D, Michael M, Rajput H, Patil RT (2012) Dietary fibre in foods: a review. J Food Sci Technol 49:255–266. https://doi.org/10.1007/s13197-011-0365-5
Dimberg LH, Theander O, Lingnert H (1992) Avenanthramides-a group of phenolic antioxidants in oats. Cereal Chem 70:478–637
Ehlers W (1989) Transpiration efficiency of oat. Agron J 81:810–817
Elshire RJ, Glaubitz JC, Sun Q, Poland JA, Kawamoto K, Buckler ES, Mitchell SE (2011) A robust, simple genotyping-by-sequencing (GBS) approach for high diversity species. PLoS ONE 6:e19379
Esvelt Klos K, Yimer BA, Babiker EM, Beattie AD, Bonman JM, Carson ML, Chong J, Harrison SA, Ibrahim AM, Kolb FL, McCartney CA (2017) Genome-wide association mapping of crown rust resistance in oat elite germplasm. Plant Genome 10(2):1–10
Esvelt Klos K, Huang YF, Bekele WA, Obert DE, Babiker E, Beattie AD, Bjørnstad Å, Bonman JM, Carson ML, Chao S, Gnanesh BN (2016) Population genomics related to adaptation in elite oat germplasm. Plant Genome 9:1–12. https://doi.org/10.3835/plantgenome2015.10.0103
FAO (2017a) FAOSTAT. In: Food agriculture organ. United Nations. http://www.fao.org/faostat/en/
FAO (2017b) FAO. In: Food agriculture organ. United Nations. http://www.fao.org/home/en/
FDA (2012) Code of federal regulations. (Title 21, Section 101.81)
Feng B, Ma LJ, Yao JJ, Fang Y, Mei YA, Wei SM (2013) Protective effect of oat bran extracts on human dermal fibroblast injury induced by hydrogen peroxide. J Zhejiang Univ Sci B 14 (2):97–105
Fetch JM, Fetch T (2011) Inheritance of resistance to oat stem rust in the cultivars Ronald and AC Gwen. Can J Plant Sci 91:419–423
Fincher GB (2009) Exploring the evolution of (1,3;1,4)-β-*D*-glucans in plant cell walls: comparative genomics can help! Curr Opin Plant Biol 12:140–147
Fominaya A, Hueros G, Loarce Y, Ferrer E (1995) Chromosomal distribution of a repeated DNA-sequence from C-genome heterochromatin and the identification of a new ribosomal DNA locus in the *Avena* genus. Genome 38:548–557
Foresman BJ, Oliver RE, Jackson EW, Chao S, Arruda MP, Kolb FL (2016) Genome-wide association mapping of barley yellow dwarf virus tolerance in spring oat (*Avena sativa* L.). PLoS One 11:e0155376. https://doi.org/10.1371/journal.pone.0155376
Frid A, Tura A, Pacini G, Ridderstrale M (2017) Effect of oral pre-meal administration of betaglucans on glycaemic control and variability in subjects with type 1 diabetes. Nutrients. https://doi.org/10.3390/nu9091004
Fu YB (2018) Oat evolution revealed in the maternal lineages of 25 *Avena* species. Sci Rep 8:4252. https://doi.org/10.1038/s41598-018-22478-4
Fu YB, Williams DJ (2008) AFLP variation in 25 *Avena* species. Theor Appl Genet 117:333–342. https://doi.org/10.1007/s00122-008-0778-3
Glaubitz JC, Casstevens TM, Lu F, Harriman J, Elshire RJ, Sun Q, Buckler ES (2014) TASSEL-GBS: a high capacity genotyping by sequencing analysis pipeline. PLoS ONE 9: e90346
Gnanesh BN, Fetch JM, Menzies JG, Beattie AD, Eckstein PE, McCartney CA (2013) Chromosome location and allele-specific PCR markers for marker-assisted selection of the oat crown rust resistance gene *Pc91*. Mol Breed 32:679–686. https://doi.org/10.1007/S11032-013-9900-6
Goddard ME, Hayes BJ (2009) Genomic selection based on dense genotypes inferred from sparse genotypes. Proc Assoc Advmt Anim Breed Genet 18:26–29
Gramza-Michałowska A, Kmiecik D, Kobus-Cisowska J, Żywica A, Dziedzic K, Brzozowska A (2018) Phytonutrients in oat (*Avena sativa* L.) Drink: effect of plant extract on antiradical capacity, nutritional value and sensory characteristics. Polish J Food Nutr Sci 68:63–71. https://doi.org/10.1515/pjfns-2017-0009

Guo L, Tong LT, Liu L, Zhong K, Qiu J, Zhou S (2014) The cholesterol-lowering effects of oat varieties based on their difference in the composition of proteins and lipids. Lipids Health Dis. https://doi.org/10.1186/1476-511x-13-182

Haas P, Machado MJ, Anton AA, Silva AS, De Francisco A (2009) Effectiveness of whole grain consumption in the prevention of colorectal cancer: meta-analysis of cohort studies. Intl J Food Sci Nutr 60:1–13. https://doi.org/10.1080/09637480802183380

Halima NB, Saad RB, Khemakhem B, Fendri I, Abdelkafi S (2015) Oat (*Avena sativa L.*): oil and nutriment compounds Valorization for potential use in industrial applications. J Oleo Sci 64:915–932. https://doi.org/10.5650/jos.ess15074

Hayes BJ, Pryce J, Chamberlain AJ, Bowman PJ, Goddard ME (2010) Genetic architecture of complex traits and accuracy of genomic prediction: coat colour, milk-fat percentage, and type in Holstein cattle as contrasting model traits. PLoS Genet. https://doi.org/10.1371/journal.pgen.1001139

He X, Skinnes H, Oliver RE, Jackson EW, Bjørnstad Å (2013) Linkage mapping and identification of QTL affecting deoxynivalenol (DON) content (Fusarium resistance) in oats (*Avena sativa* L.). Theor Appl Genet 126:2655–2670. https://doi.org/10.1007/s00122-013-2163-0

Herrmann M, Roderick HW (1996) Characterisation of new oat germplasm for resistance to powdery mildew. Euphytica 89:405–410

Holland J, Portyanko V, Hoffman D, Lee M (2002) Genomic regions controlling vernalization and photoperiod responses in oat. Theor Appl Genet 105:113–126

Holland JB, Moser HS, O'donoughue LS LS, Lee M (1997) QTLs and epistasis associated with vernalization responses in oat. Crop Sci 37:1306–1316

Hongyu K, García-Peña M, Araújo LB, Dias CTS (2014) Statistical analysis of yield trials by AMMI analysis of genotype × environment interaction. Biometrical Lett 51:89–102

Hsam SLK, Mohler V, Zeller FJ (2014) The genetics of resistance to powdery mildew in cultivated oats (*Avena sativa* L.): current status of major genes. J Appl Genet 55:155–162

Hsam SLK, Zeller FJ (1998) Chromosomal location of genes for resistance to powdery mildew in cultivated oat (Avena sativa L.). 1. Gene Eg -3 in the cultivar 'Mostyn'. Plant Breed 117:177–178. https://doi.org/10.1111/j.1439-0523.1998.tb01474.x

Huang BE, George AW (2011) R/mpMap: a computational platform for the genetic analysis of multiparent recombinant inbred lines. Bioinformatics 27:727–729

Huang BE, George AW, Forrest KL, Kilian A, Hayden MJ, Morell MK, Cavanagh CR (2012) A multiparent advanced generation inter-cross population for genetic analysis in wheat. Plant Biotechnol J 10:826–839

Ishihara A, Kojima K, Fujita T, Yamamoto Y, Nakajima H (2014) New series of avenanthramides in oat seed. Biosci Biotechnol Biochem 78:1975–1983. https://doi.org/10.1080/09168451.2014.946390

Isidro-Sánchez J, Akdemir D, Burke J (2017) Genomic selection. In: The world wheat book, pp 1001–1019

Jackson EW, Obert DE, Menz M, Hu G, Avant JB, Chong J, Bonman JM (2007) Characterization and mapping of oat crown rust resistance genes using three assessment methods. Phytopathology 97:1063–1070. https://doi.org/10.1094/PHYTO-97-9-1063

Janatuinen EK, Kemppainen TA, Julkunen RJ, Kosma VM, Mäki M, Heikkinen M, Uusitupa MI (2002) No harm from five year ingestion of oats in coeliac disease. Gut 50:332–335. https://doi.org/10.1136/gut.50.3.332

Jellen E, Leggett J (2006) Cytogenetic manipulation in oat improvement. Genet Resour Chromosom Eng Crop Improv 2:199–231

Jellen EN, Gill BS, Cox TS (1994) Genomic in-situ hybridization differentiates between a/*D*-genome and C-genome chromatin and detects intergenomic translocations in polyploid oat species (Genus *Avena*). Genome 37:613–618

Jensen S, D'Arcy C (1995) Barley yellow dwarf: 40 years of progress. Phytopathology 97:55–74

Jin H, Domier LL, Kolb FL, Brown CM (1998) Identification of quantitative loci for tolerance to barley yellow dwarf virus in oat. Phytopathology 88:410–415. https://doi.org/10.1094/PHYTO.1998.88.5.410

Jones IT (1986) Inheritance of adult plant resistance to mildew in oats. Ann Appl Biol 109:187–192. https://doi.org/10.1111/j.1744-7348.1986.tb03198.x
Jones IT (1977) Effect on grain-yield of adult plant resistance to mildew in oats. Ann Appl Biol 86:267–277. https://doi.org/10.1111/j.1744-7348.1977.tb01840.x
Jones IT, Jones ERL (1979) Mildew of oats. UK Cereal Pathog Virulence Surv 1978 Annu Rep, pp 59–63
Jones JM, Engleson J (2010) Whole grains: Benefits and challenges. Annu Rev Food Sci Technol 1:19–40
Kalbasiashtari A, Hammond EG (1977) Oat oil: refining and stability. J Am Oil Chem Soc 54:305–307. https://doi.org/10.1007/bf02672430
Kasum CM, Jacobs DR, Nicodemus K, Folsom AR (2002) Dietary risk factors for upper aerodigestive tract cancers. Int J Cancer 99:267–272. https://doi.org/10.1002/ijc.10341
Kebede AZ, Friesen-Enns J, Gnanesh BN, Menzies JG, Fetch JW, Chong J, Beattie AD, Paczos-Grzęda E, McCartney CA (2018) Mapping oat crown rust resistance gene *Pc45* confirms association with PcKM. G3 (Bethesda) g3.200757.2018. https://doi.org/10.1534/g3.118.200757
Kellogg EA, Camara PE, Rudall PJ, Ladd P, Malcomber ST, Whipple C, Doust AN (2013) Early inflorescence development in the grasses (Poaceae). Front Plant Sci 4:250. https://doi.org/10.3389/fpls.2013.00250
Khan K, Jovanovski E, Ho HV, Marques AC, Zurbau A, Mejia SB, Sievenpiper JL, Vuksan V (2018) The effect of viscous soluble fiber on blood pressure: a systematic review and meta-analysis of randomized controlled trials. Nutr Metab Cardiovasc Dis 28:3–13. https://doi.org/10.1016/j.numecd.2017.09.007
Kianian SF, Phillips RL, Rines HW, Fulcher RG, Webster FH, Stuthman DD (2000) Quantitative trait loci influencing β-glucan content in oat (*Avena sativa*, 2n = 6x = 42). Theor Appl Genet 101:1039–1048. https://doi.org/10.1007/s001220051578
Koprivnjak O, Majetić V, Staver MM, Lovrić A, Blagović B (2010) Effect of phospholipids on extraction of hydrophilic phenols from virgin olive oils. Food Chem 119:698–702. https://doi.org/10.1016/j.foodchem.2009.07.016
Kulcheski FR, Graichen FA, Martinelli JA, Locatelli AB, Federizzi LC, Delatorre CA (2010) Molecular mapping of Pc68, a crown rust resistance gene in *Avena sativa*. Euphytica 175:423–432. https://doi.org/10.1007/s10681-010-0198-8
Ladizinsky G (1998) A new species of *Avena* from Sicily, possibly the tetraploid progenitor of hexaploid oats. Genet Resour Crop Evol 45:263–269
Ladizinsky G (2012) Studies in oat evolution : a man's life with *Avena*. Springer
Lande R, Thompson R (1990) Efficiency of marker-assisted selection in the improvement of quantitative traits. Genetics 124:743–756
Langseth W, Elen O (1996) Differences between barley, oats and wheat in the occurrence of deoxynivalenol and other trichothecenes in Norwegian grain. J Phytopathol Zeitschrift 144:113–118
Leonard KJ (2002) Oat lines with effective adult plant resistance to crown rust. Plant Dis 86:593–598. https://doi.org/10.1094/pdis.2002.86.6.593
Li Z, Chen Y, Meesapyodsuk D, Qiu X (2019) The biosynthetic pathway of major avenanthramides in oat. Metabolites. https://doi.org/10.3390/metabo9080163
Lin Y, Gnanesh BN, Chong J, Chen G, Beattie AD, Fetch JW, Kutcher HR, Eckstein PE, Menzies JG, Jackson EW, McCartney CA (2014) A major quantitative trait locus conferring adult plant partial resistance to crown rust in oat. BMC Plant Biol 14:250. https://doi.org/10.1186/s12870-014-0250-2
Linares C, Ferrer E, Fominaya A (1998) Discrimination of the closely related A and D genomes of the hexaploid oat *Avena sativa* L. Proc Natl Acad Sci USA 95:12450–12455. https://doi.org/10.1073/pnas.95.21.12450
Lippi MM, Foggi B, Aranguren B, Ronchitelli A, Revedin A (2015) Multistep food plant processing at Grotta Paglicci (Southern Italy) around 32,600 cal B.P. Proc Natl Acad Sci USA 112:12075–12080. https://doi.org/10.1073/pnas.1505213112

Liu H, Chen G-B (2017) A fast genomic selection approach for large genomic data. Theor Appl Genet. https://doi.org/10.1007/s00122-017-2887-3

Liu Q, Li X, Zhou X, Li M, Zhang F, Schwarzacher T, Heslop-Harrison JS (2019) The repetitive DNA landscape in *Avena* (Poaceae): chromosome and genome evolution defined by major repeat classes in whole-genome sequence reads. BMC Plant Biol 19(1):226–236

Liu RH (2007) Whole grain phytochemicals and health. J Cereal Sci 46:207–219

Liu S, Manson JE, Stampfer MJ, Hu FB, Giovannucci E, Colditz GA, Hennekens CH, Willett WC (2000) A prospective study of whole-grain intake and risk of type 2 diabetes mellitus in US women. Amer J Public Health 90:1409–1415

Locatelli AB, Federizzi LC, Milach SC, Wight CP, Molnar SJ, Chapados JT, Tinker NA (2006) Loci affecting flowering time in oat under short-day conditions. Genome 49:1528–1538. https://doi.org/10.1139/g06-108

Løes AK, Henriksen TM, Eltun R (2007) N supply in stockless organic cereal production under northern temperate conditions. Undersown legumes or whole season green manure? In: 3rd QLIF congress improving sustainability in organic and low input food production system, University of Hohenheim, p 230

Londono DM, Van't Westende WP, Goryunova S, Salentijn EM, Van den Broeck HC, Van der Meer IM, Visser RG, Gilissen LJ, Smulders MJ (2013) Avenin diversity analysis of the genus *Avena* (oat). Relevance for people with celiac disease. J Cereal Sci 58:170–177

Loskutov I, Rines H (2011) *Avena*. In: Kole C (ed) Wild crop relatives: genomic and breeding resources. Springer, Berlin, Heidelberg, pp 109–183

Loskutov IG (2008) On evolutionary pathways of *Avena* species. Genet Resour Crop Evol 55:211–220. https://doi.org/10.1007/s10722-007-9229-2

Lu F, Lipka AE, Glaubitz J, Elshire R, Cherney JH, Casler MD, Buckler ES, Costich DE (2013) Switchgrass genomic diversity, ploidy, and evolution: novel insights from a network-based SNP discovery protocol. PLoS Genet 9:e1003215

Lüpken T, Stein N, Perovic D, Habekuß A, Krämer I, Hähnel U, Steuernagel B, Scholz U, Zhou R, Ariyadasa R, Taudien S (2013) Genomics-based high-resolution mapping of the BaMMV/BaYMV resistance gene *rym11* in barley (*Hordeum vulgare* L.). Theor Appl Genet 126:1201–1212. https://doi.org/10.1007/s00122-013-2047-3

MacArthur-Grant LA (1986) Sugars and nonstarchy polysaccharides in oats. Oat Chemistry and Technology. F. H. Webst. American Association of Cereal Chemists, St. Paul, Minnesota, USA, pp 75–92

Mandakova T, Lysak MA (2018) Post-polyploid diploidization and diversification through dysploid changes. Curr Opin Plant Biol 42:55–65

Marshall GC, Ohm HW (1987) Yield responses of 16 winter wheat cultivars to row spacing and seeding rate. Agron J 79:1027–1030

Marshall HG (1992) Breeding oat for resistance to environmental stress. Oat Sci Technol:699–749

Martens JW (1985) Oat Stem rust. Dis Distrib Epidemiol Control. https://doi.org/10.1016/B978-0-12-148402-6.50012-2

Martinelli JA, Chaves MS, Graichen FA, Federizzi LC, Dresch LF (2014) Impact of Fusarium head blight in reducing the weight of oat grains. J Agri Sci 6:188–198

Martinez-Villaluenga C, Penas E (2017) Health benefits of oat: current evidence and molecular mechanisms. Curr Opin Food Sci 14:26–31. https://doi.org/10.1016/j.cofs.2017.01.004

Mattila P, Pihlava JM, Hellstrom J (2005) Contents of phenolic acids, alkyl- and alkenylresorcinols, and avenanthramides in commercial grain products. J Agric Food Chem 53:8290–8295. https://doi.org/10.1021/jf051437z

Mauler-Machnik A, Zahn K (1994) Ear fusarioses in wheat: new findings on their epidemiology and control with Folicur (tebuconazole). Pflanzenschutz-Nachrichten Bayer 47:129–155

McCartney CA, Stonehouse RG, Rossnagel BG, Eckstein PE, Scoles GJ, Zatorski T, Beattie AD, Chong J (2011) Mapping of the oat crown rust resistance gene *Pc91*. Theor Appl Genet 122:317–325. https://doi.org/10.1007/s00122-010-1448-9

McKirdy SJ, Jones RAC, Nutter FW (2002) Quantification of yield losses caused by barley yellow dwarf virus in wheat and oats. Plant Dis 86:769–773

Mesterházy Á, Bartók T (1996) Control of Fusarium head blight of wheat by fungicides and its effect on the toxin contamination of the grains. Pflanzenschutz-Nachrichten Bayer 49:181–198

Meuwissen THE, Hayes BJ, Goddard ME (2001) Prediction of total genetic value using genome-wide Dense marker maps. Genetics 157:1819–1829

Meydani M (2009) Potential health benefits of avenanthramides of oats. Nutr Rev 67:731–735

Meyer KA, Kushi LH, Jacobs DR Jr, Slavin J, Sellers TA, Folsom AR (2000) Carbohydrates, dietary fiber, and incident type 2 diabetes in older women. Am J Clin Nutr 71:921–930. https://doi.org/10.1093/ajcn/71.4.921

Molnar SJ, Chapados JT, Satheeskumar S, Wight CP, Bancroft B, Orr W, Luckert DE, Kibite S (2012) Comparative mapping of the oat Dw6/dw6 dwarfing locus using NILs and association with vacuolar proton ATPase subunit H. Theor Appl Genet 124:1115–1125. https://doi.org/10.1007/s00122-011-1773-7

Montilla-Bascón G, Armstrong PR, Rongkui H, Mark S (2017) Quantification of betaglucans, lipid and protein contents in whole oat groats (*Avena sativa* L.) using near infrared reflectance spectroscopy. J Near Infrared Spectrosc 25:172–179. https://doi.org/10.1177/0967033517709615

Montilla-Bascón G, Rispail N, Sánchez-Martín J, Rubiales D, Mur LA, Langdon T, Howarth CJ, Prats E (2015) Genome-wide association study for crown rust (*Puccinia coronata* f. sp. avenae) and powdery mildew (*Blumeria graminis* f. sp. avenae) resistance in an oat (*Avena sativa*) collection of commercial varieties and landraces. Front Plant Sci 6:103. https://doi.org/10.3389/fpls.2015.00103

Morikawa T, Leggett JM (1996) Cytological and morphological variations in wild populations of *Avena canariensis* from the Canary Islands. Genes Genet Syst 71:15–21

Morikawa T, Leggett JM (2008) Isozyme polymorphism and genetic differentiation in natural populations of an endemic tetraploid species *Avena maroccana* in Morocco. Genet Resour Crop Evol 55:1313–1321. https://doi.org/10.1007/s10722-008-9330-1

Nava IC, Wight CP, Pacheco MT, Federizzi LC, Tinker NA (2012) Tagging and mapping candidate loci for vernalization and flower initiation in hexaploid oat. Mol Breed 30:1295–1312. https://doi.org/10.1007/s11032-012-9715-x

Nazareno ES, Li F, Smith M, Park RF, Kianian SF, Figueroa M (2018) *Puccinia coronata* f. sp *avenae*: a threat to global oat production. Mol Plant Pathol 19:1047–1060. https://doi.org/10.1111/mpp.12608

Nebus J, Nystrand G, Fowler J, Wallo W (2009) A daily oat-based skin care regimen for atopic skin. J Am Acad Dermatol 60:AB67–AB67

Newell MA, Asoro FG, Scott MP, White PJ, Beavis WD, Jannink JL (2012) Genome-wide association study for oat (*Avena sativa L.*) beta-glucan concentration using germplasm of worldwide origin. Theor Appl Genet 125:1687–1696

Nutzmann HW, Huang AC, Osbourn A (2016) Plant metabolic clusters—from genetics to genomics. New Phytol 211:771–789

Nutzmann HW, Scazzocchio C, Osbourn A (2018) Metabolic gene clusters in eukaryotes. Annu Rev Genet 52(52):159–183

O'Donoughue LS, Wang Z, Röder M, Kneen B, Leggett M, Sorrells ME, Tanksley SD (1997) An RFLP-based linkage map of oats based on a cross between two diploid taxa (*Avena atlantica* × *A. hirtula*). Genome 35:765–771 http://dx.doi.org/101139/g92-117

Okarter N, Liu RH (2010) Health benefits of whole grain phytochemicals. Crit Rev Food Sci Nutr 50:193–208. https://doi.org/10.1080/10408390802248734

Okon SM (2015) Effectiveness of resistance genes to powdery mildew in oat. Crop Protec 74:48–50. https://doi.org/10.1016/j.cropro.2015.04.004

Okoń SM, Ociepa T (2018) Effectiveness of new sources of resistance against oat powdery mildew identified in A. sterilis. J Plant Dis Protec 125:505–510. https://doi.org/10.1007/s41348-018-0171-7

Oliver RE, Lazo GR, Lutz JD, Rubenfield MJ, Tinker NA, Anderson JM, Morehead NH, Adhikary D, Jellen EN, Maughan PJ, Guedira GL (2011) Model SNP development for complex genomes based on hexaploid oat using high-throughput 454 sequencing technology. BMC Genom 12:77

Oliver RE, Tinker NA, Lazo GR, Chao S, Jellen EN, Carson ML, Rines HW, Obert DE, Lutz JD, Shackelford I, Korol AB (2013) SNP discovery and chromosome anchoring provide the first physically-anchored hexaploid oat map and reveal synteny with model species. PLoS ONE 8:58068

Panisson E, Reis EM, Boller W (2003) Quantificação de danos causados pela Giberela em cereais de inverno, na safra 2000. Fitopatol Bras 28:189–192

Pavan S, Schiavulli A, Appiano M, Marcotrigiano AR, Cillo F, Visser RG, Bai Y, Lotti C, Ricciardi (2011) Pea powdery mildew er1 resistance is associated to loss-of-function mutations at a MLO homologous locus. Theor Appl Genet 123:1425–1431. https://doi.org/10.1007/s00122-011-1677-6

Peng Y, Zhou P, Zhao J, Li J, Lai S, Tinker NA, Liao S, Yan H (2018) Phylogenetic relationships in the genus *Avena* based on the nuclear *Pgk1* gene. PLoS ONE 13:e0200047. https://doi.org/10.1371/journal.pone.0200047

Peng YY, Wei YM, Baum BR, Yan ZH, Lan XJ, Dai SF, Zheng YL (2010) Phylogenetic inferences in *Avena* based on analysis of FL intron2 sequences. Theor Appl Genet 121:985–1000. https://doi.org/10.1007/s00122-010-1367-9

Peng YY, Wei YM, Baum BR, Zheng YL (2008) Molecular diversity of the *5S rRNA* gene and genomic relationships in the genus *Avena (Poaceae: Aveneae)*. Genome 51:137–154. https://doi.org/10.1139/g07-111

Penner GA, Chong J, Wight CP, Molnar SJ, Fedak G (1993) Identification of an RAPD marker for the crown rust resistance gene *Pc68* in oats. Genome 36:818–820

Peräaho M, Kaukinen K, Mustalahti K, Vuolteenaho N, Mäki M, Laippala P, Collin P (2004) Effect of an oats-containing gluten-free diet on symptoms and quality of life in coeliac disease. A randomized study. Scand J Gastroenterol 39:27–31. https://doi.org/10.1080/00365520310007783

Peterson DM (2001) Oat antioxidants. J Cereal Sci 33:115–129. https://doi.org/10.1006/jcrs.2000.0349

Peterson DM, Qureshi A (1993) Genotype and environmental effects on tocols of barley and oats. Cereal Chem 70:157–162

Peterson DM, Saigo RH, Holy J (1985) Development of oat aleurone cells and their protein bodies. Cereal Chem 62:366–371

Peterson DM, Wood DF (1997) Composition and structure of high-oil oat. J Cereal Sci 26:121–128. https://doi.org/10.1006/JCRS.1996.0111

Pirgozliev SR, Edwards SG, Hare MC, Jenkinson P (2003) Strategies for the control of Fusarium head blight in cereals. Eur J Plant Pathol 109:731–742. https://doi.org/10.1023/A:1026034509247

Poland J, Endelman J, Dawson J, Rutkoski J, Wu S, Manes Y, Dreisigacker S, Crossa J, Sánchez-Villeda H, Sorrells M, Jannink JL (2012) Genomic selection in wheat breeding using genotyping-by-sequencing. Plant Genome 5:103–113. https://doi.org/10.3835/Plantgenome2012.06.0006

Poland JA, Rife T (2013) Genotyping-by-Sequencing for Plant Breeding and Genetics. Plant Genome 5:92–98

Prats E, Llamas MJ, Rubiales D (2007) Characterization of resistance mechanisms to Erysiphe pisi in Medicago truncatula. Phytopathology 97:1049–1053. https://doi.org/10.1094/phyto-97-9-1049

Qi X, Bakht S, Leggett M, Maxwell C, Melton R, Osbourn AA (2004) A gene cluster for secondary metabolism in oat: Implications for the evolution of metabolic diversity in plants. Proc Natl Acad Sci USA 101:8233–8238

Qin F, Shinozaki K, Yamaguchi-Shinozaki K (2011) Achievements and challenges in understanding plant abiotic stress responses and tolerance. Plant Cell Physiol 52:1569–1582

Rajhathy T, Thomas H (1974) Cytogenetics of oats (*Avena* L.). Genet Soc Canada 2:86–90

Rasane P, Jha A, Sabikhi L, Kumar A, Unnikrishnan VS (2015) Nutritional advantages of oats and opportunities for its processing as value added foods—a review. J Food Sci Technol 52:662–675. https://doi.org/10.1007/s13197-013-1072-1

Ren CZ, Ma BL, Burrows V, Zhou J, Hu YG, Guo L, Wei L, Sha L, Deng L (2007) Evaluation of early mature naked oat varieties as a summer-seeded crop in dryland northern climate regions. Field Crop Res 103:248–254. https://doi.org/10.1016/j.fcr.2007.07.001

Rines HW, Miller ME, Carson M, Chao S, Tiede T, Wiersma J, Kianian SF (2018) Identification, introgression, and molecular marker genetic analysis and selection of a highly effective novel oat crown rust resistance from diploid oat, *Avena strigosa*. Theor Appl Genet 131:721–733

Rines HW, Molnar SJ, Tinker NA, Phillips RL (2006) Oat. In: Kole C (ed) Genome maping and molecular breeding in plants, vol 1. Cereals and Millets. Springer, Berlin, Heidelberg, pp 211–242

Roderick HW, Jones ERL, Sebesta J (2000) Resistance to oat powdery mildew in Britain and Europe: a review. Ann Appl Biol 136:85–91. https://doi.org/10.1111/j.1744-7348.2000.tb00012.x

Rondanelli M, Opizzi A, Monteferrario F (2009) The biological activity of beta-glucans. Minerva Med 100:237–245

Rooney WL, Rines HW, Phillips RL (1994) Identification of RFLP Markers linked to crown rust resistance genes *Pc 91* and *Pc* in oat. Crop Sci 34:940. https://doi.org/10.2135/cropsci1994.0011183X003400040019x

Sánchez-Martín J, Rispail N, Flores F, Emeran AA, Sillero JC, Rubiales D, Prats E (2017) Higher rust resistance and similar yield of oat landraces versus cultivars under high temperature and drought. Agron Sustain Dev. https://doi.org/10.1007/s13593-016-0407-5

Sánchez-Martín J, Rubiales D, Flores F, Emeran AA, Shtaya MJ, Sillero JC, Allagui MB, Prats E (2014) Adaptation of oat (*Avena sativa*) cultivars to autumn sowings in Mediterranean environments. Field Crop Res 156:111–122. https://doi.org/10.1016/j.fcr.2013.10.018

Sánchez-Martín J, Rubiales D, Sillero JC, Prats E (2012) Identification and characterization of sources of resistance in *Avena sativa, A. byzantina and A. strigosa* germplasm against a pathotype of *Puccinia coronata* f.sp. avenae with virulence against the *Pc94* resistance gene. Plant Pathol. https://doi.org/10.1111/j.1365-3059.2011.02514.x

Sanz MJ, Jellen EN, Loarce Y, Irigoyen ML, Ferrer E, Fominaya A (2010) A new chromosome nomenclature system for oat (*Avena sativa* L. and *A. byzantina* C. Koch) based on FISH analysis of monosomic lines. Theor Appl Genet 121:1541–1552. https://doi.org/10.1007/s00122-010-1409-3

Satheeskumar S, Sharp PJ, Lagudah ES, McIntosh RA, Molnar SJ (2011) Genetic association of crown rust resistance gene *Pc68*, storage protein loci, and resistance gene analogues in oats. Genome 54:484–497. https://doi.org/10.1139/g11-014

Schubert M, Marcussen T, Meseguer AS, Fjellheim S (2019) The grass subfamily Pooideae: cretaceous-palaeocene origin and climate-driven cenozoic diversification. Glob Ecol Biogeogr 28:1168–1182

Sengupta S, Muir JG, Gibson PR (2006) Does butyrate protect from colorectal cancer? J Gastroenterol Hepatol 21:209–218. https://doi.org/10.1111/j.1440-1746.2006.04213.x

Shewry PR, Casey R (1999) Seed proteins. Seed proteins. Springer, Netherlands, Dordrecht, pp 1–10

Simons MD (1985) Crown rust. In: Roelfs AP (ed) The cereal rusts: diseases, distribution, epidemiology and control. Academic Press, New York, NY, USA, pp 132–172

Singh R, De S, Belkheir A (2013) *Avena sativa* (Oat), a potential neutraceutical and therapeutic agent: an overview. Crit Rev Food Sci Nutr 53:126–144. https://doi.org/10.1080/10408398.2010.526725

Siripoonwiwat W, O'Donoughue LS, Wesenberg D, Hoffman DL, Barbosa-Neto JF, Sorrells ME (1996) Chromosomal regions associated with quantitative traits in oat. J Quant Trait Loci 2 (3):1–10

Slavin JL, Martini MC, Jacobs DR, Marquart L (1999) Plausible mechanisms for the protectiveness of whole grains. Am J Clin Nutr 70:459s–463s. https://doi.org/10.1093/ajcn/70.3.459s

Smith S (2017) Consensus mapping and association mapping in hexaploid oat. North Carolina

Snowdon RJ, Friedt W (2004) Molecular markers in Brassica oilseed breeding: current status and future possibilities. Plant Breed 123:1–8

Stevens EJ, Armstrong KW, Bezar HJ, Griffin WB, Hampton JG(2004) Fodder oats an overview. In: Suttie JM, Reynolds SG (eds) Fodder oats: a world overview, vol 33, pp 11–18

Stewart D, McDougall G (2014) Oat agriculture, cultivation and breeding targets: implications for human nutrition and health. Br J Nutr 112:S50–S57

Suarna C, Hood RL, Dean RT, Stocker R (1993) Comparative antioxidant activity of tocotrienols and other natural lipid-soluble antioxidants in a homogeneous system, and in rat and human lipoproteins. Biochim Biophys Acta 1166:163–170. https://doi.org/10.1016/0005-2760(93)90092-n

Sur R, Nigam A, Grote D, Liebel F, Southall MD (2008) Avenanthramides, polyphenols from oats, exhibit anti-inflamatory and anti-itch activity. Arch Dermatol Res 300:564–574

Surampudi P, Enkhmaa B, Anuurad E, Berglund L (2016) Lipid lowering with soluble dietary fiber. Curr Atheroscler Rep. https://doi.org/10.1007/s11883-016-0624-z

Szydlowska-Czerniak A, Karlovits G, Hellner G, Szlyk E (2010) Effect of enzymatic and hydrothermal treatments of rapeseeds on quality of the pressed rapeseed oils: part II. Oil yield and oxidative stability. Process Biochem 45:247–258. https://doi.org/10.1016/j.procbio.2009.09.014

Tanhuanpää P, Kalendar R, Schulman AH, Kiviharju E (2008) The first doubled haploid linkage map for cultivated oat. Genome 51:560–569

Tapola N, Karvonen H, Niskanen L, Mikola M, Sarkkinen E (2005) Glycemic responses of oat bran products in type 2 diabetic patients. Nutr Metab Cardiovasc Dis 15:255–261. https://doi.org/10.1016/j.numecd.2004.09.003

Tekauz A, McCallum B, Ames N, Fetch JM (2004) Fusarium head blight of oat—current status in western Canada. Can Plant Dis Surv 26(4):473–479

Tekle S, Skinnes H, Bjornstad A (2013) The germination problem of oat seed lots affected by Fusarium head blight. Eur J Plant Pathol 135:147–158. https://doi.org/10.1007/s10658-012-0074-6

Thies F, Masson LF, Boffetta P, Kris-Etherton P (2014a) Oats and CVD risk markers: a systematic literature review. Br J Nutr 112:S19–S30

Thies F, Masson LF, Boffetta P, Kris-Etherton P (2014b) Oats and bowel disease: a systematic literature review. Br J Nutr 112:S31–S43

Thomas H, Aung T (1978) The transfer of mildew resistance from the tetraploid wild oat *Avena barbata* into the cultivated oat. In: Proceedings of the 8th EUCARPIA congress. II. Interspecific gene transfer. Madrid, pp 109–112

Tinker NA, Bekele WA, Hattori J (2016) Haplotag: software for haplotype-based genotyping-by-sequencing analysis. G3 (Bethesda) 6:857–863

Tinker NA, Chao S, Lazo GR, Oliver RE, Huang YF, Poland JA, Jellen EN, Maughan PJ, Kilian A, Jackson EW (2014) A SNP genotyping array for hexaploid oat. Plant Genome. https://doi.org/10.3835/plantgenome2014.03.0010

Tripathi V, Singh A, Ashraf MT (2018) Avenanthramides of oats: Medicinal importance and future perspectives. Pharmacogn Rev 12:66. https://doi.org/10.4103/phrev.phrev_34_17

Tumino G, Voorrips RE, Morcia C, Ghizzoni R, Germeier CU, Paulo MJ, Terzi V, Smulders MJ (2017) Genome-wide association analysis for lodging tolerance and plant height in a diverse European hexaploid oat collection. Euphytica 213:163. https://doi.org/10.1007/s10681-017-1939-8

Valentine J (1984) Accelerated pedigree selection: an alternative to individual plant selection in the normal pedigree breeding method in the self-pollinated cereals. Euphytica 33:943–951

Valentine J, Cowan A, Marshall A (2011) Oat breeding. In: Webster F, Wood P (eds) Oats chemistry and technology. AACC International, St. Paul Minnesota, USA, pp 11–30

Visscher PM, Wray NR, Zhang Q, Sklar P, McCarthy MI, Brown MA, Yang J (2017) 10 years of GWAS discovery: biology, function, and translation. Am J Hum Genet 101:5–22
Vitaglione P, Napolitano A, Foliano V (2008) Cereal dietary fiber, a natural functional ingredient to deliver phenolic compounds in the gut. Trends Food Sci Technol 19:541–561
Watkins JE, Lane LC (2004) Barley yellow dwarf disease of barley, oats, and wheat. In: Nebguide, University of Nebraska. http://ianrpubs.unl.edu/plantdisease/g906.htm
Watson A, Ghosh S, Williams MJ, Cuddy WS, Simmonds J, Rey MD, Hatta MA, Hinchliffe A, Steed A, Reynolds D, Adamski NM (2018) Speed breeding is a powerful tool to accelerate crop research and breeding. Nat Plants 4:23–29
Webster FH, Wood PJ (2011) OATS: chemistry and technology. AACC International, St. Paul, Minnesota, USA
Welch RW (1995) The chemical composition of oats. The Oat Crop. Springer, Netherlands, Dordrecht, pp 279–320
Wight CP, O'Donoughue LS, Chong J, Tinker NA, Molnar SJ (2004) Discovery, localization, and sequence characterization of molecular markers for the crown rust resistance genes *Pc38*, *Pc39*, and *Pc48* in cultivated oat (*Avena sativa* L.). Mol Breed 14:349–361. https://doi.org/10.1007/s11032-004-0148-z
Wight CP, Penner GA, O'Donoughue LS, Burrows VD, Molnar SJ, Fedak G (1994) The identification of random amplified polymorphic DNA markers for daylength insensitivity in oat. Genome 37:910–914
Wilson WA, McMullen MS (1997) Dosage dependent genetic suppression of oat crown rust resistance gene *Pc-62*. Crop Sci 37:1699. https://doi.org/10.2135/cropsci1997.0011183X003700060004x
Winterfeld G, Doring E, Roser M (2009) Chromosome evolution in wild oat grasses (Aveneae) revealed by molecular phylogeny. Genome 52:361–380. https://doi.org/10.1139/g09-012
Wooten DR, Livingston DP, Lyerly HJ, Holland JB, Jellen EN, Marshall DS, Murphy JP (2009) Quantitative trait loci and epistasis for oat winter-hardiness component traits. Crop Sci 49:1989
Yan H, Bekele WA, Wight CP, Peng Y, Langdon T, Latta RG, Fu YB, Diederichsen A, Howarth CJ, Jellen EN, Boyle B (2016a) High-density marker profiling confirms ancestral genomes of *Avena* species and identifies D-genome chromosomes of hexaploid oat. Theor Appl Genet 129:2133–2149. https://doi.org/10.1007/s00122-016-2762-7
Yan H, Martin SL, Bekele WA, Latta RG, Diederichsen A, Peng Y, Tinker NA (2016b) Genome size variation in the genus *Avena*. Genome 59:209–220
Yao J, Zhao D, Chen X, Zhang Y, Wang J (2018) Use of genomic selection and breeding simulation in cross prediction for improvement of yield and quality in wheat (*Triticum aestivum* L.). Crop J 6:353–365. https://doi.org/10.1016/j.cj.2018.05.003
Youngs VL (1986) Oat lipids and lipid-related enzymes. In: Webst FH (ed) Oats: chemistry and technology. American Association of Cereal Chemists, St. Paul. MN, pp 205–226
Yu J, Herrmann M (2006) Inheritance and mapping of a powdery mildew resistance gene introgressed from *Avena* macrostachya in cultivated oat. Theor Appl Genet 113:429–437
Zhou MX, Robards K, Glennie-Holmes M, Helliwell S (1999a) Oat lipids. J Am Oil Chem Soc 76:159–169. https://doi.org/10.1007/s11746-999-0213-1
Zhou X, Jellen EN, Murphy JP (1999b) Progenitor germplasm of domesticated hexaploid oat. Crop Sci 39:1208–1214
Zhu S, Kaeppler HF (2003) A genetic linkage map for hexaploid, cultivated oat (*Avena sativa* L.) based on an intraspecific cross "Ogle/MAM17-5". Theor Appl Genet 107:26–35

Chapter 5
Genomic Designing for Climate Smart Sorghum

Ganapathy Kuyyamudi Nanaiah and Sujay Rakshit

Abstract Sorghum is a model crop among grass species to study stress response and ensuring food security for millions of poor masses living in the most impoverished drought-prone regions of the world. Low productivity in sorghum is mainly due to their cultivation in marginal soils where soil moisture is a factor limiting its productivity. Drought and other climate-resilient traits has been the major driver for crop production especially in arid and semiarid regions. Sorghum possesses C_4 photosynthetic system and stands tall compared to other major cereals due to inherent ability to overcome drought. Possible effects of drought on productivity especially due to climate change, mechanisms and genetics underlying drought tolerance and modern genomic strategies to overcome drought and other climate resilience traits are well studied and are discussed in this chapter. It is now imperative that sorghum scientists should formulate or strategically revise their programs, physiologists and molecular scientists to explore novel traits to better understand the drivers of heat stress, crop modelers to refine their predictions for future climatic conditions, and economists to identify the most profitable and sustainable adaptation paths for producers.

Keywords Millets · Drought · Mechanisms · Genetics · Genomics

5.1 Introduction

Among the cereal crops, Sorghum [*Sorghum bicolor* (L.) Moench] stands tall and is the fifth major cereal of the world following maize, paddy, wheat and barley as per Food and Agriculture Organization (FAO) production data of 2014. The crop contributes significantly to food, nutritional and livelihood security in many of the

G. Kuyyamudi Nanaiah (✉)
ICAR-Indian Institute of Millets Research, Rajendranagar, Hyderabad 500030, India
e-mail: ganapathy@millets.res.in

S. Rakshit
ICAR-Indian Institute of Maize Research, PAU Campus, Ludhiana 141004, Punjab, India

C. Kole (ed.), *Genomic Designing of Climate-Smart Cereal Crops*,
https://doi.org/10.1007/978-3-319-93381-8_5

world's poorest regions of both Africa and Asia. The crop is valued for its grain and stover and is widely grown for both food and feed. In the Asian and African countries, the crop is predominantly grown for food purposes, while in the developed countries like United States and Australia it is grown for livestock feed and fodder purposes. India and China are among the top ten producers of sorghum across the world. In India, sorghum is predominantly grown in southern and southwestern dry land parts of the country (Aruna et al. 2018). In Asia, specifically India, the crop is grown in two seasons, viz. rainy and post-rainy seasons. The rainy-season grain produce has diverse, uses both for human consumption and livestock feed, industrial uses, etc., while post-rainy-season produce is used primarily for human consumption.

5.1.1 Sorghum as a Crop of the Future

The crop is predominantly grown in the semi-arid tropics of the world and it has inherent tolerance to various stresses such as drought, heat and salinity. The crop has the potential to produce substantial yields even under low input conditions (Ashok Kumar et al. 2011). Sorghum being a C_4 species makes more efficient use of CO_2 and has a higher photosynthetic ability and an ability to utilize nitrogen and water more efficiently. Hence, the crop is more efficient to be grown under conditions with low rainfall and drought where conditions do not favor for growing crops such as rice and wheat. As food, the crop serves as dietary staple for millions of people living in about 30 countries in the subtropical and semi-arid regions of Africa and Asia. It serves as source of food and fodder, especially for the smallholder farming sector. The crop also finds place in the high-input farming as a feed crop, and is fast emerging as a biofuel crop (Hariprasanna and Rakshit 2017).

Compared to fine cereals like rice, wheat and maize, sorghum crop received less attention for genetic and genomics studies in the past and this is the mainly due to its lesser economic importance. During last two and a half decade significant progress has been made in this area. Sorghum and rice together turned out to complement each other in understanding the genome complexicity of the grass genomes. The crop shows close proximity not only to cereals but also commercial crops including sugarcane due to this, sorghum emerged as model crop to intiate comparartive genomics research through syntenic studies. With publication of the sorghum genome sequence in 2009 (Paterson et al. 2009a, b) the scenario was revolutionized and this neglected crop started receiving prominence in genomics studies. Stress tolerance of the crop proved to be an added advantage for its popularity. Over the period of a few decades many reports on sorghum genomics as well as transgenic research have come into the public domain, which deals with almost all traits related to the crop. These studies have exhibited promise to improve the crop further in terms of stress tolerance and yielding ability.

5.1.2 Area and Production Trends

Sorghum is cultivated in 105 countries (Rakshit et al. 2015). Out of these, 37 countries have more than 0.1 m ha sorghum harvested area and eight countries (In decreasing order—Sudan, India, Nigeria, Niger, United States, Ethiopia, Burkina Faso, and Mexico) have more than 1 m ha area under sorghum, which together contribute 71% of world sorghum harvested area. More than 80% of sorghum area globally, i.e., 42.12 m ha is occupied by the developing countries in the Africa and Asia continents where sorghum is grown mainly as dietary source for low income farmers. In western and central Africa, sorghum is grown between the Sahara desert in the north and the equatorial forests in the south. In southern and eastern Africa it is grown predominantly in drier regions (FAO and ICRISAT 1996). Sudan has the largest area under sorghum in northern Africa and the area has increased more than four times during 2011–2013 as compared to 1961–1963. The average sorghum area (2011–2013) in different growing countries is—about 35% of the area is occupied by western and central Africa, 17% by southern Asia, 16% by northern Africa, 12% by eastern and southern Africa, 5% each by Central America and South America, and less than 5% by countries in eastern Asia, Europe and Oceania. In the Americas, the United States has the largest area followed by Mexico. In South America, Argentina and Brazil have some appreciable acreage under sorghum. Australia grows sorghum on more than 0.6 m ha (Hariprasanna and Rakshit 2017).

Developed countries produce about one-third of the world's sorghum and the rest comes from the remaining parts of the world where more than 70% of the sorghum area is geographically located. With regard to production from 2011 to 2013 average about 23% is occupied by western and central Africa, 13% by South America, 12% each by Central America and North America, 11% by southern Asia, 8% by northern Africa, 4% each by eastern Asia and Oceania, and about 2% by Europe (Hariprasanna and Rakshit 2017). In Asia, geographically only India and China are important sorghum-growing countries. United States is the world's largest producer of sorghum. In the United States, sorghum cultivation is concentrated in the central and southern Great Plains where rainfall is low and variable. Nearly 90% of the grain sorghum in the United States is produced in five states including Kansas, Oklahoma, Texas, Nebraska, and Missouri. Sorghum production in Central America and the Caribbean is dominated by Mexico, which accounts for about 94% of the total production. In South America production is concentrated in Argentina (56% of the region's total) and in the dry areas of Brazil (27% of the region's total). Production in Europe is limited to small areas in France, Italy, and Ukraine. In Oceania, Australia is the only significant producer. Production in Asia is far more concentrated in just two countries, China and India, which together contribute more than 85% of the regional total (Hariprasanna and Rakshit 2017). In India, the main sorghum-producing states are Maharashtra, Karnataka, Telangana, Madhya Pradesh, and Gujarat. In the recent past, sorghum has been gaining increased popularity in coastal Andhra Pradesh under a rice-fallow situation (Chapke et al. 2017). Sorghum production in China is concentrated in the drier regions of the north and especially the northeast. However, it is distributed from Taiwan in the east, Xinjiang in the west, to Aihui county in Heilongjiang in the northeast, and to Sisha Island in the south (Gao et al. 2010). In northern Africa, Sudan is the largest sorghum producer, and production levels have nearly tripled compared to production during 1961–1963. Nigeria is the major sorghum producer in western Africa and production has increased there by more than 60% in the period 2011–2013 over 1961–1963. However, a lot of variation has been observed over the years both in area harvested and production. In central Africa, Cameroon is the largest producer whereas in eastern Africa, Ethiopia is the biggest sorghum producer with all other countries far behind with respect to quantity of grain produced (Hariprasanna and Rakshit 2017).

5.1.3 Production Challenges

The greatest challenge facing agricultural scientists is to feed the ever increasing global human population, which is expected to reach about 9.7 billion by 2050. The production demand of cereals both for food and feed are projected to about 3 billion tons by 2050 from the current 2.1 billion tons (www.fao.org). The greatest challenge will be to increase production in spite of decreasing availability of land resources, deteriorating soil fertility, and exposure of global climate changes. Under this context, agriculturists need to look for crops which are climate resilient and can contribute to the food and nutritional security in the days to come. With regard to sorghum area and production, 80% of the global sorghum area experiences low yield levels, and maximum production comes from the developed world with high yield levels. Consumption of sorghum for food purposes is declining because of a change in food habits and consumer preference brought about by economic status, whereas use for animal feed and other industrial purposes is increasing. Under a changing climate regime sorghum would assume renewed importance as a food and industrial crop, and therefore concerted focus is necessary on such marginalized crops to ensure food and nutritional security in a sustainable manner in the years to come. Although the crop is reported to be grown in more than 100 countries, major contribution is from eight countries having over 1 million ha area that contributes to about 60% of world sorghum grain production. In Africa, the crop is mostly cultivated in a few countries and contributes to a major share of area and is a major staple food grain in large parts of the continent. In spite of its economic importance, sorghum cropped area around the world has declined over the last four decades at a rate of over 0.15 million ha per year. However, in some countries including Brazil, Ethiopia, Sudan, Australia, Mexico, Nigeria, and Burkina Faso it is expanding, mainly because of new land brought under sorghum cultivation or diversion of a portion of area planted to other crops such as maize and wheat. Global sorghum production peaked during the mid-1980s, and thereafter it declined by about 13–15%, but not steadily.

In almost all the sorghum growing regions except Africa yield levels have been enhanced over the years as a result of improved cultivars, higher input use, better resources, and better crop management. Most of the sorghum is consumed in the countries where it is produced and world trade is mainly linked to demand for livestock products, which is governed by the feed requirements and prices in developed countries.

Various measures that can be taken up to improve competiveness of sorghum are through grain quality improvement and development of end-product specific cultivar development and breeding of nutrient-enriched varieties, Reduction in cost of production by providing low-cost cultivation practices to the farmers, enhance of the export competitiveness of both Indian and Asian sorghum. Value addition can greatly benefit sorghum through improvement in the export competitiveness. Creation of value addition chains across the country through public–private partnership would enhance the effectiveness of the value addition network, increasing diversification by developing niche export markets such as gluten-free products, sweet sorghum syrup, alcoholic beer, stalk-based ethanol/grain-based potable spirit, value-added ready-to-eat (RTE)/ready-to-cook (RTC) foods (Aruna et al. 2018).

5.2 Climate Change Predictions and Implications on Sorghum Production

Globally, an estimated 500 million people rely on sorghum as a primary staple (Morris et al. 2013), making grain nutritional quality very important. Impacts of global climate change are predicted to have negative impacts on crop productivity in the coming years (Lobell et al. 2008; Rosenzweig et al. 2014; Kole et al. 2015). Sorghum regions are expected to experience warmer climate during the second half of the century. Both minimum and the maximum temperature of the day is expected to rise at all the growing locations.

5.2.1 Climate Change Predictions

Global warming can reduce the net carbon gain by increasing plant respiration rates, which in turn would result in the invasion of weed, pathogens, and pests (Högy and Fangmeier 2008; Asseng et al. 2011) in addition to decreasing yield of the crops. While increased CO_2 may result in higher photosynthesis rates, foliage quality may reduce as change in the concentration of plant defensive compounds in foliages (Yao et al. 2012). Climate change impacts the insects feeding on the foliage, influences the incidence of plant diseases and also higher order interactions of predation and parasitism (Porter et al. 2014). The effects of climate change are expected to propagate throughout food webs. Maintaining diverse germplasm in global repositories as well as diversifying breeding programs can help protecting against widespread sorghum loss against possible impact arising from pests and diseases. Identifying host resistance sources against current biotic threats would reduce possible impact due to potential crop epidemics, such as stem rust outbreak Ug99 in wheat (Salcedo et al. 2017). In sorghum stalk rot disease that affects plant root and pith tissues is believed to significantly reduce the amount of macronutrients and minerals in mature grain (Bandara et al. 2017). The severity of charcoal rot caused by *Macrophomina* spp. is more severe when sorghum is exposed to long periods of drought during grain filling stage (Adeyanju et al. 2015). Higher temperature and moisture stress would possible increase pathogen virulence to disease and will certainly impact negatively on sorghum grain yield and quality. Location specific management options can be adopted to mitigate the negative impacts of the change in climate in future projected scenarios, as they are found beneficial (Sandeep et al. 2018).

5.2.2 *Effects of Climate Change in Sorghum*

Although sorghum has relative advantage over other crops for resisting extreme heat exposure, warming temperatures are still associated with large yield reductions in major growing regions, thereby suggesting need for investigating the potential for enhanced heat resilience. The regions that are predicted to have negative impacts include Sub-saharan Africa and southern Asia, where sorghum is grown as a major staple food. Sorghum is widely grown in Sahel African regions where Battisti and Naylor (2009) reported child deaths from hunger-related causes. This large region in Africa is at considerable risk of rising temperatures (Battisti and Naylor 2009). Other sorghum production regions projected to have reduced crop productivity in another eight decade (2080) include regions of Australia, North America, and South America (Cline 2007). Sandeep et al. (2017) have reported increasing trend in the water requirement of sorghum over majority of sorghum growing regions both in kharif and rabi season during 2050–2080 in spite of increasing trend in rainfall. Decline in sorghum productivity in future climate change scenarios at different locations of India was primarily attributed to reduction in crop growth period with increase in temperature (Boomiraj et al. 2011).

Climate change impact studies involving post-rainy sorghum projected a decline in yield up to 7% by 2020, up to 11% by 2050 and up to 32% by 2080 (Srivastava et al. 2010) over the base period of 1978–1999. Fu et al. (2016) used CERES-sorghum model and simulated the effects of climate change on sorghum production at the University of Arizona, Maricopa, Arizona, USA. The validated model was subsequently applied to understand the possible effects of climate change on grain yield of sorghum and water use efficiency in western North America for the years 2080 -2100. The study inferred that temperature appears to be a dominant driver of the global climate change influencing future sorghum productivity. Gebrekiros et al. (2016) used agricultural production system simulator (APSIM) model was conducted from research data (2006–2009) and they inferred that the implication of this analysis was that adoption of shorter duration rather than longer duration sorghum cultivars seems an important response in dealing with the main effects of climate change. A detailed crop modeling done using data from sorghum and millets from 35 stations across Senegal, Mali, Burkina Faso, and Niger showed negative impacts associated with a 2 °C increase in temperature and that was not compensated by additional precipitation.

5.2.3 *Effect of Drought on Growth and Development*

In arid and semi-arid regions where crops are grown on light and marginal soil conditions, drought is a regular phenomenon which occurs at various stages during the growing season. Drought involves combination of stress and the effects are caused by high temperatures (Prasad et al. 2008) and a lack of soil moisture (Campos et al. 2004). Evapo-transpiration is the major force affecting soil, plant, and atmospheric continuum of the hydrologic cycle. Most of the earlier studies, drought predictions were based on amount and precipitation distribution (Blum 2011). Recent studies indicate the soil characteristics and the soil moisture balance are included for drought assessment. Soil moisture deficit affects plant growth and development (Moussa and Abdel-Aziz 2008). Drought or moisture stress can bring change in normal growth and development, yield by affecting various morphological, physiological and biochemical related processes (Simpson 1981). Low yield and total crop failure are the main effects. The probability of moisture stress are high both during the start and end of the season in the tropics.

Drought occurs at various stages and are classified as pre- and post-flowering stages of crop development, adversely affecting yield (Tuinstra et al. 1997; Kebede et al. 2001). Seedling stage drought drastically effects the plant stand and establishment (Baalbaki et al. 1999). Moisture at flowering and grain filling stages causes reduced yield and sometimes complete crop failure (Blum 1996). Based on the nature of stress, genotype responses vary and are controlled by genetic mechanisms. Pre-anthesis moisture stress effects components of yield such as plant stand, tillering, number of earheads, seeds per plant, while post-anthesis stress affects transpiration efficiency, CO_2 fixation and carbohydrate translocation (Thomas and Howarth 2000; Xin et al. 2008).

5.2.3.1 Effect of Drought on Grain Yield

As with all crops, sorghum grain yield is dependent on water supply (soil water at planting and in-season precipitation). A summary of 30 years of data from Tribune,

KS, indicated that every millimeter of water above 100 mm resulted in an additional 16.6 kg of grain (Stone and Schlegel 2006). However, the relationship between grain yield and water is complex because yield is more sensitive to water deficits at certain growth stages (Garrity et al. 1982). Therefore, grain yield is more dependent on rainfall or irrigation well distributed over the growing season depending on demand at each stage than on total water available through the growing season. Howell and Hiler (1975) reported that yield response of grain sorghum was not strongly correlated to seasonal evapotranspiration but was highly dependent on timing of the evapotranspiration deficit. Sorghum can tolerate short periods of less severe water deficit. However, long-term and severe stress can affect sorghum growth and the final yield. Eck and Musick (1979) studied the effect of various periods of water stress on irrigated grain sorghum at early boot, heading, and early grain filling stages. Their report indicated that 13–15 days of stress did not affect grain yield. A 27- to 28-day stress, however, reduced yield at early boot, heading, and early grain fill by 27, 27, and 12%, respectively. Stress period of 35 and 42 days beginning at boot stage reduced yield by 43 and 54%, respectively. Lewis et al. (1974) showed that a soil water potential drop to −13 bars from late vegetative to boot stage reduced grain sorghum yield by 17%. The same water potential drop from boot to bloom and milk through soft dough stages caused 34 and 10% yield reductions, respectively. Inuyama et al. (1976) reported 16 and 36% yield reduction due to 16 days and 28 days of water, respectively, deficit during the vegetative stage of sorghum. In same study, 12 days of water deficit during boot stage resulted in 36% yield reduction. Withholding 100 mm of irrigation water at the early 6- to 8-leaf stage and at heading and bloom reduced sorghum grain yield by about 10 and 50%, respectively (Jordan and Sweeten 1987).

5.2.3.2 Effect of Water Stress on Seed Emergence, Plant Stand and Vegetative Growth

Moisture is important for seedling germination (Arau et al. 2001). The embryo in the seed is dormant and highly tolerant to desiccation (water stress). After seeds start to germinate and emerge, however, it is susceptible to moisture stress (Blum 1996). Seedling stage moisture stress could be due to drought, high soil temperature, and or higher salt concentration in soil. In Sorghum, seed stage stress can affect seed germination and emergence. Studies have shown that water stress at seeding will reduce weight of endosperm of the planted seed as well as growth of coleoptile, mesocotyl, radicle, shoot, and root of sorghum (Jafar et al. 2004; Bayu et al. 2005). In sorghum the plant stand is dependent on seed germination and emergence and therefore drought can cause loss in a yield even before crop establishment (Blum 1996). The effects of water stress are reduction in the rate of cell expansion and ultimately, cell size and consequently, growth rate, stem elongation, and leaf expansion (Hale and Orcutt 1987). Sorghum leaf area is also reduced because of water stress. Garrity et al. (1982) reported that the 14–26% reduction in photosynthesis of water-stressed sorghum accounted for a decrease in leaf area. Blum and Arkin (1984) also observed a significant leaf area reduction due to drought before there is a decrease in stomatal conductance. In most cases, the sorghum root-to-shoot ratio has been reported to increase under water stress (Wright et al. 1983; Salih et al. 1999). Increase in sorghum root-to-shoot ratio is due to the decrease in growth of shoot rather than absolute increase in root growth under moisture stress. This increase is a result of diversion of significant amounts of assimilates to root growth; these assimilates could be used to produce grain under non-water-stressed conditions (Wright et al. 1983).

5.2.3.3 Effect of Water Stress on Reproductive Growth

The sensitivity of sorghum to drought stress is greater during reproductive stages than during the vegetative stage (Doorenbos and Kassam 1979). Moisture stress from boot stage to approximately 10 days after anthesis will severely affect yield. Water stress during reproductive stages can stop the pollen and ovule development, prevent fertilization, and also induce premature abortion of fertilized ovules (McWilliams 2003; Saini 1997). Sorghum yield is the product of harvested panicles, seeds per panicle, and test weight. All of which can be affected by duration and severity of drought during reproductive stages. Eck and Musick (1979) reported that decrease in yield due to water stress at the early stage is due to both reduced seed size and seed number and that yield reduction due to water stress at heading or later stages was due to reduced seed size.

5.2.3.4 Effect of Water Stress on Physiology and Biochemical Traits

Severe water stress can cause stomata closure which results in low stomatal conductance and low transpiration rates (Salih et al. 1999; Reddy et al. 2014). It is also believed that CO assimilation by leaves is mainly reduced because of stomata closure in drought stress conditions (Farooq et al. 2009). Vinita et al. (1998) reported a reduction in photochemical efficiency of photosystem II (PSII), activities of phosphoenolpyruvate carboxylase (PEPcase) and ribulose-1,5-bisphosphate carboxylase/oxygenase (Rubisco) for severe water stress conditions. In addition to stomata closure, a reduction in PEPcase activity, a reduced Rubisco regeneration and functionality, and inhibited functional activity of PSII lowers the net photosynthetic rate (Shangguan et al. 1999). An increase in the rate of photorespiration under severe drought conditions can be attributed to stomata closure, an increase in the internal oxygen concentration, and a decrease in internal CO concentration. An increase in reactive oxygen in cells results in injury to sorghum cells due to peroxidation (Farooq et al. 2009). A reduction in leaf area and net photosynthesis and an increase in photorespiration rates eventually reduce total dry matter production in drought conditions (Perry et al. 1983; Terbea et al. 1995).

5.2.3.5 Effects on Grain Quality

In sorghum, Murray et al. (2008) found that the macronutrients concentration in the grain (starch, protein, and crude fat) as well as calcium, phosphorus has major environment and G × E effects. Factors such as nutrient and moisture availability during the post-flowering can significantly affect the grain quality. It is during this growth stage when sorghum is filling grain with starch, protein, and other health-promoting compounds that are important for human nutrition and health (Yang et al. 2009). While research on climate change impact on quality of sorghum grain is limited, data from studies in wheat suggest that elevated temperature and CO_2 levels will have negative impacts on grain quality, especially when vital soil nutrients such as N are restricting (Kimball et al. 2001).

5.2.3.6 Strategies to Overcome Effects of Climate Changes

Sorghum is known for its intrinsic ability to grow in dry and semi-arid environments and this can pave the way for genetic improvement in sorghum. The carbon-nitrogen ratio would increase and this would impact C_3 crops (wheat, soybeans, rice) more than C4 crops such as corn, sorghum, sugarcane. Moreover, the greater diversity within the genus *Sorghum* can be exploited to develop improved sorghum genotypes for target environments. Understanding the genotype by environment interaction in different crop growing regions will be of greater interest to improve sorghum productivity under changing climatic conditions. Along with increasing the yield potential in the target environments, it is important to identify stable genotypes under contrasting growing conditions where and when weather patterns are more unpredictable. Moreover, crop modeling may enhance selection accuracy in the targeted environments so that the efficiency of breeding can be increased. Chapman et al. (2000a, b) implemented crop simulation modeling using historical environment and yield data to predict genotype performance across targeted production environments. Using multi-environment trial data Chapman et al. (2002) estimated the genotype performance under water-limited conditions by simple additive genetic model. The results although is limited genetically and geographically, crop modeling holds promise for breeding sorghum with improved performance within target locations (i.e., yield potential) and adaptation across production environments (i.e., yield stability). Crop growth simulation models (CGSMs) are widely used to study the impact of climate change on crop productivity.

The expected negative effects on climate change can best be analyzed using data from large sample of cultivars, which typically is not possible when using aggregate regional yield data, controlled experimentation in growth chambers, or crop simulation models. It is now imperative that sorghum scientists should now formulate or strategically revise their programs, physiologists and molecular scientists to explore responses or traits to better understand the drivers of heat stress, crop modelers to refine their predictions for future climatic conditions, and economists to identify the most profitable and sustainable adaptation paths for producers.

5.3 Sorghum and Millets as the Model for Stress Biology

Sorghum and Millets possess several morphological, physiological, molecular, and biochemical characteristics that confer tolerance to abiotic stresses compared to major cereals. In small millets, the short life-cycle of millets is responsible escaping from stress as they require 12–14 weeks to complete their life-cycle (seed to seed) whereas rice and wheat requires a maximum of 20–24 weeks. However, the prevalence of stress conditions and their consequences are circumvented by several traits such as short stature, small leaf area, thickened cell walls, and the capability to form dense root system (Li and Brutnell 2011). Also, the C_4 photosynthetic trait is highly advantageous to millets. In the C_4 system, carbon dioxide (CO_2) is concentrated around ribulose-1,5-bisphosphate carboxylase/oxygenase (RuBisCO), which in turn suppresses ribulose 1,5-bisphosphate (RuBP) oxygenation and photorespiration (Aubry et al. 2011). Thus, C_4 photosynthetic pathway enhances the concentration of carbon dioxide in bundle sheath. Depending on the temperature, photorespiration is suppressed (around 80%) and also increases the *in planta* catalytic activity of RuBisCO (Sage et al. 2011). Since RuBisCO of C_4 crops works at higher CO_2 levels, millets have enhanced photosynthetic rates during warm conditions and confers immediate water use efficiency (WUE) and nitrogen use efficiency (NUE) which are $\sim$1.5–4-fold higher than C_3 photosynthesis (Sage and Zhu 2011).

Foxtail millet requires just 257 g of water to produce 1 g of dry biomass, whereas maize and wheat require 470 and 510 g, respectively (Li and Brutnell 2011). In addition to higher WUE and NUE, C_4 mechanism provides other benefit to millets including improved growth and ecological enactment in warm temperatures, enhanced flexible allocation patterns of biomass, and reduced hydraulic conductivity per unit leaf area (Sage and Zhu 2011). These attributes of millets make them ideal for next-generation crops holding the potential for research to explore the climate-resilient traits and for further utilization of the information for genetic improvement of major cereals. One significant effort made under this direction is the engineering of C_4 traits in rice using millet as models; however, through the realization of stress tolerance potential of millets it is imperative for increased progress in developing climate-resilient crop species (Bandyopadhyay et al. 2017).

5.3.1 *Sorghum as a Drought Tolerant Cereal*

At the end of this century, scientists expect that average temperature across the globe is set to rise due to the increase in emission of greenhouse gases to the atmosphere. It is expected that the change has major effect on agricultural production, resulting in food and nutritional insecurity. Estimates show the climate change impacts on food grain production along with the added pressure of feeding huge population which is about 9 billion by 2050 could lead to 2–3 billion people starving from hunger, food and nutritional insecurities (Hawkesford et al. 2013; Ray et al. 2013). Potential crop improvement interventions for minimizing the influence of climate change on agricultural production can be obtained by breeding food crops that are more tolerant to conditions caused by climate change (Wang et al. 2018).

It is believed that parts of regions across globe could benefit from climate change due to increased productivity and yields. Larger scientific fraternity agrees that the current rates of global warming and emissions of greenhouse gas would significantly reduce the overall crop productivity. Moreover, agriculture is one of major contributors to greenhouse gas emissions into the atmosphere. Increase the agricultural production to next level by intensive agricultural practices would further increase the emission of green house gases. Comparative evaluation across from predictive models inferred that climate change would reduce production of major cereal crops except crops like sorghum and millets mainly due to their ability to grow in variable climatic conditions (Wang et al. 2018). Moreover, millets are less resource-intensive crops and release fewer green house gases compared to major cereals. Therefore, in addition to addressing nutritional security, millets have an enormous potential use for reducing the impact of agriculture on global warming and should be grown on a global scale as an alternative to major cereals and grains (Bandyopadhyay et al. 2017).

Thus, sorghum is a model crop among grass species to study stress response and ensuring food security for millions of poor masses living in the most impoverished drought-prone regions of the world. Sorghum not only provides food and feed but also serves as an important source of fodder for large cattle with its dry stover. Green plants are also a source of forage for cattle. In recent years, sweet sorghum has turned out to be a source of ethanol production and second-generation lingo cellulose-based biofuels. Thus, sorghum has the potential to provide food, feed, fodder, and fuel.

5.3.2 *C_4 Photosynthetic System*

Sorghum and most other millets are C_4 crops. Among the different C_4 crops, Sorghum is most suited to drought prone environments. During the C_4 system of photosynthesis, the CO_2 gets concentrated around RUBisCo and this in turn suppresses the RUBP oxygenation and photorespiration (Aubry et al. 2011). The leaves of the C_4 plants posses more nitrogen and higher water use efficiency compared to C_3 plants. The C_4 mechanism enhances the concentration of CO_2 in bundle sheath, which suppresses photorespiration (around 80%) depending on the temperature and increases the *in planta* catalytic activity of RuBisCO (Sage et al. 2011). Since RuBisCO of C_4 plants functions at elevated CO_2 levels, sorghum and millets have enhanced photosynthetic rates at warmer conditions and confers higher water use efficiency (WUE) and nitrogen use efficiency (NUE) which is almost 1.5- to 4-fold higher than C_3 photosynthesis (Sage and Zhu 2011). Although the modifications to leaves of C_4 plants are complex, their faster growth led to the proposal that C_4 photosynthesis should be installed in C_3 crops in order to increase yield potential. Relatively high genetic variability among genome sorghum and relatively small size of its genome make sorghum ideal model for the dissection of drought-tolerance genomic regions and genes valuable to unravel the high complexity of drought tolerance related traits (Sanchez et al. 2002; Paterson et al. 2009a, b).

5.3.3 *Mechanisms Favoring Stress Tolerance in Sorghum and Millets*

The exceptional tolerance of sorghum and millets toward diverse abiotic stresses including drought, salinity, light and heat makes them a tractable system to study their stress-responsive traits at the cellular, molecular and physiological levels. Several morpho-physiological and biochemical studies in millets have shown their stress adaptation strategies (Bandyopadhyay et al. 2017). For example, Bidinger et al. (2007) have reported that pearl millet adjusts flowering phenology based on the pattern of rainfall. Balsamo et al. (2006) observed increase in leaf tensile strength during drought in teff, and in Panicum spp, an increase in root length was reported by Ajithkumar and Panneerselvam (2014). Changes in biochemical activities such as enhanced levels of antioxidants, reactive oxygen species and scavenging enzymes, catalase enzyme activity and superoxide, and osmolytes synthesis and production of stress-related proteins has been reported in response to abiotic stresses in foxtail millet (Lata et al. 2011), little millet (Ajithkumar and Panneerselvam 2014) and teff (Smirnoff and Colombe 1988). van der Weerd et al. (2001) reported changes in membrane permeability for water in pearl millet in comparison to maize for achieving better water status during osmotic stress. In addition, several novel genes, alleles and quantitative trait loci (QTLs) have been identified in millets whose functional characterization has revealed their roles in conferring stress tolerance. Compared to other millets, foxtail millet has been studied extensively, and several genetic and genomic resources have been developed (Muthamilarasan and Prasad 2015).

Whole genome sequencing of foxtail millet and comparison of gene families among 15 sequenced plant genomes showed that 1,517 genes were specific only to foxtail millet (Zhang et al. 2012). Among this, 586 genes were annotated as "response to water," which could be playing significant roles in conferring drought and dehydration stresses, thus facilitating the adaptation of this crop to arid and semi-arid zones. The genes involved in C_4 pathway namely, carbonic anhydrase (CAH), malate dehydrogenase (MDH), malic enzyme (ME), phosphoenolpyruvate carboxylase (PEPC), phosphoenolpyruvate carboxylase kinase (PPCK), and pyruvate orthophosphate dikinase (PPDK) were also identified and compared to that of sorghum, maize, rice and *Brachypodium*. The study showed that foxtail millet has a higher number of MDH (7 genes) and PPDK (3 genes) than other crops. Zhang et al. (2012) have also performed phylogenetic and evolutionary analysis of CAH homologs among all the five grass genomes, which showed that *Ft_CA*1 was highly expressed in the mesophyll, and this could be a potential candidate for studying C_4 pathway in foxtail millet. Although most millet crops are known for climate resilience mechanisms, studies providing detailed insights into the molecular machinery underlying stress tolerance is largely lacking in small millet crops such as little millet, barnyard millet, proso millet, and kodo millet. Very limited knowledge base has been generated on genetic determinants of stress tolerance from biparental and association mapping approaches. In this context, extensive phenotypic screening to observe the natural genetic variations in stress tolerance across diverse small millet germplasm is greatly needed to fully harness the underlying genetic potential through conventional/molecular breeding approaches and transgenic technologies. Systemic research efforts should be carried out to facilitate crop improvements in millets and in general field crops as a whole in the wake of increased desertification and salinity of the crop fields due to ill effects of climate change. Global researchers should work collectively to understand the complexity of moisture stress through improved crop phenotyping and modeling, combined with effective sharing of knowledge, facilities, and data, will boost the cost effectiveness and facilitate genetic gains of all staple crops which will spill over to less studied crops.

5.3.4 Drought General Mechanisms

Drought stress expresses wide range of responses in plants (Chaves et al. 2003). The response include reduced photosynthesis (Kyparissis et al. 2000a, b; Havaux and Tardy 1999), include increase in oxidative damage of chloroplasts (Munné-Bosch et al. 2001; Jubany-Marí et al. 2010), alterations in metabolic reactions (Beck et al. 2007), triggers sugar catabolism, in order to provide osmotically active compound and signal molecules (Pelleschi et al. 1997; Pinheiro et al. 2001; Yang et al. 2001) and modifies cellular lipid composition (Toumi et al. 2008). Crops have variety of inbuilt mechanisms such as larger and deeper root systems (Passioura 2004), modification in stomatal closure to reduce water loss (Cornic 2000), accumulation of compatible solutes and protective proteins (Chen and Murata 2002), and an increase in the level of antioxidants (Zhang and Kirkham 1996). The tolerance to drought is a due to morphological and anatomical characters such as thick leaf wax, deep root system, and physiological traits such as osmotic adjustment, stay green, quiescence (Dugas et al. 2011; Vadez et al. 2011). Crop plants have various inbuilt mechanisms to overcome moisture stress and can be classified as morphological, physiological, biochemical, cellular, and molecular. The major mechanism of are classified into four basic types such as drought avoidance, drought tolerance, drought escape, and drought recovery. Drought avoidance and drought tolerance are the mechanisms mostly operating in crop plants.

5.3.4.1 Drought Avoidance

During drought avoidance mechanism, the plants maintain cell turgor and high cell water content under water limiting conditions. Plants do adopt to one of the strategies such as leaf rolling or shedding and reducing leaf area in order to reduce water loss by transpiration, and increase in stomatal and cuticular resistance (Morgan 1984; Turner 1986). Most sorghum genotypes have a thick waxy cuticle that controls water loss during moisture. Another strategy plant adopt is by increasing the water uptake ability through a well-developed and efficient root system especially by longer depth of roots, root biomass or root/shoot ratio, etc., and enhancing the ability of plants in storing the water in tissues (Tardieu 2013; Sawidis et al. 2005; Ogbum and Edwards 2010). A deep and efficient root system can explore deep soil layers is associated to the conservation of water supply during period of water stress. The cell architecture of mesophyll tissue of C_4 plants allow them to accumulate CO_2 in the bundle sheath cells, reducing photorespiration, reducing stomatal conductance to preserve water without decreasing carbon fixation rates. Mechanisms such as leaf abscission, dormancy, and any other mechanisms that reduce loss of water by transpiration are considered drought avoidance mechanisms. One different strategy plants may adopt include accelerating or slowing down the switchover of growth stages from vegetative to reproductive to avoid complete abortion at the severe drought stress stage (Mitra 2001; Luo 2010). Greater leaf senescence of in moisture stress plants is also a major drought avoidance mechanism used by sorghum to decrease transpiration requirements. Leaf diffusive resistance measurements indicated open stomata in both irrigated and non-irrigated plants so that under the imposed water stress conditions stomatal closure was not affecting the transpiration requirement (Stout and Simpson 1978).

5.3.4.2 Drought Tolerance

Drought tolerance refers to the ability of plants to maintain a certain level of physiological mechanisms under moisture stress conditions through the regulation of thousands of genes and involving metabolic pathway mechanisms to reduce or repair the resulting stress damage (Passioura 1997a, b; Mitra 2001; Luo 2010). Certain plants exert protoplasmic tolerance by increasing osmo-regulatory molecules in the cells to maintain the cell turgor pressure, and adjusting the activities of cell defense enzymes to reduce the accumulation of hazardous substances. In order to lower the osmotic potential and maintain turgor, the plants accumulate compatible solutes including organic acids, sugars, amino acids, sugar, ions, etc. Osmotic adjustment is important in the drought tolerance of many C_4 species including sorghum growing in arid environments and allows the growth of sorghum when leaf water potential is low (Girma and Krieg 1992; Jones and Turner 1978). In particular Sorghum bicolor accumulates glycine betaine and proline in response to water deficit (Buchanan et al. 2005). On the other hand, the antioxidant capacity is by the plant's ability to detoxify reactive oxygen species that may cause cell damage by lipid peroxidation or protein and nucleic acid modification (Scandalios 2005). Some plants are able to prevent ROS damage by using superoxide dismutase, catalases, or peroxidase and by using free radical scavengers as carotenoids, ascorbate, proline, tocopherols, and glutathione (Mundree et al. 2002). In sorghum, evidence also indicated that leaf osmotic potential decreased due to osmo-regulation, a drought avoidance mechanism, resulting in formation of more osmotically active cellular solutes. Studies performed by (Girma and Selassie-Theses) observed partitioning of the total diurnal change in osmotic potential indicated that dehydration and net solute accumulation were the major mechanisms for diurnal change in Osmotic potential. Osmotic adjustment maintained turgor pressure over a wide range of Osmotic potential Despite the maintenance of turgor pressure by osmotic adjustment, stomatal conductance, and photosynthesis declined as Osmotic potential decreased. Results from this study indicate that osmotic adjustment does not maintain stomatal conductance and photosynthesis in grain sorghum. Fracasso et al. (2016) reported dehydrin (DHN) compound in sorghum could be used for screening drought tolerance both in durra and in caudatum races. NADP-Malic Enzyme, Carbonic Anhydrase (CA) and Plasma membrane Intrinsic Protein (PIP2-5), being up-regulated by drought stress only in durra race, have a more limited, though nonetheless useful application. In the tolerant durra genotype IS22330 in particular, the regulation of stomatal openings was strongly related to NADP-Malic Enzyme expression.

5.3.4.3 Drought Escape

The term Drought escape refers to natural or artificial adjustment of the growth period, life cycle, or planting time of plants to prevent the growing season from encountering local seasonal or climatic drought (Mitra 2001; Manavalan et al. 2009). Many farmers prefer short duration crop varieties. These varieties short life cycles complete their life cycle by avoiding the seasonal drought stress in agricultural production. In sorghum some varieties adopt this strategy to avoid water deficit periods that could occur during the growing season in some regions.

From a molecular point of view, drought induces the expression of genes encoding for proteins involved in protection and signal transduction (Mundree et al. 2002). Several drought responsive genes are induced by external treatment the hormone involved in water deficit signalling, i.e., abscisic acid, but an additional gene set also gets induced by drought in an ABA-independent signal transduction pathway (Mundree et al. 2002). The promoters fragments of these genes contain sequence specific ABA-responsive cis-elements (called ABRE). The same cis-elements were found in sorghum genes responding to ABA (Buchanan et al. 2005). ABA-independent pathway genes contain other nine characteristic base pair defined as dehydration responsive elements (DRE). Proteins that can bind DRE include the ethylene-responsive element binding proteins, AP2 proteins, and DRE binding factor 1 and 2 that can activate the transcription of genes containing the DRE sequence (Mundree et al. 2002).

5.3.4.4 Drought Recovery

This mechanism refers to the plant capability to resume growth and gain yield (for crops) after exposure to severe drought stress which causes a complete loss of turgor pressure and leaf dehydration. Liu et al. (2018) reported relative water content, water potential, osmotic potential, chlorophyll content and photosynthetic parameters decreased after imposition of drought but other plant characters recovered fully after re-watering. Their studies from correlation analysis revealed that drought adaption is more closely related to drought recovery ($r = 0.85$) than drought resistance ($r = 0.46$). The potential of the plant to maintain high photosynthetic rate and lower transpiration contributes to drought resistance while high-relative water content improves drought recovery mechanisms. These results reveal that drought recovery ability of the crop plants is more important for drought adaption than drought resistance. The physiological mechanisms of drought resistance and drought recovery are governed differently. Parameters such as high relative water content and high photosynthetic rate could be used to select drought adapted cultivars in sorghum.

5.4 Target Traits for Sorghum Adaptation and Improvement

Development and cultivation of climate smart crops and varieties with desired traits is more critical to safeguard against global food scarcity and malnutrition (Mickelbart et al. 2015) especially during the global climate change scenario. The rich diversity for adaptation in sorghum is one of the reasons for these crops to be grown for food, feed, fuel in African and Asian continents. Sorghum as the name suggests camel crop, stands tall for its ability to thrive under harsh climates across major growing conditions. The genetic mechanisms underlying adaptive traits responsible for its ability to tolerate harsh growing environments will have positive implications in breeding programs in sorghum as well as other agronomically important crops. Here we focus on major traits that are critical for broadening adaptation and enhancing production especially under the climate change scenario.

5.4.1 Yield Potential and Yield Stability

Sorghum has shown to be more productive and profitable in drought prone environments (Staggenborg et al. 2008). Selection methods for assessing yield for wide adaptation to climate change are classified as *direct selection* for yield or performance in the target stress environments and *indirect selection* methods through which selection is made for morphological, phonological, or physiological traits in the target stress environments. Adaptive traits are in general the traits that enhance phenotypic plasticity and/or tolerance to particular stresses. Photoperiod-sensitive flowering guinea-race sorghum of West African is an example for a trait conferring phenotypic plasticity or individual buffering. Photoperiod sensitive lines flowers at a more fixed calendar date despite highly variable planting dates of sowing. Selection for extremely early maturity in the wake of drought periods in the 1970s and 1980s resulted in loss of photoperiodic sensitivity from much of the breeding material in pearl millet (Niangado 2001), as well as sorghum. Photoperiod sensitive types are more prevalent in Guinea-race sorghums which predominate in West Africa (Curtis 1968), and a similar strong relationship was reported between the duration of vegetative phase in an early sowing date and level of photoperiod sensitivity in sorghum (Clerget et al. 2007). However it is cautioned that by adding additional traits to a selection protocol reduces the selection intensity, and thus expected gain for yield performance, unless the indirect selection criterion targets yielding ability under stress conditions Stability in yield is a result of multiple performance characteristics such as nutrient use efficiency, abiotic stress resistance, photosynthetic efficiency, etc., and is often considered and measured as a standalone trait. Commercial companies are interested in hybrids that produce high yields across a wide range of growing environments, making yield stability an important trait for selection. Yield stability, a measure of change in productivity of a genotype over growing season and environment is observed to be higher in hybrids than for inbred lines in sorghum (Reich and Atkins 1970). Because yield stability is highly quantitative and is a function of several adaptive traits, dissecting its genetic basis is a difficult task. Targeting the stack of traits especially components of stability (e.g., drought resistance and disease resistance) is likely the best mechanism to stabilize sorghum yield across environments (Haussman et al. 2012).

5.4.2 Drought Tolerance

The majority of sorghum production is located in semi-arid environments, which makes drought resistance a target trait for genetic dissection and improvement. Genotypes with post-flowering drought resistance in sorghum exhibit higher chlorophyll and photosynthetic capability under drought compared to wild type lines, a phenomenon commonly recognized as '*stay-green*'. Borrell et al. (2000) reported that the stay-green genotypes significantly improves grain yield under water stress (post-anthesis). Stay-green sorghum is widely used in breeding programs to develop hybrids with increased drought resistance (Henzell et al. 2001). The primary source for the stay-green phenotype has been BTx642 (also known as B35), which was derived from a durra sorghum from Ethiopia (Harris et al. 2007). The grain yield benefit of the stay-green phenotype under drought was found to be a result of reduced vegetative biomass and water uptake during the pre-flowering growth stages (Borrell et al. 2014a, b). Under artificial conditions, sorghum root length during the seedling stage was observed to be a major factor associated with drought tolerance (Bibi et al. 2012). More recently, Mace et al. (2012) found a relationship between drought adaptation and nodal root angle to further support the role of below-ground biomass traits in sorghum production under water stress.

5.4.3 Heat Tolerance

Sorghum domestication and diffusion in the semi-arid tropics lead to the natural tolerance to high temperatures. Heat stress mainly during the reproductive stage, can negatively impact grain yield. During the vegetative stage, persistent high temperatures resulted in decrease in photosynthetic rate by damaging chloroplast structures (Djanaguiraman et al. 2014) in sorghum. Sorghum is one of the cereal crops vulnerable to heat stress during anthesis, as pollen viability is reduced and subsequently leads to a decrease in fertilization (Djanaguiraman et al. 2014; Prasad et al. 2015; Djanaguiraman et al. 2018). Few studies indicated wide genetic variation for high temperatures and this can be used to develop genotypes with improved heat tolerance (Singh et al. 2015; 2016). Among popular lines used in commercial production, BTx642 showed poor heat tolerance while BTx623 was reported to be one of the most tolerant (Singh et al. 2015). Sorghum yield can also be reduced when exposed to high temperatures during the post anthesis stage by shortening grain fill duration (Prasad et al. 2015). Reduced grain filling period as a result of heat stress reduces grain weight, which is a significantly associated with final yield (Boyles et al. 2016).

5.4.4 Cold Tolerance

A current limiting factor in temperate sorghum production is limited cold tolerance among commercial hybrids and elite breeding lines compared to other cereals (Yu et al. 2004). Persistent temperatures below 15 °C can reduce seed germination and plant emergence to lower plant density, which as an important component of grain yield (Maulana et al. 2017). While many sorghums, originating in the semi-arid tropics, are susceptible to low temperatures, Franks et al. (2006) observed cold tolerance in germplasm from China. Based on seedling vigor, additional landraces from temperate regions were found to have higher tolerance levels than the standard check for cold tolerance, Shan Qui Red (Maulana et al. 2017). It is unknown if genetic resistance to cold tolerance in sorghum causes yield drag when low temperatures are absent during growth.

5.4.5 Nutrient Use Efficiency

Interest in developing nutrient use efficient crops has risen with growing concerns on environmental quality, which has been relegated as a result of nitrate leaching and dwindling of phosphorus reserves. Moreover, increasing uptake and utilization of vital nutrients for growth will have positive impacts on crop productivity and profitability (Hirel et al. 2007) as well as improve global food security (Hufnagel et al. 2014). Nitrogen is a common limiting nutrient for cereal crop production, including sorghum (Gelli et al. 2014). It takes approximately 2.1 kg N to produce 100 kg of grain for sorghum, which is slightly less N use efficient compared to maize (Muchow 1988). Conversely, photoperiod-sensitive sorghum grown for ethanol production had the highest N use efficiency among other major bioenergy crops (Ra et al. 2012). Additionally, protein levels in grain or silage is important for nutritional quality for food and feed use. Levels of crude protein in vegetative and reproductive tissues are dependent on soil N levels in sorghum. There is genetic variation for nitrogen stress tolerance, and Chinese kaoliang sorghum San Chi San observed as the most N use efficient lines found to date (Gelli et al. 2014). Similar to other crops, phosphorus is often supplied to soils for sorghum production. Phosphorous deficiency is commonly observed in sorghum grown on clay soils, under low pH, P bioavailability is low (Hufnagel et al. 2014). Similar to nitrogen use efficiency, phosphorous content is heritable, and phenotypic variation exists in uptake and utilization.

5.4.6 Resistance to Disease and Insect Pests

Sorghum is a one of the preferred host for various diseases and insects resulting in lower crop yield and quality significantly. The biotic threats are major concern, especially when much of sorghum production occurs in tropical latitudes. In temperate climates, diseases can overwinter on crop residues in the soil or on the noxious perennial Johnson grass (*Sorghum halepense*). Anthracnose caused by *Colletotrichum graminicola* is a fungal disease that regularly causes significant crop loss. This pathogen invades all major tissues and can reduce yield and quality by lowering photosynthetic efficiency when present on the leaves or by blocking assimilate movement into developing grain when present in the stem. Multiple sources of host resistance are available among the germplasm lines and are frequently used in breeding programs (Mehta et al. 2005).

Insects that commonly invade sorghum demonstrate overwintering capability but also can move seasonally in an aggressive manner. Recently, sorghum production in North America became a preferred host for the sugarcane aphid (*Melanaphis sacchari*), which like other aphid pests, penetrate outer tissues to feed on sap moving throughout the phloem. Since 2013, several incidence of aphid outbreaks were reported in Mexico resulting in complete crop loss (Bowling et al. 2016).

Fortunately, cross-resistance from greenbug was identified and these sources of resistance to sugarcane aphid have been extensively used to develop aphid tolerant commercial hybrids (Armstrong et al. 2015). With the introduction of aphid tolerant hybrids in commercial production and an increase in beneficial insects, the aphid population significantly declined during 2018 growing season. Maintaining these greenbug-resistant lines in repositories were instrumental in developing sugarcane aphid tolerant sorghum. This example epitomizes the benefits of protecting genetic diversity as it can have major implications in short- and long-term global agricultural production and food security. Other insects that pose threats to production include sorghum midge (Jordan et al. 1998), headworms (Starks and Burton 1979), and other species of aphids (Michels and Burd 2007).

5.4.6.1 Grain Quality and Nutritional Improvement

Diverse varieties of sorghum possess range of health promoting compounds in grains. Various components of sorghum grain including proteins (kafirins) (Duodu et al. 2003; Belton et al. 2006), phytonutrients (Awika and Rooney 2004), phenols (Dykes and Rooney 2006; Rhodes et al. 2014), starch (Zhu 2014), micronutrients (Shakoor et al. 2016), and general grain composition (Bean et al. 2016; Boyles et al. 2017) have been widely studied over the years. Several efforts have been made towards improving quality and nutrition in sorghum grain by developing and introduction of improved cultivars (Miller et al. 1996; Rooney et al. 2013) or identifying mutants with unique compositional characteristics (Pederson et al. 2005; Tesso et al. 2006). It has been demonstrated that grain nutrients are significantly dependent on environment and G × E effects (Murray et al. 2008), and thus improving grain quality under stressful environments may be challenging but necessary.

5.5 Genetics of Drought Tolerance

Drought tolerance is a quantitative trait. The performance of sorghum under water stress, the high genetic variance among genotypes and the relatively small size of sorghum genome, make this cereal an ideal crop for the identification of drought related genomic regions and genes necessary to unravel the highly complex drought-tolerance trait (Paterson et al. 2009a, b; Sanchez et al. 2002). Several sorghum linkage maps were built using different type of DNA markers (Mace et al. 2009; Rami et al. 1998) reaching high density level (Ashraf 2010). Different genomic regions related to drought tolerance in pre-flowering and post-flowering stage were identified recently (Kebede et al. 2001).

Genetic advances in improvement of crops under water-limited conditions are only possible if drought resistance traits are identified and selected for in addition to yield (Borrell et al. 2000; Sanchez et al. 2002). In plants variety of genes get expressed during drought associated with morphological and physiological traits and many of these are believed to express not only in stress tolerance but also they play major role during gene regulation and signal transduction during stress response (Bohnert et al. 2006; Nguyen et al. 1997; Shinozaki and Yamaguchi-Shinozaki 2000). Many traits associated with drought resistance in sorghum have been identified and mapped but the most widely studied trait stay-green trait is considered to have major role in drought resistance trait in sorghum. Stay green trait in sorghum is well characterized and is used to describe the post flowering drought tolerance in sorghum (Rosenow et al. 1987). The trait refers to plant ability to retain for functional photosynthesis during grain filling under post flowering drought stress. QTL associated with pre and post flowering drought tolerance have been mapped and markers associated with the QTLs were identified (Tuinstra et al. 1996, 1997, 1998; Crasta et al. 1999; Ejeta et al. 2000; Kebede et al. 2001; Hasussmann et al. 2002; Sanchez et al. 2002; Hash et al. 2002; Harris et al. 2007; Kassahun et al. 2010). Tuinstra et al. (1996) reported six different genomic regions associated with post flowering drought tolerance in RIL population derived from cross TX7078 x B 35. The genomic regions explained about 40% of the variation for grain yield under preflowering drought and were stable across environments. Kedebe et al. (2001) used RIL population SC56 × TX 7000 and identified four QTLs for preflowering drought tolerance and the QTLs were not stable across environments. He also noticed strong association of preflowering drought tolerance with days to flowering.

5.5.1 Stay-Green Is an Integrated Drought-Adaptation Trait in Sorghum

Delayed leaf senescence during grain filling stage of the crop is an consequence of dynamics occurring in the early crop growth stages and is due to improved balance between the water supply and demand, also the efficiency with which the crop convert water to biomass and grain yield (Borrell et al. 2009; Jordan et al. 2012). The supply side factor, i.e., crop water use during grain filling can be increased by enhancing the availability of water at anthesis stage and/or during grain filling (van Oosterom et al. 2011). The demand side factor i.e., crop water use can be reduced by lessening the leaf area and/or transpiration per unit leaf area. Further, leaf area can be reduced by bringing down tillering (Kim et al. 2010), leaf number per culm, and/or individual leaf size (Borrell et al. 2000). Transpiration per unit leaf area can be brought down by modifying stomatal density, stomatal opening time, and hydraulic factors or combination of factors.

There can be five mechanisms operating for a plant to stay green (Thomas and Howarth 2000). The stay green may arise if the onset of senescence is delayed referred as (type A), the rate with which the senescence is reduced (type B), photosynthesis declines but still the chlorophyll is retained (type C), greenness retained due to rapid death of plant at harvest (type D), or the phenotype is greener to begin with (type E). These classifications indicate that stay-green may be functional or cosmetic (Borrell et al. 2014a, b). Functional stay-green is due to the maintenance of leaf photosynthesis during grain filling (types A, B, and E), while cosmetic stay-green occurs when photosynthetic capacity is disconnected from leaf greenness (types C and D). Enhanced crop productivity in water-limited environments is dependent on functional stay-green (Borrell et al. 2014a, b). However, not all functional stay-green is necessarily productive (Fig 5.1).

For example, low sink demand compared to source, resulted by a small panicle or low grain number in sorghum, will possess a stay-green phenotype since there is little demand to translocate carbon and nitrogen from leaves to grain (Henzell and Gillieron 1973; Rosenow et al. 1987). Therefore, selection for both stay-green and grain yield should be undertaken simultaneously in plant breeding programs to ensure that delayed senescence is not due to low sink demand (Borrell et al. 2014a, b).

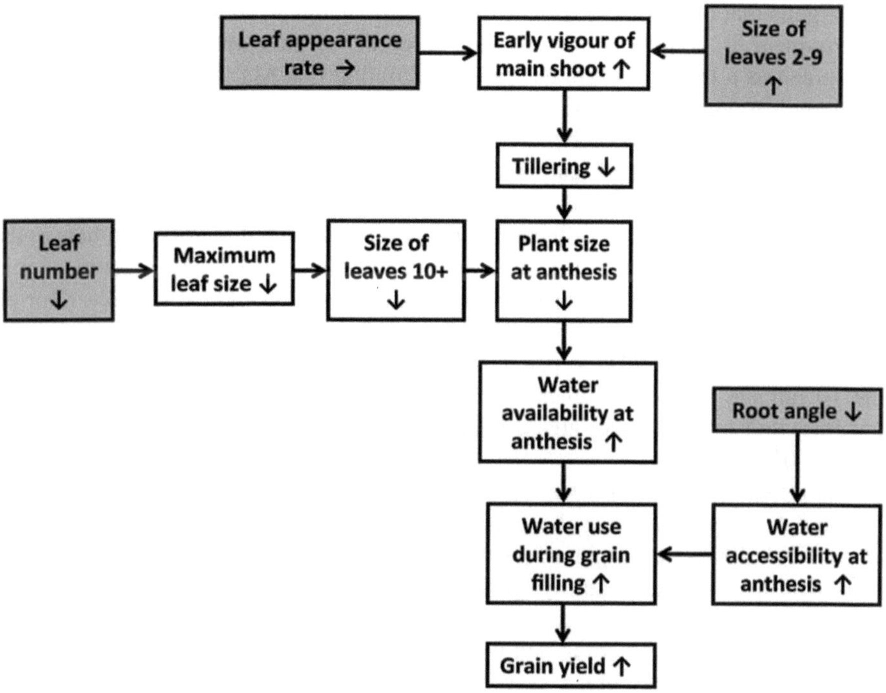

Fig. 5.1 Flowchart of physiological processes that determine plant size and crop water use at anthesis, with consequences for water uptake during grain filling. (Grey boxes indicate traits that are directly affected by *Stg*QTLs and white boxes indicate traits for which the effect is an emergent consequence of the effect on the grey box. Up arrows indicate increase; down arrows indicate decrease; side arrows indicate no change.) (*Source* Borrell et al. (2014a, b)

Tuinstra et al. (1997) identified thirteen genomic regions related with post-flowering drought tolerance of which four QTLs were identified for yield traits and stability; seven were identified for grain development and grain weight, and two QTLs for the stay-green trait. There are three major stay-green gene sources (B 35, SC 56, and E 36-1) from which QTLs that have been mapped onto 10 linkage groups on sorghum. Crasta et al. (1999) and Xu et al. (2000) identified four stay-green QTLs and mapped two of the QTLs (*Stg1* and *Stg2*) on linkage group A, and the other two, *Stg3* and *Stg4* onto linkage group D and J, respectively. Tao et al. (2000) mapped two stay-green QTLs on linkage group B and I. Xu et al. (2000) mapped QTLs *Chl1*, *Chl2* and *Chl3* for chlorophyll content in sorghum and also reported that the map position of chlorophyll content coincides with the stay-green QTLs. The stay-green QTLs were ordered based on their merits to the stay-green phenotype as Stg2, Stg1, Stg3, and Stg4 [18]. Most of the reports highlight the potential of the stay-green trait in developing drought tolerant sorghum, but the magnitude of their contribution of the stay-green QTLs to the yield under post-flowering moisture stress largely depends on the genetic background and the target environments

Walulu et al. (1994) reported that stay green trait in sorghum is controlled by a major gene that expresses varying levels of dominant gene action, depending on the environment it is grown. van Oosterom et al. (1996) observed and reported stay-green as a function of green leaf area duration (GLAD) and this is dependent on green leaf area at flowering, time of onset of senescence, and subsequent rate of senescence and these components appeared to be inherited independently. The inheritance of the onset of leaf senescence is additive, and the senescence rate is dominant and therefore, GLAD was found to be partially dominant. The expression of these three factors is also affected by many environmental factors, and hence, the combined genetic effects of the three factors and the environmental factors should be considered when designing breeding programs for drought resistance (Borrell et al. 2000; Mahalakshmi and Bidinger 2002). In addition to detection of Quantitative trait loci (QTLs) per se for drought QTLs were also been mapped on the 10 linkage groups for other controlling traits related to yield and yield components, root systems, stay-green, plant height, flowering, and maturity (Sanchez et al. 2002).

Variations in flowering time, reproductive sink strength along with other environmental factors influence the expression of stay-green trait (Tao et al. 2000, Harris et al. 2007). In Sorghum, six maturity genes (Ma1-Ma6) have been mapped and identified. These maturity genes while dominant cause extreme lateness (Morgan et al. 2002). Two maturity QTLs were identified to be located near a stay-green QTL linkage group and the major effect maturity QTL were reported to be correlated with stay-green score (Xu et al. 2000). Sorghum genotypes from tropical climates were found to be dominant for all four maturity loci (Ma1-Ma4) which control the flowering time (Quinby 1974). However, replacing the Ma1 dominant maturity gene to recessive ma1 alters the tropical sorghum to a temperate type which can flower even at high altitudes (Major et al. 1990). Tuinstra et al. (1998) identified two QTLs that conditioned the expression of the stay-green trait.

Goldsbrough and Wood (1997) reported expression of dehydrin gene, dhn1 in sorghum as drought response mechanism and expression of the gene in seedlings and pre-flowering sorghum was found to be identical among different genotypes, but time of expression of the gene varied among the genotypes. This suggested that the expression of dehydrins is an important drought adaptation mechanism in sorghum.

5.5.2 Root Traits in Drought Tolerance

Many researchers have indicated roots traits significant role in drought tolerance in sorghum. The root characters are reported to be governed by both additive and dominant genetic effects (Ekanayake et al. 1985). Mace et al. (2012) and Rajkumar et al. (2013) reported four QTLs associated with nodal root angle (qRA1_5, qRA2_5, qRA1_8, qRA1_10), three with root dry weight (qRDW1_2, qRDW1_5, qRDW1_8) and eight for root fresh weight, root volume, and root dry weight in sorghum. One of the root angle QTL was found to be co-located with stay-green QTLs and also found to be associated with grain yield (Mace et al. 2012). Two other QTLs (qRT6 and qRT7) linked with brace roots have been identified and mapped Chromosome 6 and 7 in sorghum. Brace roots have role in anchoring, water and nutrient uptake during late growth and development and also associated with grain yield under water limited conditions (Li et al. 2011). Leaf rolling is considered as one of the major indications as response to drought stress in crops such sorghum (Matthews et al. 1990). The rolling of leaves is mainly controlled by major genes and several genes such RL1 to RL10 were identified rice. The genes RL7, RL8, and RL9 were linked with corresponding chromosomes with molecular markers. The heterozygote condition of each gene exhibited rolled leaves (Zhang et al. 2009).

The recent availability of the sorghum genome allows monitoring the genome-wide gene expression profiling at a single time in response to several abiotic stresses through microarray or RNA-Seq analysis (Buchanan et al. 2005; Dugas et al. 2011; Yazawa et al. 2013) permitting to identify drought stress responsive genes, their relationship and their regulatory elements as well as the post-transcriptional modifications due to small RNAs. These small RNAs (including microRNAs, miRNAs, and short-interfering RNAs, siRNAs) are able to silence genes by driving target mRNAs to degradation or repressing their translation (Ambros 2004; Bartel 2004). MiRNAs were found in plants, animals and other eukaryotes as well as in DNA virus. In plants, miRNAs are 20–24 nucleotides long non-coding RNAs complementing their mRNAs target inducing their cleavage and their silencing (Khraiwesh et al. 2012). Considering their importance in post-transcriptional gene silencing, their involvement in stress regulated gene expression seemed likely and was confirmed in several studies as important "players" in plant resistance to biotic and abiotic stresses (Navarro et al. 2006; Pasini et al. 2014; Sunkar et al. 2007).

5.5.3 Role of Abscisic Acid and Other Genes in Drought Tolerance

Drought tolerance genes are found to be induced by water stress, desiccation, and abscisic acid (ABA). Yamaguchi-Shinozaki et al. (2002) found wide variation in time for induction and expression of drought related genes and classified into two groups. The first set of genes are involved in responsible for proteins function directly involved in stress tolerance and the second set of genes produces protein factors required for regulation of signal transduction and gene expression functions under drought (Yamaguchi-Shinozaki et al. 2002). Although most of the drought stress genes are induced by ABA, researchers have reported ABA-dependent and independent genes and also genes involved in signal transduction cascades between the initial signal of drought stress and the expression of the genes (Shinozaki and Yamaguchi-Shinozaki 1997, 2000). Inhibition of development of lateral roots under moisture stress condition is one of the mechanisms for drought tolerance in plants (Xiong et al. 2006). The inhibition of lateral root growth due to drought tolerance is mediated by abscisic acid and genotypes that are sensitive to abscisic acid are more drought tolerant than those that are insensitive to abscisic acid (Xiong et al. 2006). It is also observed that abscisic acid insensitive plants do lose water at faster rate due to transpiration than abscisic acid sensitive types (Yamaguchi-Shinozaki et al. 2002).

5.6 Genomic Resources

During the last decade, genomics-based approaches have been extensively used to dissect the genetic make-up of drought and abiotic stress adaptation and quantitative nature of abiotic stress tolerance, QTL have been the main target of research to identify the genetic loci regulating the adaptive response of cereal crops to unfavorable environmental conditions. The traits includes drought-tolerance traits (Serraj et al. 2009; Tuberosa 2012), stay-green traits (Yang et al. 2007; Borrell et al. 2014a, b), accumulation of water-soluble carbohydrates and transport to storage organs (Landi et al. 2005; Salem et al. 2007; Rebetzke et al. 2008), Nature of root architecture (Wasson et al. 2012; Uga et al. 2013; Lynch et al. 2014), concentration of abscisic acid (Rebetzke et al. 2008; Rehman et al. 2011), canopy temperature (Lopes et al. 2014) and carbon isotope discrimination (Pinto et al. 2010).

Advances in Genomic tools offers tools and techniques to address challenges in increasing grain yield, quality by assisting conventional breeding methodologies. Application of DNA markers to facilitate marker-aided selection (MAS) for crop improvement has proved successful in crossbreeding. Recent genomic approaches provide means to improve the understanding of gene diversity at sequence level and offer DNA markers to accelerate the genetic improvement (Muthamilarasan et al. 2013, 2014). A genomics assisted breeding strategy for development of new cultivars that are "climate change ready" (Varshney et al. 2005) especially for crops of semi-arid tropics are being initiated and involves defining the stress(es) that will likely affect crop production and productivity under certain climate change scenarios. Data from multilocational testing provides opportunity for modeling of "stress-impacts" on crops. Plant breeders and germplasm scientists will search for morphological and physiological traits in available germplasm that could enhance crop adaptation under upcoming climate variability. Crop physiologists may help define and constitution of ideotypes for enhancing yield under such adaptation. Moreover, the use of geographical information systems and passport data can allow identification of accessions from stress-prone environments, whereas the available characterization, including DNA/gene sequence information, and evaluation data as well as mapping of desired genes or QTL will assist in selecting promising accessions for further screening against specific stress(es). Similarly, precise phenotypic data and appropriate biometric/statistical analysis will assist in identifying unique responses of a set of genotypes under given physiological condition influenced by variation in weather patterns. This information will be further used in genomics-assisted breeding approaches such as genome-wide selection of promising germplasm for further use in crop breeding aiming at both population improvement and cultivar identification and release.

Genetic mapping and QTL analysis, via bi-parental or association mapping populations, led to the dissection of genomic regions of economically important traits, potentially allowing MAS, QTL, and AM studies in addition to calculation of breeding values of desired lines and genomic selection (GS) of high value genotypes to be made in the context of breeding programs (Kulwal et al. 2011). Until recently, AM and GS were hampered by the need for very high marker density coverage of the genome. Advancement of next-generation sequencing (NGS) methods has facilitated the development of large numbers of genetic markers, such as single nucleotide polymorphisms (SNP), insertion-deletions (InDels), etc., even in relatively research-neglected crop species. Discovery of novel genes/alleles for any given trait could be then performed through genotyping-by-sequencing (GBS) approaches. Similarly, genome-wide association studies (GWAS) could be used to identify the genomic regions governing traits of interest by performing statistical associations between DNA polymorphisms and trait variations in diverse collection of germplasms that are genotyped and phenotyped for traits of interest. NGS coupled with GWAS increases the mapping resolution for precise location of genes/alleles/QTL (Ma et al. 2012; Liu et al. 2013; Varshney et al. 2014).

During the course of evolution, nature played role in evolution of new genes, shuffled and selected these genes or alleles in a wide range of environments producing the diversity as evidenced in wild species. In contrast, the selection and domestication of crops by humans is recent, having occurred over the last 10,000 years. During domestication and breeding processes, there has been significant reduction of genetic diversity in major crops by undergoing selections for grain yield under managed agricultural environments. Currently, breeders are shuffling relatively few alleles to generate desired combinations of alleles that provide higher yield and desired agronomic characteristics. In many large sized genomes of crop species, this reshuffling process is limited by restricted recombination patterns within species, leading to repeated inheritance of same blocks of genes, raising issues of linkage drag, which may not contain the best possible combination of alleles. Breaking the linkage drags will allow breeders to develop novel allele/gene combinations within the elite set of parents. Genomics assists in identifying the potential to increase the diversity of alleles available to breeders through mining the gene pools of crop wild relatives (CWRs). Genomics tools enables rapid identification and selection of novel beneficial genes and their introgression into novel germplasm. In the current genomics era, this technology will be used to assist crop researchers in safeguard the future through improved food security. Altogether the application of genomics for genetic enhancement offers the greatest potential to sustaining food production in the coming decades. With rapid advancement in genomic technologies, the application of genomics to identify and transfer valuable agronomic genes from related gene pools and crop relatives to elite crops will increase the pace and assist in achieving the challenges of global food and nutritional security.

Sorghum genome was sequenced in 2009 by Andrew Paterson and group involving 20 other research laboratories across USA, Germany China, Switzerland, and India. The genome of sorghum is relatively smaller ($\sim$730 Mb) making it ideal model for functional genomics study of C_4 grasses. Paterson et al. (2009a, b) revealed 34,496 sorghum genes out of which 27,640 were protein coding. With the availability of genome sequence, rapid development in genomics it is now possible to assign functional role to important genes of economically important traits especially biotic and abiotic stress tolerance. Advances in sequencing technologies such as next generation sequencing has resulted in generation of large genomic resources in sorghum as well as other crops. The greater challenge would be to effectively use the genomic resources for genetic improvement. Large scale sequencing and EST projects led to deposition of large-scale sequence information in the public databases and these online sequence information are the major sources for development of molecular markers such as simple sequence repeats (SSRs), Insertion deletions (InDels), Single Nucleotide Polymorphism (SNPs) and these are done using various bioinformatics and computational resources available in the public domain. The molecular markers developed are then used in number of crop-based applications such as assessment of diversity and population structure, gene identification and cloning, QTL mapping and marker assisted selection. Among various DNA markers, SSRs are widely used due to their hypervariable, co-dominant, robust, chromosome specific and multi-allelic in nature. Prior to genome sequencing in sorghum, researchers have developed large number of markers which were used for various genetic studies in sorghum. Large number of genomic SSR of Xtxp series by Kong et al. (2000), Bhattramakki et al. (2000), Xsb series SSRs (Taramino et al. 1997), Xgap series (Brown et al. 1996), Xcup series SSRs developed from CDNAs (Schloss et al. 2002), EST based SSR markers belonging to Xisep series by Ramu et al. (2009), Xiabt series by Arun 2006 and Reddy et al. 2008), Ungnhsbm developed from unigenes (Srinivas et al. 2009), (GATA)n intron length polymorphism based SbGM series based SSR markers (Jaikishan et al. 2013) and mSbCIR developed by CIRAD France. Sorghum diversity kit comprising of 48 markers distributed throughout the genome was developed by Billot et al. (2012). Owing to limited polymorphism by SSRs compared to SNP based markers, researchers have now opting to use single nucleotide based markers for crop improvement based applications. SNPs are considered as ultimate genetic markers as they represent the finest resolution of DNA and are abundant in population also have low mutation rate. Further due to advancement in technology as well lowered cost of sequencing, resequencing projects led to rapid discovery of SNPs and InDels in sorghum. InDel based markers stands next to SNPs resulting from large scale sequencing projects and are abundant in genome. InDels can be used in routing breeding applications as they can be resolved under gel electrophoresis. InDels are successfully used in number of applications especially in mapping of agrnomically important loci such as waxy locus (McIntyre et al. 2008) and tannin (Wu et al. 2012). Zheng and his coworkers have discovered more than 90,000 InDels of length varying from 1 bp to 10 bp from sequencing of grain and sweet sorghum genotypes. Yang et al. (2007) developed potential Intron Length Polymorphic markers (PIP) that targets both SNPs and InDels. Intron length polymorphism (ILP) can be detected easily by exon primed intron crossing PCR (EPIC-PCR) (Palumbi 1995) and the exon regions flanking the targeted intron are used for developing primers. A database on PIC was developed by Yang et al. (2007) comprising of more than 57,000 markers pertaining to 59 plant species and more than 4000 belonging to sorghum.

Several high density maps (Ashraf 2010), have been built using variety of DNA markers (Mace et al. 2009; Rami et al. 1998). Genomic regions associated with pre-flowering and post-flowering drought tolerance were identified but after the developments in sequencing of sorghum (Paterson et al. 2009a, b) that has enabled the monitoring of the genome-wide expression profiles to abiotic stresses through microarray techniques or RNA-Seq analysis (Dugas et al. 2011; Buchanan et al. 2005; Yazawa et al. 2013; Pasini et al. 2014). The advancements led to identifications of drought stress responsive genes and their regulatory elements and their role in gene function. Several transcriptomics studies were carried out on sorghum using RNA-Seq analysis to monitor gene expression in response to osmotic stress and abscisic acid (Paterson et al. 2009a, b), to provide a S. bicolor expression atlas on the dynamic genotype-specific expression profiles (Shakoor et al. 2014), or to identify genome-wide SNPs that can potentially enhance genetic analysis and the application of molecular markers in sorghum genomics and breeding (Bekele et al. 2013).

Next generation sequencing technologies led to rapid identification of genes and complex traits for drought traits in sorghum, complementing the use of unique genetic materials such as Near Isogenic lines, which are commonly used to identify complex quantitative traits (Tuinstra et al. 1998). However, the library of drought-regulated expressed sequence tags (DRESTs) also provided a more defined view of the sorghum transcriptome (Pontius et al. 2003), also referred to "UniGenes" that represent putative unique genes. The UniGene database include collection of non-redundant unified view of transcriptome that comprise expressed sequence tags (ESTs) that are derived from differentially expressed cDNA libraries (Pontius et al. 2003). Presently, the sorghum represents about 14,000 UniGene clusters in more than 90 diverse libraries from several genotypes (Kresovich et al. 2005). The UniGene gene transcripts which are expressed under drought along with their genomic location information represent a collection of candidate genes for drought tolerance.

5.7 Genomic-Assisted Breeding Approaches

Advancements made in the science of genome sequencing technologies and the availability of next-generation sequencing technologies enabled identification of SNP throughout the genome at relatively lesser costs for many crops and these NGS technologies has greatly helped to undergo genomic selection. Among various sequencing methods genotyping by sequencing methodology provides a quicker and lower cost tool (Poland and Rife 2012). The genomic region coverage is good with the genotyping-by-sequencing but generally leads to large amounts of missing data and this is one of the disadvantages of this technique (Davey et al. 2011; Beissinger et al. 2013). In crops like sorghum being predominantly self pollinated crops where homozygote lines are used for sequencing and compared with available reference genome, genotype imputation allows use of GBS without losses in predictive ability (Habier et al. 2009; Weigel et al. 2010; Dassonneville et al. 2011; Mulder et al. 2012).

Marker assisted selection are mostly applied in areas where the breeding programs for traits with high heritability and are governed by monogene or few major QTL that explains large portion of the phenotypic variability. However use of these techniques for improvement of complex traits have been difficult with complex traits such as yield and abiotic stress tolerance, etc., due to complex genetics based on the interaction of multiple QTLs with minor effects on traits. Nevertheless, marker-assisted selection tools are now gradually emerging into 'genomics-assisted breeding' especially to hasten the genetic gains in the breeding program as well to improve precisions in breeding program.

5.7.1 Genomic Selection

Genomic selection as a technology has gained importance in recent years compared to the use of marker-assisted selection in plant breeding, and this different greatly from QTL mapping and the genome wide association mapping.

Genomic selection (GS) is latest approach in crop improvement for improving quantitative traits in large plant breeding populations which uses whole genome high density markers and high throughput genotyping. Genomic prediction combines molecular marker data with phenotypic and pedigree data (if available) in an attempt to increase the prediction accuracy of breeding/genotypic values. Genomic selection is applied in a breeding population and this is different from reference population in which the marker effects were estimated. Genomic selection uses two types of phenotypic datasets (a) training dataset (b) validation dataset. The training set is the reference or source population in which the marker effects will be estimated; it contains: (1) phenotypic information from the breeding germplasm evaluated diverse environmental conditions (2) molecular marker scores and (3) pedigree or kinship information of the germplasm. Hence, marker effects are estimated using the training set by certain statistical methods to incorporate this information; the genetic values of new genotypes are predicted based on the marker effects (Goddard and Hayes 2007; Heffner et al. 2011). The validation set includes the selection candidates (derived from the reference population) that have been genotyped (but not phenotyped) and are selected based on marker effects estimated in the training set (Fig. 5.2).

Meuwissen et al. (2001) explained advanced breeding methodology called genomic selection (GS) that is believed to solve problems associated with marker

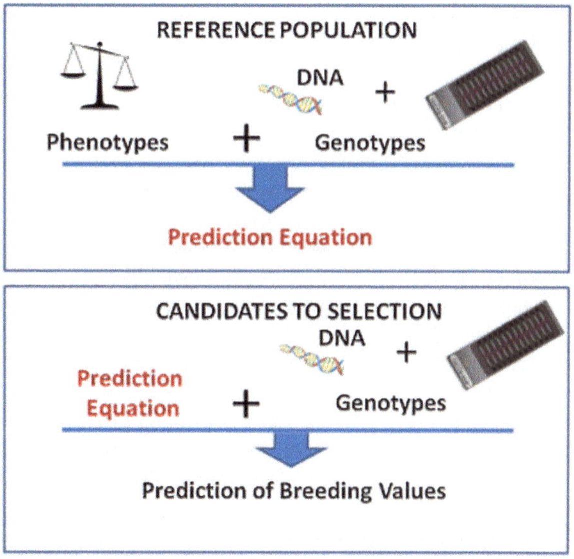

Fig. 5.2 Principles of genomic selection (Top: A prediction equation is obtained from a reference population with phenotypes and genotypes, bottom: this prediction equation is used on candidates with genotypic information only). (*Source* Boichard et al. 2016)

assisted selection of complex traits. In genomic assisted breeding high-density markers coverage are required to potentially have all QTL to be in linkage disequilibrium with at least one marker. The information on all possible loci, haplotypes and their effects across the whole genome is used to estimate the genomic breeding value of a particular genotype in the population. Eventually, knowledge of the relative values of alleles at all loci segregating in a population could allow the breeder to design a genotype in silico and to practice whole genome selection (Table 5.1).

QTL mapping aims to map chromosomal regions affecting phenotypic traits of interest, and thereby enabling the linked markers to use for selection of genomic regions (Bernardo 2008; Lorenz et al. 2011). However, the use bi-parental populations and use of QTL mapping are more specific to specific breeding population or programs. The statistical models used for the study are not largely applicable for breeding of polygenic traits controlled by many loci of small effects (Meuwissen et al. 2001; Goddard and Hayes 2007; Heffner et al. 2011). Association mapping involves identification of chromosomal regions associated with a particular trait, using a diversity panel instead of a breeding population (Ingvarsson and Street 2011; Huang and Han 2014).

Table 5.1 Sorghum drought tolerance/avoidance mechanisms

Drought tolerance/ avoidance in sorghum	How the mechanism help	Limits of the mechanism
Deep root system	Increases water extraction depth	Up to 2.50 m
Higher root density (secondary roots)	Increases water extraction area	Root density of about 4.1 cm per cm soil
Stomata remain open at wide range of leaf turgor	Maintain CO_2 exchange (photosynthesis)	from 11 bars to 1 bar
Stomata closing at higher level of stress	Avoids further water loss	About −14 bar to −15 bar is lowest leaf water potential where stomata closes
Leaf roll	Avoids further water loss by decreasing surface area of leaf contact with radiation	Starts after about 10–14 days of water stress
Forming small vacuoles from large central vacuoles	Avoids cell rapture by maintaining tonoplast integrity	At about a leaf water potential of −37 bar
Production of antioxidant	Protect from lipid peroxidation	Until late in the drought stage
Cuticle and epicuticular wax (waxy bloom substance)	Checks transpiration (decreases water lose from leaves by obstructing the path)	It can check up to 30% of transpiration loss

Source Assefa et al. (2010)

Genomic selection differs from association mapping in the way that genomic selection does not intend to test the significance of markers/genes associated with target trait. Genomic selection leverages or calculates the genomic estimated breeding value using the genome-wide markers. This makes this methodology directly applicable to plant breeding (Bernardo and Yu 2007; Dekkers 2007; Goddard and Hayes 2007; Habier et al. 2007). Nevertheless, the markers located near genes associated with biological functions present effects of greater magnitude. Thus, the predicted effects of markers in genomic selection may be used functional enrichment analysis to identify important functional groups. One of the major advantages of genomic selection in breeding programs is the reduced time needed to develop new materials provided the predictive abilities are large. Genomic selection reduces breeding cycle through prediction of phenotypic performance in the set of genotypes for various traits of interest (Meuwissen et al. 2001; Bernardo and Yu 2007; Bernardo 2008; Crossa et al. 2011). Genomics assisted breeding is especially more useful for improvement of complex traits due to its advantages of precision, direct improvement, short breeding cycle, and high selection efficiency (Varshney et al. 2005). The ultimate goal of genomic assisted selection is to find the best combinations of alleles (or haplotypes), optimal gene networks, and specific genomic regions to facilitate crop improvement (Xu et al. 2012). These tools will enhance, but not replace, the conventional breeding (Varshney et al. 2005).

5.7.2 Genomic Selection Models and Other Factors

A number of genomic selection models are available and models are adjusted and selection can be made based on the best model for the target trait. However it is better to have a simple model as complex models have difficulties in access of computational resources. The genomic selection depends on the different assumptions and the effects of QTLs however the difference in the prediction abilities is quite small (de Oliveira et al. 2018). Bernardo and Yu (2007) reported that the Bayesian based models exhibited less advantage in prediction in comparison to RR-BLUP model and suggested as an alternative in the breeding programs (Hofheinz and Frisch 2014). Selection of models for genomic selection depends on the population or breeding cycles and may affect the outcome with different assumptions. It is better to have comparative evaluation of different models (Habier et al. 2007) and adopt the best-suited models for selections of lines.

Different combination of training datasets maximizes the use of phenotypic and genotypic information available in the genetic material, generating large training populations and increasing the predictive ability (De Roos et al. 2009; Hayes et al. 2009; Asoro et al. 2011; Lorenz et al. 2012; Technow et al. 2013). Both size and the constitution of the training population are important factors (Habier et al. 2010; Rincent et al. 2012). Predictions are better when selections are made for traits with high heritability (Combs and Bernardo 2013; Lorenz 2013). The major factors affecting the predictive abilities are effective population size, training population size, nature of trait, linkage disequilibrium, number of markers, choice of model for prediction and the relationship between training and breeding population (Daetwyler et al. 2008; Grattapaglia and Resende 2010; Asoro et al. 2011; Nakaya and Isobe 2012; de los Campos et al. 2013).

The major attribute is that the training and test populations must be highly related to ensure an effective selection. The interaction of trait architecture and population structure plays an important role in creating a training population. In the study conducted by de Oliveira et al. (2018) they showed that the genetic structure of the 200 genotypes reflected their subdivision in two population sub-panels using principal component analysis. The sub-panel I mostly constituted the saccharine and high biomass genotypes from CIRAD and ICRISAT institutes, while their sub-panel II mostly composed of genotypes from the Embrapa sorghum germplasm bank and from their breeding program. Therefore in joint analysis, the similarity between genotypes in training and test populations was large. This partially explained higher predictive abilities found in the joint analysis than those obtained in the cross analysis between sub-panels. The influence of marker density on the predictive ability of the trait showed that the best predictions were obtained when the marker density are large. The median predictive ability remained constant with roughly four thousand markers. This indicates that a reduced number of markers explained most of the genetic variation, from the practical point of view opening new opportunities for breeders with relatively small subset of SNPs in sorghum, by constructing genotyping chips (Matukumalli et al. 2009; Yu et al. 2013; Wang et al. 2014).

Phenotyping of traits is as important as genotyping in genomic selection since it influences the prediction of marker effects in the process of selection of superior genotypes. The genotyping cost is expected to go down drastically in future and the phenotyping cost to go up. It is also expected that the precision of phenotypic measurements taken for trait influences the predictive abilities of the models (Heslot et al. 2015). The ultimate economic value of a test line depends on its performance in the target environment.

5.7.2.1 Genomic Selection in Sorghum

Yu et al. (2016) applied genomic selection tools for studying the genetics of biomass related traits such as dry biomass yield and plant height. They also cited that genomic selection can be successfully applied for direct selection in sorghum biomass breeding programs. Using a cross-validation approach they found the predictive abilities ranging from 0.35 to 0.78 which can be applicable directly for selection of traits and breeding. The study demonstrated that the methodology can potentially reduce the time required for the launching of new cultivars of biomass sorghum.

Fernandes et al. (2018) used different genomic selection methods which use single and correlated traits to predict biomass yield. Genomic selection is mostly performed on a single trait, but correlated traits can also aid in selection of the target trait through multi-trait genomic selection. In their study, they used a population of high biomass sorghum to compare different strategies that use correlated traits to improve prediction of biomass yield. The correlated traits used were moisture, height of the plant measured at monthly intervals starting from planting and harvest, and the area under the growth progress curve. They also tested new strategy called trait-assisted genomic selection, in which correlated traits are used along with marker data in the validation population to predict the target trait. Their results showed that single and multi-trait genomic selection recorded similar results, indicating that correlated traits did not improve prediction of biomass yield. However, trait assisted genomic selection method increased prediction accuracy when plant height was used in both the training and validation populations to help predict yield in the validation population. Coincidence between selected genotypes in phenotypic and genomic selection was also highest in trait-assisted genomic selection. Overall, these results suggest that trait-assisted GS can be an efficient strategy when correlated traits are obtained earlier or more inexpensively than a focal trait.

de Oliveira et al. (2018) predicted genomic estimated breeding value using genomic selection models from a set of 200 sorghum lines comprising landraces and breeding lines from high biomass and saccharine groups. They evaluated for days to flowering, plant height, fresh and dry biomass yield and quality parameters such as fiber, cellulose, hemi-cellulose and lignin. From genotyping by sequencing they obtained about 2,58,000 SNP markers. Their study reported predictive abilities in order as 0.39, 0.38, 0.35, 0.28, 0.26, 0.26, and 0.21 for the different traits acid detergent fiber, lignin, dry matter yield, cellulose, neutral detergent fiber, fresh matter yield and days to flowering, respectively while their study reported plant height showed predictive ability of 0.65. They then fitted different models and compared genomic models Bayes A, Bayes B, Bayes Cπ, Bayes Lasso, Bayes Ridge Regression model, and random regression best linear unbiased predictor. The resulting predictive abilities varied little between the different models but substantially between traits.

5.7.3 Future Prospects

Sorghum is a climate-smart crop providing food–fodder-feed–fuel to most of the resource-poor farming communities in arid and semiarid regions of the world where it is a staple crop. During the last decade, molecular markers have played a major role in indirect selection of traits of economic importance through marker assisted breeding. Genetic mechanisms underlying drought and other related traits responsible for its ability to tolerate harsh growing environments will have positive implications in breeding programs in sorghum as well as other agronomically important crops. High-throughput sequencing led to increasing amount of sequence data especially SNPs in the public domain and this greatly contributed to science of genomics research for crop improvement. These genomic tools and resources led to evolution of new breeding technologies such as genomic selection and genomic assisted breeding tools, as well as new knowledge about statistics that could increase the efficiency and precision of crop improvement. The integration of these new tools along with new-generation breeding populations will further help accumulate the required information to develop the framework for implementing genomic selection for complex traits such as drought tolerance in sorghum thus the gap between breeding and genomics is expected to narrow down significantly.

References

Adeyanju A, Little C, Yu J, Tesso T (2015) Genome-wide association study on resistance to stalk rot diseases in grain sorghum. G3 Genes Genomes Genet 5(6):1165–1175

Ajithkumar IP, Panneerselvam R (2014) ROS scavenging system, osmotic maintenance, pigment and growth status of *Panicum sumatrense* Roth under drought stress. Cell Biochem Biophys 68:587–595

Ambros V (2004) The functions of animal microRNAs. Nature 431(7006):350–355

Arau MI, Valades-Cerda MC, Maiti RK, Hernandez-Pingero JN, Valades-Cerda MC (2001) Variability among non-glossy and glossy genotypes-in seedling and callus growth in response to 2,4-D treatments. Indian J Plant Physiol 6:19–23

Armstrong JS, Rooney WL, Peterson GC, Villenueva RT, Brewer MJ, Sekula -Ortiz D (2015) Sugarcane aphid (hemiptera: Aphididae): host range and sorghum resistance including cross—resistance from greenbug sources. J Econ Entomol 108(2):576–582

Arun SS (2006) In silico EST data mining for elucidation of repeats biology and functional annotation in sorghum [Sorghum bicolor (L.) Moench.]. MSc thesis. University of Agricultural Sciences, Dharwad

Aruna C, Dayakar Rao B, Vilas Tonapi, Nageshwar Rao TG (2018) Achieving sustainable cultivation of sorghum. In: Rooney W (ed) Volume 2. Sorghum utilization around the world. Burleigh Dodds Science Publishing, Cambridge, UK, pp 1–24. https://doi.org/10.19103/as.2017.0015.28

Ashok Kumar A, Reddy BV, Sharma HC, Hash CT, Srinivasa Rao P, Ramaiah B, Reddy P (2011) Recent advances in sorghum genetic enhancement research at ICRISAT. Amer J Plant Sci 2 (4):589–600

Ashraf M (2010) Inducing drought tolerance in plants: Recent advances. Biotechnol Adv 28: 169–183

Asoro FG, Newell MA, Beavis WD, Scott MP, Jannink JL (2011) Accuracy and training population design for genomic selection on quantitative traits in elite north American oats. Plant Genome J 4:132–144

Assefa Y, Staggenborg SA, Prasad PVV (2010) Grain sorghum water requirement and responses to drought stress: A review. Crop Manag 10:1109–1116

Asseng S, Foster IA, Turner NC (2011) The impact of temperature variability on wheat yields. Glob Change Biol 17(2):997–1012

Aubry S, Brown NJ, Hibberd JM (2011) The role of proteins in C3 plants prior to their recruitment into the C4 pathway. J Exp Bot 62(9):3049–3059

Awika JM, Rooney LW (2004) Sorghum phytochemicals and their potential impact on human health. Phytochemistry 65:1199–1221

Baalbaki RZ, Zurayk RA, Bleik MM, Saxena NS (1999) Germination and seedling development of drought tolerant and susceptible wheat under moisture stress. Seed Sci Technol 27:291–302

Balsamo RA, Willigen CV, Bauer AM, Farrant J (2006) Drought tolerance of selected Eragrostis species correlates with leaf tensile properties. Ann Bot 97(6):985–991

Bandara YM, Weerasooriya DK, Tesso TT, Prasad PV, Little CR (2017) Stalk rot fungi affect grain sorghum yield components in an inoculation stage-specific manner. Crop Protec 94: 97–105

Bandyopadhyay T, Muthamilarasan M, Prasad M (2017) Millets for next generation climate-smart agriculture. Front Plant Sci 8:1266

Bartel DP (2004) MicroRNAs: genomics, biogenesis, mechanism, and function genomics: the miRNA genes. Cell 116:281–297

Battisti DS, Naylor RL (2009) Historical warnings of future food insecurity with unprecedented seasonal heat. Science 323:240–244

Bayu W, Rethman NF, Hammes PS, Pieterse PA, Grimbeek J, Linde M (2005) Water stress affects the germination, emergence and growth of different sorghum cultivars. Eth J Sci 28:119–128

Bean SR, Wilson JD, Moreau RA, Galant A, Awika JM, Kaufman RC, Adrianos SL, Ioerger BP (2016) Structure and composition of the sorghum grain. In: Ciampitti I, Prasad V (eds) Sorghum: state of the art and future perspectives. American Society of Agronomy and Crop Science Society of America, Inc

Beck PA, Hutchison S, Gunter SA, Losi TC, Stewart CB, Capps PK, Phillips JM (2007) Chemical composition and in situ dry matter and fiber disappearance of sorghum × Sudangrass hybrids. J Anim Sci 85(2):545–555

Beissinger TM, Hirsch CN, Sekhon RS, Foerster JM, Johnson JM, Muttoni G, Vaillancourt B, Buell CR, Kaeppler SM, de Leon N (2013) Marker density and read depth for genotyping populations using genotyping-by-sequencing. Genetics 193:1073–1081

Bekele WA, Wieckhorst S, Friedt W, Snowdon RJ (2013) High-throughput genomics in sorghum: from whole-genome resequencing to a SNP screening array. Plant Biotech J. 11(9):1112–1125

Belton PS, Delgadillo I, Halford NG, Shewry PR (2006) Kafirin structure and functionality. J Cereal Sci 44:272–286

Bernardo R (2008) Molecular markers and selection for complex traits in plants: learning from the last 20 years. Crop Sci 48:1649

Bernardo R, Yu J (2007) Prospects for genomewide selection for quantitative traits in maize. Crop Sci 47:1082–1090

Bhattramakki D, Dong J, Chhabra AK, Hart GE (2000) An integrated SSR and RFLP linkage map of Sorghum bicolor (L.) Moench. Genome 43:988–1002

Bibi A, Sadaqat HA, Tahir MHN, Akram HM (2012) Screening of sorghum (Sorghum bicolor var Moench) for drought tolerance at seedling stage in polyethylene glycol. J Anim Plant Sci 22 (3):671–678

Bidinger FR, Napolean T, Hash CT (2007). Quantitative trait loci for grain yield in pearl millet under variable post flowering moisture conditions. Crop Sci 47:969–980

Billot C, Rivallan R, Sall MN, Fonceka D, Deu M, Glaszmann JC, Noyer JL, Rami JF, Risterucci AM, Wincker P, Ramu P, Hash CT (2012) A reference microsatellite kit to assess for genetic diversity of Sorghum bicolor (Poaceae). American J Bot 99:245–250

Blum A (1996) Crop responses to drought and the interpretation of adaptation. Plant Growth Reg 20:135–148

Blum A (2011) Drought resistance and its improvement. In: Blum A (ed) Plant breeding for water-limited environments. Springer Science and Business Media, NY, pp 53–137

Blum A, Arkin GF (1984) Sorghum root growth and water-use as affected by water supply and growth duration. Field Crops Res 9:131–142

Bohnert HJ, Gong Q, Li P, Ma S (2006) Unraveling abiotic stress tolerance mechanisms—getting genomics going. Curr Opin Plant Biol 9:180–188

Boichard D, Ducrocq V, Croiseau P, Fritz S (2016) Genomic selection in domestic animals: principles, applications and perspectives. CR Biol 339(7–8):274–277

Boomiraj K, Wani SP, Agrawal PK (2011) Impact of climate change on dryland Sorghum in India. https://doi.org/10.13140/2.1.2528.7689

Borrell AK, Hammer GL, Henzell RG (2000) Does maintaining green leaf area in sorghum improve yield under drought? II Dry matter production and yield. Crop Sci 40(4):1037–1048

Borrell AK, Jordan DR, George-Jaeggli B, Hammer GL, Van Oosterom E, Klein P, Mullet J (2009) Fine-mapping candidate genes for 'stay-green' in sorghum: are we there yet? 3rd International Conference on Integrated Approaches to Improve Crop Production under Drought-Prone Environments (InterDrought-III). Shanghai, China: Shanghai Academy of Agricultural Science. L 5.03

Borrell AK, Mullet JE, George-Jaeggli B, van Oosterom EJ, Hammer GL, Klein PE, Jordan DR (2014a) Drought adaptation of stay-green sorghum is associated with canopy development, leaf anatomy, root growth, and water uptake. J Exp Bot 65(21):6251–6263

Borrell AK, van Oosterom EJ, Mullet JE, George-Jaeggli B, Jordan DR, Klein PE, Hammer GL (2014b) Stay-green alleles individually enhance grain yield in sorghum under drought by modifying canopy development and water uptake patterns. New Phytol 203:817–830

Bowling RD, Brewer MJ, Kerns DL, Gordy J, Seiter N, Elliott NE, Buntin GD, Way MO, Royer TA, Biles S, Maxson E (2016) Sugarcane aphid (hemiptera: Aphididae): a new pest on sorghum in North America. J Integr Pest Manag 7(1):12

Boyles RE, Cooper EA, Myers MT, Brenton Z, Rauh BL, Morris GP, Kresovich S (2016) Genome -wide association studies of grain yield components in diverse sorghum germplasm. Plant Genome 9

Boyles RE, Pfeiffer BK, Cooper EA, Rauh BL, Zielinski KJ, Myers MT, Brenton Z, Rooney WL, Kresovich S (2017) Genetic dissection of sorghum grain quality traits using diverse and segregating populations. Theor Appl Genet 130:697–716

Brown SM, Hopkins MS, Mitchell SE, Senior ML, Wang TY, Duncan RR, Gonzalez-Candelas F, Kresovich S (1996) Multiple methods for the identification of polymorphic simple sequence repeats (SSRs) in sorghum [Sorghum bicolor (L.) Moench]. Theor Appl Genet 93:190–198

Buchanan CD, Lim S, Salzman RA, Kagiampakis I, Morishige DT, Weers BD, Klein RR, Pratt LH, Cordonnier-Pratt MM, Klein PE, Mullet JE (2005) Sorghum bicolor's transcriptome response to dehydration, high salinity and ABA. Plant molr biol 58(5):699–720

Campos H, Cooper M, Habben JE, Edmeades GO, Schussler JR (2004) Improving drought tolerance in maize: a view from industry. Field Crops Res 90:19–34

Chapke RR, Babu S, Subbarayudu B, Tonapi VA (2017) Growing popularity of sorghum in rice fallows: an IIMR case study. Bulletin, ICAR-Indian Institute of Millets Research, Hyderabad 500030, India. 40 ISBN: 81-89335-58-8

Chapman SC, Cooper M, Butler D, Henzell RG (2000a) Genotype by environment interactions affecting yield of grain sorghum I Characteristics that confound interpretation of hybrid yield. Aust J Agric Res 51:197–207

Chapman SC, Cooper M, Hammer GL, Butler D (2000b) Genotype by environment interactions affecting yield of grain sorghum II Frequencies of different seasonal patterns of drought stress are related to location effects on hybrid yields. Aust J Agric Res 51:209–222

Chapman SC, Hammer GL, Podlich DW, Cooper M (2002) Linking biophysical and genetic models to integrate physiology, molecular biology and plant breeding. In: Kang M (ed) Quantitative genetics, genomics, and plant breeding, vol 1, pp 167–187

Chaves MM, Maroco JP, Pereira JS (2003) Understanding plant response to drought-from genes to the whole plant. Funct Plant Biol 30:239–264

Chen TH, Murata N (2002) Enhancement of tolerance of abiotic stress by metabolic engineering of betaines and other compatible solutes. Curr Opin Plant Biol 5:250–257

Clerget B, Rattunde HF, Dagnoko S, Chantereau J (2007) An easy way to assess photoperiod sensitivity in sorghum: relationships of the vegetative-phase duration and photoperiod sensitivity. J SAT Agric Res 3(1):1–3

Cline W (2007) Global Warming and Agriculture: Impact Estimates by Country. Peterson Institute, Washington DC

Combs E, Bernardo R (2013) Accuracy of genomewide selection for different traits with constant population size, heritability, and number of markers. Plant Genome 6:1–7

Cornic G (2000) Drought stress inhibits photosynthesis by decreasing stomatal aperture—not by affecting ATP synthesis. Trends Plant Sci 5(5):187–188

Crasta OR, Xu W, Rosenow DT, Mullet J, Nguyen HT (1999) Mapping of post-flowering drought resistance traits in grain sorghum: association between QTLs influencing premature senescence and maturity. Mol Gen Genet 262:579–588

Crossa J, Pérez P, de los Campos G, Mahuku G, Dreisigacker S, Magorokosho C (2011) Genomic selection and prediction in plant breeding. J Crop Improv 25:239–261

Curtis DL (1968) The relation between yield and date of heading of Nigerian sorghums. Expl Agric 4(1):93–101

Daetwyler HD, Villanueva B, Woolliams JA (2008) Accuracy of predicting the genetic risk of disease using a genome-wide approach. PLoS ONE 3:e3395

Dassonneville R, Brondum RF, Druet T, Fritz S, Guillaume F, Guldbrandtsen B, Lund MS, Ducrocq V, Su G (2011) Effect of imputing markers from a low-density chip on the reliability of genomic breeding values in Holstein populations. J Dairy Sci 94:3679–3686

Davey JW, Hohenlohe PA, Etter PD, Boone JQ, Catchen JM, Blaxter ML (2011) Genome-wide genetic marker discovery and genotyping using next-generation sequencing. Nat Rev Genet 12:499–510

de los Campos G, Hickey JM, Pong-Wong R, Daetwyler HD, Calus MPL (2013) Whole-genome regression and prediction methods applied to plant and animal breeding. Genetics 193:327–345

de Oliveira AA, Pastina MM, de Souza VF (2018) Mol Breeding 38:49

De Roos AP, Hayes BJ, Goddard ME (2009) Reliability of genomic predictions across multiple populations. Genetics 183:1545–1553

Dekkers JC (2007) Prediction of response to marker-assisted and genomic selection using selection index theory. J Anim Breed Genet 124:331–341

Djanaguiraman M, Perumal R, Jagadish SV, Ciampitti IA, Welti R, Prasad PVV (2018) Sensitivity of sorghum pollen and pistil to high-temperature stress. Plant, Cell Environ 41:1065–1082

Djanaguiraman M, Vara Prasad PV, Murugan M, Perumal R, Reddy UK (2014) Physiological differences among sorghum (Sorghum bicolor L. Moench) genotypes under high temperature stress. Environ Exp Bot 100:43–54

Doorenbos J, Kassam AH (1979) Yield response to water Irrig and Drain Page 33. Food and Agriculture Organization of the United Nations, Rome, Italy

Dugas DV, Monaco MK, Olsen A, Klein RR, Kumari S, Ware D, Klein PE (2011) Functional annotation of the transcriptome of Sorghum bicolor in response to osmotic stress and abscisic acid. BMC Genom 12:514

Duodu K, Taylor JR, Belton P, Hamaker B (2003) Factors affecting sorghum protein digestibility. J Cereal Sci 38:117–131

Dykes L, Rooney LW (2006) Sorghum and millet phenols and antioxidants. J Cereal Sci 44: 236–251

Eck HV, Musick JC (1979) Plant water stress effect on irrigated sorghum I. Effect on yield. Crop Sci 19:586–592

Ejeta G, Goldsbrough MR, Tunistra EM, Grote A, Menkir Y, Ibrahim Y, Cisse N, Weerasuriya Y, Melake-Berhan A, Shaner CA (2000) Application of molecular markers in plant breeding, Training manual for a seminar held at IITA, Ibadan, Nigeria, pp 16–17

Ekanayake IJ, O'Toole JC, Garrity DP, Masajo TM (1985) Inheritance of root characters and their relations to drought resistance in rice. Crop Sci 25:927–993

Farooq M, Wahid A, Kobayashi N, Fujita D, Basra SM (2009) Plant drought stress: effects, mechanisms and management. Agron Sustain Dev 29:185–212

Fernandes SB, Dias KOG, Ferreira DF, Patrick JB (2018) Efficiency of multi-trait, indirect, and trait-assisted genomic selection for improvement of biomass sorghum. Theor Appl Genet 131:747

Fracasso A, Trindade L, Amaducci S (2016) Drought tolerance strategies highlighted by two Sorghum bicolor races in a dry-down experiment. J Plant Phys 190:1–14

Franks CD, Burow GB, Burke JJ (2006) A comparison of U.S. and Chinese sorghum germplasm for early season cold tolerance. Crop Sci 46:1371–1376

Fu T, Ko J, Wall GW, Pinter PJ, Kimball BA, Ottman MJ, Kim HY (2016) Simulation of climate change impacts on grain sorghum production grown under free air CO2 enrichment. Int Agrophys 30(3):311–322

Gao S, Wang Y, Li G (2010) Sorghum breeding and production in China. In: He Z, Alain Bonjean PA (eds) Cereals in China, D.F. CIMMYT, Cereals in China, pp 97–107

Garrity DP, Darrel GW, Charles YS, James RG (1982) Moisture deficits and grain sorghum performance: evapotranspiration-yield relationships. Agron J 74:815–820

Gebrekiros G, Araya A, Yemane T (2016) Modeling impact of climate change and variability on sorghum production in Southern Zone of Tigray, Ethiopia. J Earth Sci Clim Change 7(2):322

Gelli M, Duo Y, Konda AR, Zhang C, Holding D, Dweikat I (2014) Identification of differentially expressed genes between sorghum genotypes with contrasting nitrogen stress tolerance by genome-wide transcriptional profiling. BMC Genom 15:179

Girma FS, Krieg DR (1992) Osmotic adjustment in sorghum: II. Relationship to gas exchange rates. Plant Physiol 99(2):583–588

Goddard ME, BJ Hayes (2007) Genomic selection. J Anim Breed Genet 124:323–330

Goldsbrough PB, Wood AJ (1997) Characterization and expression of dehydrins in water-stressed Sorghum bicolor. Physiol Plant 99:144–152

Grattapaglia D, Resende MD (2010) Genomic selection in forest tree breeding. Tree Genet Genomes 7:241–255

Habier D, Fernando RL, Dekkers JC (2007) The impact of genetic relationship information on genome-assisted breeding values. Genetics 177:2389–2397

Habier D, Fernando RL, Dekkers JCM (2009) Genomic selection using low-density marker panels. Genetics 182:343–353

Habier D, Tetens J, Seefried FR, Lichtner P, Thaller G (2010) The impact of genetic relationship information on genomic breeding values in German Holstein cattle. Genet Sel Evol 42:5

Hale MG, Orcutt DM (1987) The physiology of plant under stress. Wiley Interscience Publ, New York, NY

Hariprasanna K, Rakshit S (2017) Economic importance of sorghum. Sorghum Genome 1–25

Harris K, Subudhi PK, Borrell A, Jordan D, Rosenow D, Nguyen H, Klein P, Klein R, Mullet J (2007) Sorghum stay-green QTL individually reduce post-flowering drought-induced leaf senescence. J Exp Bot 58:327–338

Hash CT, Folkertsma RT, Ramu P, Reddy BV, Mahalakshmi V, Sharma HC, Rattunde HF, Weltzien ER, Haussmann BI, Ferguson ME, Crouch JH (2002) Marker-assisted breeding across ICRISAT for terminal drought tolerance and resistance to shoot fly and Striga in sorghum. In the wake of the double helix: from the Green Revolution to the Gene Revolution, pp 27–31

Haussmann BI, Fred Rattunde H, Weltzien-Rattunde E, Traoré PS, Vom Brocke K, Parzies HK (2012) Breeding strategies for adaptation of pearl millet and sorghum to climate variability and change in West Africa. J Agron Crop Sci 198(5):327–339

Haussmann BI, Mahalakshmi V, Reddy BV, Seetharama N, Hash CT, Geiger HH (2002) QTL mapping of stay-green in two sorghum recombinant inbred populations. Theor Appl Genet 106:133–142

Havaux M, Tardy F (1999) Loss of chlorophyll with limited reduction of photosynthesis as an adaptive response of Syrian barley landraces to highlight and heat stress. Aust J Plant Physiol 26:569–578

Hawkesford MJ, Araus JL, Park R, Calderini D, Miralles D, Shen T, Zhang J, Parry MA (2013) Prospects of doubling global wheat yields. Food Energy Secur 2:34–48

Hayes BJ, Bowman PJ, Chamberlain a J, Goddard ME (2009) Invited review: genomic selection in dairy cattle: progress and challenges. J Dairy Sci 92:433–443

Heffner EL, Jannink JL, Iwata H (2011) Genomic selection accuracy for grain quality traits in biparental wheat populations. Crop Sci 51:2597–2606

Henzell RG, Gillieron W (1973) Effects of partial and complete panicle removal on the rate of death of some Sorghum bicolor genotypes under moisture stress. Queensland J Agri Animal Sci 30:291–299

Henzell RG, Hare BR, Jordan DR, Fletcher DS, McCosker AN, Bunker G, Persley DS (2001) Sorghum breeding in Australia: public and private endeavours. In: Proceedings of the 4th Australian sorghum conference, Kooralbyn, Australia

Heslot N, Jannink JL, Sorrells ME (2015) Perspectives for genomic selection applications and research in plants. Crop Sci 55

Hirel B, Le Gouis J, Ney B, Gallais A (2007) The challenge of improving nitrogen use efficiency in crop plants: towards a more central role for genetic variability and quantitative genetics within integrated approaches. J Exp Bot 9:2369–2387

Hofheinz N, Frisch M (2014) Heteroscedastic ridge regression approaches for genome-wide prediction with a focus on computational efficiency and accurate effect estimation. G3 (Bethesda) 4:539–546

Högy P, Fangmeier A (2008) Effects of elevated atmospheric CO_2 on grain quality of wheat. J Cereal Sci 48:580–591

Howell TA, Hiler EA (1975) Optimization of water use efficiency under high frequency irrigation: I. Evapotranspiration and yield relationship. Trans ASAE 18:873–878

Huang X, Han B (2014) Natural variations and genome-wide association studies in crop plants. Annu Rev Plant Biol 65:531–551

Hufnagel B, de Sousa SM, Assis L, Guimaraes CT, Leiser W, Azevedo GC, Negri B, Larson BG, Shaff JE, Pastina MM, Barros BA, Weltzien E, Rattunde HFW, Viana JH, Clark RT, Falcao A, Gazaffi R, Garcia AAF, Schaffert RE, Kochian LV, Magalhaes JV (2014) Duplicate and conquer: multiple homologs of PHOSPHORUS-STARVATION TOLERANCE 1 enhance phosphorus acquisition and sorghum performance on low-phosphorus soils. Plant Physiol 166 (2):659–677

Ingvarsson PK, Street NR (2011) Association genetics of complex traits in plants. New Phytol 189:909–922

Inuyama S, Musick JT, Dusek DA (1976) Effect of plant water deficit at various growth stages on growth, grain yield and leaf water potential of irrigated grain sorghum. Proc Crop Sci Soc Jpn 45:298–307

Jafar MS, Nourmohammadi G, Maleki A (2004) Effect of water deficit on seedling, plantlets and compatible solutes of forage Sorghum cv Speedfeed. In: 4th International crop science conference, Brisbane, Australia

Jaikishan I, Paik GR, Madhusudhana R, Elangovan M, Rajendrakumar P (2013) Development of microsatellite markers targeting (GATA)n motifs in sorghum [Sorghum bicolor (L.) Moench]. Mol. Breed. 31:223–231

Jones MM, Turner NC (1978) Osmotic adjustment in leaves of sorghum in response to water deficits. Plant Physiol 61(1):122–126

Jordan DR, Hunt CH, Cruickshank AW, Borrell AK, Henzell RG (2012) The relationship between the stay—green trait and grain yield elite sorghum hybrid grown in a range of environments. Crop Sci 52:1153–1161

Jordan DR, Tao YZ, Godwin ID, Henzell RG, Cooper M, McIntyre CL (1998) Loss of genetic diversity associated with selection for resistance to sorghum midge in Australian sorghum. Euphytica 102:1–7

Jordan WR, Sweeten JM (1987) Irrigation water management for the Texas high plains: a research summary. Online Texas Water Resource Inst Rep No TR-139, Texas A&M Univ, College Station, TX

Jubany-Marí T, Munné-Bosch S, Alegre L (2010) Redox regulation of water stress responses in field-grown plants. Role of hydrogen peroxide and ascorbate. Plant Physiol Biochem 48 (5):351–358

Kassahun B, Bidinger FR, Hash CT, Kuruvinashetti MS (2010) Stay-green expression in early generation sorghum [Sorghum bicolor (L.) Moench] QTL introgression lines. Euphytica 172:351–362

Kebede H, Subudhi PK, Rosenow DT, Nguyen HT (2001) Quantitative trait loci influencing drought tolerance in grain sorghum (Sorghum bicolor L. Moench). Theor Appl Genet 103:266–276

Khraiwesh B, Zhu JK, Zhu J (2012) Role of miRNAs and siRNAs in biotic and abiotic stress responses of plants. Biochimica et Biophysica Acta 1819(2):137–148

Kimball BA, Morris CF, Pinter PJ Jr, Wall GW, Hunsaker DJ, Adamsen FJ, LaMorte RL, Leavitt SW, Thompson TL, Matthias AD, Brooks TJ (2001) Elevated CO2, drought and soil nitrogen effects on wheat grain quality. New Phytol 150:295–303

Kim HK, Luquet D, van Oosterom EJ, Dingkuhn M, Hammer GL (2010) Regulation of tillering in sorghum: genotypic effects. Annals of Botany 106:69–78

Kole C, Muthamilarasan M, Henry R, Edwards D, Sharma R et al (2015) Application of genomics-assisted breeding for generation of climate resilient crops: progress and prospects. Frontiers Plant Sci 6:563

Kong L, Dong J, Hart GE (2000) Characteristics linkage-map positions and allelic differentiation of Sorghum bicolor (L.) Moench DNA simple-sequence repeats (SSRs). Theor Appl Genet 101:438–448

Kresovich S, Barbazuk B, Bedell JA, Borrell A, Buell CR, Burke J (2005) Toward sequencing the sorghum genome: a US National Science Foundation-Sponsored workshop report. Plant Physiol 138:1898

Kulwal PL, Thudi M and Varshney RK (2011) Genomics interventions in crop breeding for sustainable agriculture. In: Meyers RA (ed) Encyclopedia of Sustainability Science and Technology, pp 2527–2540, Springer, New York, NY

Kyparissis A, Drilias P, Manetas Y (2000a) Seasonal fluctuations in photoprotective (xanthophyll cycle) and photoselective (chlorophylls) capacity in eight Mediterranean plant species belonging to two different growth forms. Funct Plant Biol 27(3):265–272

Kyparissis A, Petropoulou Y, Manetas Y (2000b) Summer survival of leaves in a soft-leaved shrub (Phlomis fruticosa L., Labiatae) under Mediterranean field conditions: avoidance of photoinhibitory damage through decreased chlorophyll contents. J Exp Bot 46(12):1825–1831

Landi P, Sanguineti MC, Salvi S (2005) Validation and characterization of a major QTL affecting leaf ABA concentration in maize. Mol Breed 15:291–303

Lata C, Bhutty S, Bahadur RP, Majee M, Prasad M (2011) Association of an SNP in a novel DREB2-like gene SiDREB2 with stress tolerance in foxtail millet [*Setaria italica* (L.)]. J Exp Bot 62:3387–3401

Lewis RB, Hiler EA, Jordan WR (1974) Susceptibility of grain sorghum to water deficit at three growth stages. Agron J 66:589–591

Li P, Brutnell TP (2011) Setaria viridis and Setaria italica, model genetic systems for the Panicoid grasses. J Exp Bot 62(9):3031–3037

Li YJ, Fu YR, Huang JG, Wu CA, Zheng CC (2011) Transcript profiling during the early development of the maize brace root via Solexa sequencing. FEBS J 278:156–166

Liu S, Wang X, Wang H, Xin H, Yang X, Yan J, et al (2013) Genome wide analysis of ZmDREB genes and their association with natural variation in drought tolerance at seedling stage of Zea mays L. PLoS Genet 9:e1003790

Liu T, Chen D, Wang S, Yin L, Deng X (2018) Physio-ecological responses to drought and subsequent re-watering in sorghum seedlings. Acta Prataculturae Sin 27(6):100–110

Lobell DB, Burke MB, Tebaldi C, Mastrandrea MD, Falcon WP, Naylor RL (2008) Prioritizing climate change adaptation needs for food security in 2030. Science 319(5863):607–610

Lopes MS, Rebetzke GJ, Reynolds M (2014) Integration of phenotyping and genetic platforms for a better understanding of wheat performance under drought. J Exp Bot 65(21):6167–6177

Lorenz AJ (2013) Resource allocation for maximizing prediction accuracy and genetic gain of genomic selection in plant breeding: a simulation experiment. G3 (Bethesda) 3:481–491

Lorenz AJ, Chao S, Asoro FG, Heffner EL, Hayashi T, Iwata H, Smith KP, Sorrells ME, Jannink JL (2011) Genomic selection in plant breeding: knowledge and prospects, 1st edn. Elsevier Inc

Lorenz AJ, Smith KP, Jannink JL (2012) Potential and optimization of genomic selection for Fusarium head blight resistance in six-row barley. Crop Sci 52:1609–1621

Luo LJ (2010) Breeding for water-saving and drought resistance rice (WDR) in China. J Exp Bot 61:3509–3517

Lynch JP, Chimungu JG, Brown KM (2014) Root anatomical phenes associated with water acquisition from drying soil: targets for crop improvement. J Exp Bot 65:6155–6166
Ma Y, Qin F and Tran LS (2012) Contribution of genomics to gene discovery in plant abiotic stress responses. Mol Plant 5:1176–1178
Mace ES, Rami JF, Bouchet S, Klein PE, Klein RR, Kilian A, Wenzl P, Xia L, Halloran K, Jordan DR (2009) A consensus genetic map of sorghum that integrates multiple component maps and high-throughput Diversity Array Technology (DArT) markers. BMC Plant Biol 9 (1):13–27
Mace ES, Singh V, Van Oosterom EJ, Hammer GL, Hunt CH, Jordan DR (2012) QTL for nodal root angle in sorghum (Sorghum bicolor L. Moench) co-locate with QTL for traits associated with drought adaptation. Theor Appl Genet 124(1):97–109
Mahalakshmi V, Bidinger FR (2002) Evaluation of stay-green sorghum germplasm lines at ICRISAT. Crop Sci 42:965–974
Major DJ, Rood SB, Miller FR (1990) Temperature and photo period effects mediated by the sorghum maturity genes. Crop Sci 30:305–310
Manavalan LP, Guttikonda SK, Tran L-SP, Nguyen HT (2009) Physiological and molecular approaches to improve drought resistance in soybean. Plant Cell Physiol 50:1260–1276
Matukumalli LK, Lawley CT, Schnabel RD, Taylor JF, Allan MF, Heaton MP, O'Connell J, Moore SS, Smith TPL, Sonstegard TS, van Tassell CP (2009) Development and characterization of a high density SNP genotyping assay for cattle. PLoS ONE 4:e5350
Matthews RB, Azam-Ali SN, Peacock JM (1990) Response of four sorghum lines to mid-season drought II Leaf characteristics. Field Crop Res 25(3–4):297–308
Maulana F, Weerasooriya D, Tesso T (2017) Sorghum landrace collections from cooler regions of the world exhibit magnificent genetic differentiation and early season cold tolerance. Front Plant Sci 8:756
McIntyre CL, Drenth J, Gonzalez N, Henzell RG and Jordan DR (2008) Molecular characterization of the waxy locus in sorghum. Genome 51:524–533
McWilliams D (2003) Drought strategies for corn and grain sorghum. Coop Ext Circ 580, New Mexico State Univ., Las Cruces, NM
Mehta PJ, Wiltse CC, Rooney WL, Collins SD, Frederiksen RA, Hess DE, Chisi M, TeBeest DO (2005) Classification and inheritance of genetic resistance to anthracnose in sorghum. Field Crops Res 93:1–9
Meuwissen TH, Hayes BJ, Goddard ME (2001) Prediction of total genetic value using genome-wide dense marker maps. Genetics 157:1819–1829
Michels GJ, Burd JD (2007) IPM case studies—Sorghum. In: van Emden HF, Harrington R (eds) Aphids as crop pests. CABI, Wallingford, United Kingdom, pp 627–638
Mickelbart MV, Hasegawa PM, Bailey-Serres J (2015) Genetic mechanisms of abiotic stress tolerance that translate to crop yield stability. NatReviews Genet 16(4):237
Miller FR, Prihoda KL, Rooney LW, Rosenow DT, Waniska RD (1996) Registration of a food quality sorghum restorer parent, Tx2907. Crop Sci 36:479–479
Mitra J (2001) Genetics and genetic improvement of drought resistance in crop plants. Curr Sci (Bangalore) 80:758–763
Morgan JM (1984) Osmoregulation and water stress in higher plants. Annu Rev Plant Physiol 35:299–319
Morgan PW, Finlayson SA, Childs KL, Mullet JE, Rooney WL (2002) Opportunities to improve adaptability and yield in grasses: lessons from sorghum. Crop Sci (42):1781–1799
Morris GP, Ramu P, Deshpande SP et al (2013) Population genomic and genome-wide association studies of agroclimatic traits in sorghum. Proc Natl Acad Sci USA 110:453–458
Moussa HR, Abdel-Aziz SM (2008) Comparative response of drought tolerant and drought sensitive maize genotypes to water stress. Aust J Crop Sci 1:31–36
Muchow RC (1988) Effect of nitrogen supply on the comparative productivity of maize and sorghum in a semi-arid tropical environment I Leaf growth and leaf nitrogen. Field Crops Res 18:1–16

Mulder HA, Calus MP, Druet T, Schrooten C (2012) Imputation of genotypes with low-density chips and its effect on reliability of direct genomic values in Dutch Holstein cattle. J Dairy Sci 95:876–889
Mundree SG, Baker B, Mowla S, Peters S, Marais S, Vander Willigen C, Govender K, Maredza A, Muyanga S, Farrant JM, Thomson JA (2002) Physiological and molecular insights into drought tolerance. African JJ Biotech 1(2):28–38
Munné-Bosch S, Jubany-Marí T, Alegre L (2001) Drought-induced senescence is characterized by a loss of antioxidant defences in chloroplasts. Plant Cell Environ 24(12):1319–1327
Murray SC, Sharma A, Rooney WL, Klein PE, Mullet JE, Mitchell SE, Kresovich S (2008) Genetic improvement of sorghum as a biofuel feedstock: I QTL for stem sugar and grain nonstructural carbohydrates. Crop Sci 48:2165–2179
Muthamilarasan M, Theriappan P and Prasad M (2013) Recent advances in crop genomics for ensuring food security. Curr Sci 105:155–158
Muthamilarasan M, Venkata Suresh B, Pandey G, Kumari K, Parida SK and Prasad M (2014) Development of 5123 intron-length polymorphic markers for large-scale genotyping applications in foxtail millet. DNA Res 21:41–52
Muthamilarasan M, Prasad M (2015) Advances in Setaria genomics for genetic improvement of cereals and bioenergy grasses. Theor Appl Genet 128:1–14
Nakaya A, Isobe SN (2012) Will genomic selection be a practical method for plant breeding. Ann Bot 110:1303–1316
Navarro L, Dunoyer P, Jay F, Arnold B, Dharmasiri N, Estelle M, Voinnet O (2006) A plant miRNA contributes to antibacterial resistance by repressing auxin signaling. Science 312 (5772):436–439
Nguyen HT, Babu RC, Blum A (1997) Breeding for drought resistance in rice: physiology and molecular genetics considerations. Crop Sci 37:1426–1434
Niangado O (2001) The state of millet diversity and its use in West Africa. Broadening Genet Base Crop Prod 31:147–157
Ogburn R, Edwards EJ (2010) The ecological water-use strategies of succulent plants. Adv Bot Res 55:179–225
Palumbi SR (1995) Nucleic Acids II: the Polymerase Chain Reaction. In: Hillis D, Moritz C (eds) Molecular Systematics 2nd edn. Sinauer, Sunderland, MA, pp 205–247
Pasini L, Bergonti M, Fracasso A, Marocco A, Amaducci S (2014) Microarray analysis of differentially expressed mRNAs and miRNAs in young leaves of sorghum under dry-down conditions. J Plant Physiol 171(7):537–548
Passioura J (1997) Drought and drought tolerance. Drought tolerance in higher plants: genetical, physiological and molecular biological analysis. Springer, Netherlands
Passioura J (2004) Increasing crop productivity when water is scarce—from breeding to field management. In: 'New directions for a diverse planet'. In: Proceedings of 4th international crop sciences congress, Brisbane, Australia
Paterson AH, Bowers JE, Bruggmann R, Dubchak I, Grimwood J, Gundlach H, Haberer G, Hellsten U, Mitros T, Poliakov A, Schmutz J (2009a) The Sorghum bicolor genome and the diversification of grasses. Nature 457(7229):551–556
Paterson AH, Bowers JE, Bruggmann R, Dubchak I, Grimwood J, Gundlach H, Haberer G, Hellsten U, Mitros T, Poliakov A, Schmutz J, Spannagl M, Tang H, Wang X, Wicker T, Bharti AK, Chapman J, Feltus FA, Gowik U, Grigoriev IV, Lyons E, Maher CA, Martis M, Narechania A, Otillar RP, Penning BW, Salamov AA, Wang Y, Zhang L, Carpita NC, Freeling M, Gingle AR, Hash CT, Keller B, Klein P, Kresovich S, McCann MC, Ming R, Peterson DG, Mehboob-ur-Rahman WD, Westhoff P, Mayer KF, Messing J, Rokhsar DS (2009b) The Sorghum bicolor genome and the diversification of grasses. Nature 457:551–556
Pedersen JF, Bean SR, Graybosch RA, Park SH, Tilley M (2005) Characterization of waxy grain sorghum lines in relation to granule-bound starch synthase. Euphytica 144:151–156
Pelleschi S, Rocher JP, Prioul JL (1997) Effect of water restriction on carbohydrate metabolism and photosynthesis in mature maize leaves. Plant Cell Environ 20:493–503

Perry SW, Krieg DR, Hutmacher RB (1983) Photosynthetic rate control in cotton. Plant Physiol 73:662–665

Pinheiro C, Chaves MM, Ricardo CP (2001) Alterations in carbon and nitrogen metabolism induced by water deficit in the stems and leaves of *Lupinus albus* L. J Exp Bot 52:1063–1070

Pinto RS, Reynolds MP, Mathews KL, McIntyre CL, Olivares-Villegas JJ, Chapman SC (2010) Heat and drought adaptive QTL in a wheat population designed to minimize confounding agronomic effects. Theor Appl Genet 121(6):1001–1021

Poland JA, Rife TW (2012) Genotyping-by-sequencing for plant breeding and genetics. Plant Genome J 5:92–102

Pontius JU, Wagner L, Schuler GD (2003) UniGene: a unified view of the transcriptome. PubMed Abstr Publ Full Text PubMed Cent Full Text OpenURL

Porter JR, Xie L, Challinor AJ, Cochrane K, Howden SM, Iqbal MM, Lobell DB, Travasso MI, Netra Chhetri NC, Garrett K (2014) Food security and food production systems. Geneva, Switzerland, IPCC

Prasad PV, Djanaguiraman M, Perumal R, Ciampitt IA (2015) Impact of high temperature stress on floret fertility and individual grain weight of grain sorghum: Sensitive stages and thresholds for temperature and duration. Front Plant Sci 6:820

Prasad PV, Pisipati SR, Mutava RN, Tuinstra MR (2008) Sensitivity of grain sorghum to high temperature stress during reproductive development. Crop Sci 48:1911–1917

Quantitative trait loci for grain quality, productivity, morphological and agronomical traits

Quinby JR (1974) Sorghum improvement and the genetics of growth. Texas A&M University Press, College Station

Ra K, Shiotsu F, Abe J, Morita S (2012) Biomass yield and nitrogen use efficiency of cellulosic energy crops for ethanol production. Biomass Bioenerg 37:330–334

Rajkumar, Fakrudin B, Kavil SP, Girma Y, Arun SS, Dadakhalandar D, Gurusiddesh BH, Patil AM, Thudi M, Bhairappanavar SB (2013) Molecular mapping of genomic regions harbouring QTLs for root and yield traits in sorghum (*Sorghum bicolor* L. Moench). Physiol Mol Biol Plants 19:409–419

Rakshit S, Hariprasanna K, Sunil SG, Ganapathy KN, Das IK, Ramana OV, Dhandapani A, Patil JV (2015) Changes in area, yield gains, and yield stability of sorghum in major sorghum-producing countries, 1970 to 2009. Crop Sci 54(4):1571–1584

Rami JF, Dufour P, Trouche G, Fliedel G, Mestres C, Davrieux F, Blanchard P, Hamon P (1998) Quantitative trait loci for grain quality, productivity, morphological and agronomical traits in sorghum (Sorghum bicolor L. Moench). Theor Appl Genet 97(4):605–616

Ramu P, Kassahun B, Senthilvel S, Ashok Kumar C, Jayashree B, Folkertsma RT, Ananda Reddy L, Kuruvinashetti MS, Haussmann BIG, Hash CT (2009) Exploiting rice-sorghum synteny for targeted development of EST-SSRs to enrich the sorghum genetic linkage map. Theor Appl Genet 119:1193–1204

Ray DK, Mueller ND, West PC, Foley JA (2013) Yield trends are insufficient to double global crop production by 2050 19;8(6):e66428

Rebetzke GJ, Van Herwaarden AF, Jenkins C, Weiss M, Lewis D, Ruuska S, Tabe L, Fettell NA, Richards RA (2008) Quantitative trait loci for water-soluble carbohydrates and associations with agronomic traits in wheat. Aust J Agric Res 59(10):891–905

Reddy PS, Fakrudin B, Rajkumar, Punnuri SM, Arun SS, Kuruvinashetti MS, Das IK, Seetharama N (2008) Molecular mapping of genomic regions harboring QTLs for stalk rot resistance in sorghum. Euphytica 159(1–2):191–198

Reddy NR, Ragimasalawada M, Sabbavarapu MM, Nadoor S, Patil JV (2014) Detection and validation of stay-green QTL in post-rainy sorghum involving widely adapted cultivar, M35-1 and a popular stay-green genotype B35. BMC Genom 15:909

Rehman AU, Malhotra RS, Bett K, Tar'an B, Bueckert R, Warkentin TD (2011) Mapping QTL associated with traits affecting grain yield in chickpea (Cicer arietinum L.) under terminal drought stress. Crop Sci 51(2):450–463

Reich VH, Atkins RE (1970) Yield stability of four population types of grain sorghum, Sorghum bicotot (L.) Moench, in different environments. Crop Sci 10:511–517

Rhodes DH, Hoffmann L, Rooney WL, Ramu P, Morris GP, Kresovich S (2014) Genome - wide association study of grain polyphenol concentrations in global sorghum [Sorghum bicolor (L.) Moench] germplasm. J Agric Food Chem 62:10916–10927

Rincent R, Laloë D, Nicolas S (2012) Maximizing the reliability of genomic selection by optimizing the calibration set of reference individuals: comparison of methods in two diverse groups of maize inbreds (Zea mays L.). Genetics 192:715–728

Rooney WL, Rooney LW, Awika J, Dykes L (2013) Registration of Tx3362 sorghum germplasm. J Plant Regist 7:104–107

Rosenow DT (1987) Breeding sorghum for drought resistance. In: Menyonga JM, Bezune T, Yuodeowei A (eds) Proceedings of the international drought symposium. AU/STRCSAFGRAD Coordination Office, Ouagadougou, Burkina Faso, pp 19–23

Rosenzweig C, Elliott J, Deryng D, Ruane AC, Müller C, Arneth A, Boote KJ, Folberth C, Glotter M, Khabarov N, Neumann K, Piontek TAMP, Schmid E, Stehfest H, Yang, Jones JW (2014) Assessing agricultural risks of climate change in the 21st century in a global gridded crop model intercomparison. Proc Natl Acad Sci 111(9):3268–3273

Sage RF, Christin PA, Edwards EA (2011) The lineages of C4 photosynthesis on planet earth. J Exp Bot 62:3155–3169

Sage RF, Zhu XG (2011) Exploiting the engine of C4 photosynthesis. J Exp Bot 62(9):2989–3000

Saini HS (1997) Effects of water stress on male gametophyte development in plants. Sex Plant Reprod 10:67–73

Salcedo A, Rutter W, Wang S, Akhunova A, Bolus S, Chao S, Anderson N, De Soto MF, Rouse M, Szabo L, Bowden RL (2017) Variation in the AvrSr35 gene determines Sr35 resistance against wheat stem rust race Ug99. Science 358(6370):1604–1606

Salem K, Röder M, Börner A (2007) Identification and mapping quantitative trait loci for stem reserve mobilisation in wheat (Triticum aestivum L.). Cereal Res Commun 35(3):1367–1374

Salih AA, Ali IA, Lux A, Luxova M, Cohen Y, Sugimoto Y, Inanaga S (1999) Rooting, water uptake, and xylem structure adaptation to drought of two sorghum cultivars. Crop Sci 39: 168–173

Sanchez AC, Subudhi PK, Rosenow DT, Nguyen HT (2002) Mapping QTLs associated with drought resistance in sorghum (Sorghum bicolor L. Moench). Plant Mol Biol 48:713–726

Sandeep VM, Rao VU, Rao BB, Bharathi G, Pramod VP, Chowdary PS, Patel NR, Mukesh P, Kumar PV (2018) Impact of climate change on sorghum productivity in India and its adaptation strategies. J. Agrometeorol 20(2):89–96

Sandeep VM, Bapuji Rao B, Bharati G, Rao VUM, Pramod VP, Chowdary PS, Patel NR, Vijayakumar P (2017) Projected future change in water requirement of grain sorghum in India. J Agrometeorol 19(3):217–225

Sawidis T, Kalyva S, Delivopoulos S (2005) The root-tuber anatomy of Asphodelus aestivus. Flora Morphol Distrib Funct Ecol Plants 200:332–338

Scandalios JG (2005) Oxidative stress: molecular perception and transduction of signals triggering antioxidant gene defenses. Braz J Med Biol Res 38:995–1014

Schloss SJ, Mitchell SE, White GM, Kukatla R, Bowers JE, Paterson AH, Kresovich S (2002) Characterization of RFLP probe sequences for gene discovery and SSR development in Sorghum bicolor (L.) Moench. Theor Appl Gen 105:912–920

Serraj R, Kumar A, McNally KL, Slamet-Loedin I, Bruskiewich R, Mauleon R et al (2009) Improvement of drought resistance in rice. Adv Agron 103:41–99

Shakoor N, Nair R, Crasta O, Morris G, Feltus A, Kresovich S (2014) A Sorghum bicolor expression atlas reveals dynamic genotype-specific expression profiles for vegetative tissues of grain, sweet and bioenergy sorghums. BMC Plant Biol 14:35–43

Shakoor N, Ziegler G, Dilkes BP, Brenton Z, Boyles R, Connolly EL, Kresovich S, Baxter I (2016) Integration of experiments across diverse environments identifies the genetic determinants of variation in Sorghum bicolor seed element composition. Plant Physiol 170:1989–1998

Shangguan Z, Shao M, Dyckmans J (1999) Interaction of osmotic adjustment and photosynthesis in winter wheat under soil drought. J Plant Physiol 154:753–758

Shinozaki K, Yamaguchi-Shinozaki K (1997) Gene expression and signal transduction in water-stress response. Plant Physiol 115:327–334

Shinozaki K, Yamaguchi-Shinozaki K (2000) Molecular responses to dehydration and low temperature: difference and cross-talk between two stress signaling pathways. Curr Opin Plant Biol 3:217–223

Simpson GM (1981) Water stress in plants. Praeger, N

Singh V, Nguyen CT, van Oosterom EJ, Chapman SC, Jordan DR, Hammer GL (2015) Sorghum genotypes differ in high temperature responses for seed set. Field Crops Research 171:32–40

Singh V, Nguyen CT, Yang Z, Chapman SC, van Oosterom EJ, Hammer GL (2016) Genotypic differences in effects of short episodes of high - temperature stress during reproductive development in sorghum. Crop Sci 56:1561–1572

Smirnoff N, Colombe SV (1988) Drought influences the activity of enzymes of the chloroplast hydrogen-peroxide scavenging system. J Exp Bot 39:1097–1108

Srinivas GK, Satish R, Madhusudhana RN, Reddy SM, Mohan and Seetharama N (2009) Identification of quantitative trait loci for agronomically important traits and their association with genic-microsatellite markers in sorghum. Theor Appl Genet 118:1439–1454

Srivastava A, Kumar SN, Aggarwal PK (2010) Assessment on vulnerability of sorghum to climate change in India. Agr Ecosyst Environ 138(3–4):160–169

Staggenborg SA, Dhuyvetter KC, Gordon WB (2008) Grain sorghum and corn comparisons: yield, economic, and environmental responses. Agron J 100(6):1600–1604

Starks KJ, Burton RL (1979) Damage to grain sorghum by fall armyworm and corn earworm. J Econ Entomol 72:576–578

Stone LR, Schlegel AJ (2006) Yield-water supply relationship of grain sorghum and winter wheat. Agron J 98:1359–1366

Stout DG and Simpson GM (1978) Drought resistance of Sorghum bicolor. 1. Drought avoidance mechanisms related to leaf water status. Can J Plant Sci 58(1):213–224

Sunkar R, Chinnusamy V, Zhu J, Zhu JK (2007) Small RNAs as big players in plant abiotic stress responses and nutrient deprivation. Trends Plant Sci 12(7):301–309

Tao YZ, Henzell RG, Jordan DR, Butler DG, Kelly AM, McIntyre CL (2000) Identification of genomic regions associated with stay green in sorghum by testing RILs in multiple environments. Theor Appl Genet 100:1225–1232

Taramino G, Tarchini R, Ferrario S, Lee M, Pe ME (1997) Characterization and mapping of simple sequence repeats (SSRs) in Sorghum bicolor. Theor Appl Genet 95:66–72

Tardieu F (2013) Plant response to environmental conditions: assessing potential production, water demand, and negative effects of water deficit. Front Physiol 4:17

Technow F, Bürger A, Melchinger AE (2013) Genomic prediction of northern corn leaf blight resistance in maize with combined or separated training sets for heterotic groups. G3 (Bethesda) 3:197–203

Terbea M, Vranceanu AV, Petcu E, Craiciu DS, Micut G (1995) Physiological response of sunflower plants to drought. Rom Agric Res 3:61–67

Tesso T, Ejeta G, Chandrashekar A, Huang C-P, Tandjung A, Lewamy M, Axtell JD, Hamaker BR (2006) A novel modified endosperm texture in a mutant high—protein digestibility/ high - lysine grain sorghum (Sorghum bicolor (L.) Moench). Cereal Chem J 83:194–201

Thomas H, Howarth CJ (2000) Five ways to stay green. J Exp Bot 51:329–337

Toumi I, Gargouri M, Nouairi I, Moschou PN, Salem-Fnayou AB, Mliki A, Zarrouk M, Ghorbel A (2008) Water stress induced changes in the leaf lipid composition of four grapevine genotypes with different drought tolerance. Biol Plant 52(1):161–164

Tuberosa R (2012) Phenotyping for drought tolerance of crops in the genomics era. Front Physiol 3:347

Tuinstra MR, Ejeta G, Goldsbrough P (1998) Evaluation of near-isogenic sorghum lines contrasting for QTL markers associated with drought tolerance. Crop Sci 38:835–842

Tuinstra MR, Grote EM, Goldsbrough PB, Ejeta G (1996) Identification of quantitative trait loci associated with pre-flowering drought tolerance in sorghum. Crop Sci 36:1337–1344

Tuinstra MR, Grote EM, Goldsbrough PB, Ejeta G (1997) Genetic analysis of post-flowering drought tolerance and components of grain development of Sorghum bicolor (L.) Moench. Mol Breed 3:439–448

Turner NC (1986) Adaptation to water deficits: a changing perspective. Aust J Plant Physiol 13:175–190

Uga Y, Sugimoto K, Ogawa S, Rane J, Ishitani M, Hara N, et al (2013) Control of root system architecture by deeper rooting 1 increases rice yield under drought conditions. Nat Genet 45:1097–1102

Vadez V, Deshpande SP, Kholova J, Hammer GL, Borrell AK, Talwar HS, Hash CT (2011) Stay-green quantitative trait loci's effects on water extraction, transpiration efficiency and seed yield depend on recipient parent background. Funct Plant Biol 38:553–566

van der Weerd L, Claessens MM, Ruttink T, Vergeldt FJ, Schaafsma TJ, Van As H (2001) Quantitative NMR microscopy of osmotic stress responses in maize and pearl millet. J Exp Bot 52(365):2333–2343

van Oosterom EJ, Whitaker ML, Weltzien RE (1996) Integrating genotype by environment analysis, characterisation of drought patterns, and farmer preferences to identify adaptive plant traits for pearl millet. In: Adaptation Plant, Improvement Crop (eds) m. Cooper and G.L. Hammer), pp 383–402

van Oosterom EJ, Borrell AK, Deifel KS, Hammer GL (2011) Does increased leaf appearance rate enhance adaptation to postanthesis drought stress in sorghum? Crop Science 51:2728– 2740

Varshney RK, Graner A and Sorrells ME (2005) Genomics-assisted breeding for crop improvement. Trends Plant Sci 10:621–630

Varshney RK, Terauchi R and McCouch SR (2014) Harvesting the promising fruits of genomics: applying genome sequencing technologies to crop breeding. PLoS Biol. 12:e1001883. https://doi.org/10.1371/journal.pbio.1001883

Vinita J, Bhargava S, Streb P, Feierabend J (1998) Comparative effect of water, heat and light stresses on photosynthetic reactions in Sorghum bicolor (L.) Moench. J Exp Botany 49:327:1715–1721

Walulu RS, Rosenow DT, Wester DB, Nguyen HT (1994) Inheritance of the stay-green trait in sorghum. Crop Sci 34:970–972

Wang J, Vanga SK, Saxena R, Orsat V, Raghavan V (2018) Effect of climate change on the yield of cereal crops: a review. Climate 6(2):41

Wang S, Wong D, Forrest K, Allen A, Chao S, Huang BE, Maccaferri M, Salvi S, Milner SG, Cattivelli L, Mastrangelo AM, Whan A, Stephen S, Barker G, Wieseke R, Plieske J, International Wheat Genome Sequencing Consortium, Lillemo M, Mather D, Appels R, Dolferus R, Brown-Guedira G, Korol A, Akhunova AR, Feuillet C, Salse J, Morgante M, Pozniak C, Luo MC, Dvorak J, Morell M, Dubcovsky J, Ganal M, Tuberosa R, Lawley C, Mikoulitch I, Cavanagh C, Edwards KJ, Hayden M, Akhunov E (2014) Characterization of polyploid wheat genomic diversity using a high-density 90000 single nucleotide polymorphism array. Plant Biotechnol J 12:787–796

Wasson AP, Richards RA, Chatrath R, Misra SC, Prasad SV, Rebetzke GJ et al (2012) Traits and selection strategies to improve root systems and water uptake in water-limited wheat crops. J Exp Bot 63:3485–3498

Weigel KA, de Los Campos G, Vazquez AI, Rosa GJM, Gianola D, Van Tassel CP (2010) Accuracy of direct genomic values derived from imputed single nucleotide polymorphism genotypes in Jersey cattle. J Dairy Sci 93:5423–5435

Wright GC, Smit RC, McWilliams JR (1983) Differences between two grain sorghum genotypes in adaptation to drought stress I crop growth and yield responses. Aust J Agric Res 34:615–626

Wu Y, Li X, Xiang W et al (2012) Presence of tannins in sorghum grains is conditioned by different natural alleles of Tannin1. Proc Natl Acad Sci USA 109 (26):10281–10286

Xin Z, Aiken R, Burke J (2008) Genetic diversity of transpiration efficiency in sorghum. Field Crops Res 111:74–80

Xiong L, Wang RG, Mao G, Koczan JM (2006) Identification of drought tolerance determinants by genetic analysis of root response to drought stress and abscisic acid. Plant Physiol 142:1065–1074
Xu W, Subudhi PK, Crasta OR, Rosenow DT, Mullet JE, Nguyen HT (2000) Molecular mapping of QTLs conferring stay-green in grain sorghum (Sorghum bicolor L. Moench) Genome 43 (3):461–469
Xu Y, Lu Y, Xie C, Gao S, Wan J, Prasanna BM (2012) Whole genome strategies for marker-assisted plant breeding. Mol Breed 29:833–854
Yamaguchi-Shinozaki K, Kasuga M, Liu Q, Nakashima K, Sakuma Y, Abe H, Shinwari ZK, Seki M, Shinozaki K (2002) Biological mechanisms of drought stress response. JIRCAS working report, pp 1–8
Yang J, Zhang J, Liu K, Wang Z, Liu L (2007) Abscisic acid and ethylene interact in rice spikelets in response to water stress during meiosis. J Plant Growth Regul 26:318–328
Yang J, Zhang J, Wang Z, Zhu Q, Wang W (2001) Remobilization of carbon reserves in response to water-deficit during grain filling of rice. Field Crops Res 71:47–55
Yang L, Browning JD, Awika JM (2009) Sorghum 3-deoxyanthocyanins possess strong phase II enzyme inducer activity and cancer cell growth inhibition properties. J Agric Food Chem 57:1797–1804
Yao F, Huang J, Cui K, Nie L, Xiang J, Liu X, Wu W, Chen M, Peng S (2012) Agronomic performance of high-yielding rice variety grown under alternate wetting and drying irrigation. Field crops research 126:16–22
Yazawa T, Kawahigashi H, Matsumoto T, Mizuno H (2013) Simultaneous transcriptome analysis of Sorghum and Bipolaris sorghicola by using RNA-seq in combination with de novo transcriptome assembly. PloSone 8(4):e62460
Younis ME, El-Shahaby OA, Abo-Hamed SA, Ibrahim AH (2000) Effect of water stress on growth, pigments and CO Assimilation in three Sorghum Cultivars. J Agron Crop Sci 185: 73–82
Yu J, Tuinstra M, Claassen M, Gordon W, Witt M (2004) Analysis of cold tolerance in sorghum under controlled environment conditions. Field Crop Res 85:21–30
Yu H, Xie W, Li J, Zhou F, Zhang Q (2013) A whole-genome SNP array (RICE6K) for genomic breeding in rice. Plant Biotechnol J 12:28–37
Yu X, Li X, Guo T, Zhu C, Wu Y, Mitchell SE, Roozeboom KL, Wang D, Wang ML, Pederson GA, Tesso TT, Schnable PS, Bernardo R, Yu J (2016) Global strategy to turbo charge gene banks. Nat Plants 2(10):1–7
Zhang G, Liu X, Quan Z, Cheng S, Xu X, Pan S, Xie M, Zeng P, Yue Z, Wang W, Tao Y (2012) Genome sequence of foxtail millet (Setaria italica) provides insights into grass evolution and biofuel potential. Nat Biotechnol 30(6):549–554
Zhang J, Kirkham B (1996) Antioxidant responses to drought in sunflower and sorghum seedlings. New Phytol 132:361–373
Zhang GH, Xu Q, Zhu XD, Qian Q, Hong WX (2009) Shallot-like1 Is a kanadi transcription factor that modulates rice leaf rolling by regulating leaf abaxial cell development. Plant Cell 21: 719–735
Zhu F (2014) Structure, physicochemical properties, modifications, and uses of sorghum starch: sorghum starch review. Compr Rev Food Sci Food Saf 13:597–610

Chapter 6
Genomic Designing of Pearl Millet: A Resilient Crop for Arid and Semi-arid Environments

Desalegn D. Serba, Rattan S. Yadav, Rajeev K. Varshney, S. K. Gupta, Govindaraj Mahalingam, Rakesh K. Srivastava, Rajeev Gupta, Ramasamy Perumal and Tesfaye T. Tesso

Abstract Pearl millet [*Pennisetum glaucum* (L.) R. Br.; Syn. *Cenchrus americanus* (L.) Morrone] is the sixth most important cereal in the world. Today, pearl millet is grown on more than 30 million ha mainly in West and Central Africa and the Indian sub-continent as a staple food for more than 90 million people in agriculturally marginal areas. It is rich in proteins and minerals and has numerous health benefits such as being gluten-free and having slow-digesting starch. It is grown as a forage crop in temperate areas. It is drought and heat tolerant, and a climate-smart crop that can withstand unpredictable variability in climate. However, research on pearl millet improvement is lagging behind other major cereals mainly due to limited investment in terms of man and money power. So far breeding achievements include the development of cytoplasmic male sterility (CMS), maintenance counterparts (*rf*) system and nuclear fertility restoration genes (*Rf*) for hybrid breeding, dwarfing genes for reduced height, improved input responsiveness, photoperiod neutrality for short growing season, and resistance to important diseases. Further improvement of pearl millet for genetic yield potential, stress tolerance, and nutritional quality traits would enhance food and nutrition security for people living in agriculturally dissolute environments. Application of molecular technology in the pearl millet breeding program has a promise in enhancing the selection efficiency while shortening the lengthy phenotypic selection process

D. D. Serba (✉) · R. Perumal
Agricultural Research Center-Hays, Kansas State University, Hays, KS, USA
e-mail: ddserba@ksu.edu

R. S. Yadav
Institute of Biological, Environmental and Rural Sciences, Aberystwyth University, Aberystwyth, UK

R. K. Varshney · S. K. Gupta · G. Mahalingam · R. K. Srivastava · R. Gupta
International Crops Research Institute for the Semi-Arid Tropics (ICRISAT), Hyderabad, Telangana, India

T. T. Tesso
Department of Agronomy, Kansas State University, Manhattan, KS, USA

C. Kole (ed.), *Genomic Designing of Climate-Smart Cereal Crops*,
https://doi.org/10.1007/978-3-319-93381-8_6

ultimately improving the rate of genetic gains. Linkage analysis and genome-wide association studies based on different marker systems in detecting quantitative trait loci (QTLs) for important agronomic traits are well demonstrated. Genetic resources including wild relatives have been categorized into primary, secondary and tertiary gene pools based on the level of genetic barriers and ease of gene introgression into pearl millet. A draft on pearl millet whole genome sequence was recently published with an estimated 38,579 genes annotated to establish genomic-assisted breeding. Resequencing a large number of germplasm lines and several population genomic studies provided a valuable insight into population structure, genetic diversity and domestication history of the crop. Successful improvement in combination with modern genomic/genetic resources, tools and technologies and adoption of pearl millet will not only improve the resilience of global food system through on-farm diversification but also dietary intake which depends on diminishingly fewer crops.

Keywords Biofortification · Climate resilient · Cytoplasmic male sterility · Dwarfing gene · Gene pool · Genomic-assisted breeding · *Pennisetum glaucum*

6.1 Introduction

The world population is predicted to be more than 9.6 billion by the middle of twenty-first century with the highest increase projected in Sub-Saharan Africa (SSA) from the current 0.9 billion to 2.2 billion by 2050 (prb.org; fao.org). On the contrary, in the last five decades, the SSA region has recorded only an ~25% increase in cereals yield per unit land as compared to >300% in the developed countries. There could be multiple reasons for the large yield-gap but one of the major reasons is slow or no adoption of modern tools and technologies. The most suboptimal crop production in arid and semi-arid environments covering >40% of the world's land area is characterized by dry and hot weather conditions (Safriel et al. 2005). Food production in such dissolute environments is challenged by low rainfall with erratic distribution and high temperature. The climatic variability experienced in the last few decades is projected to hit the arid and semi-arid regions hard to exacerbate the biophysical and socio-economic stresses. The looming global climate change will further aggravate the drought and high temperature stresses and will adversely impact crop production.

The combined effect of dwindling water resources and competition for water among industries limit the availability of water for irrigation in dry areas. Moreover, the available water for irrigation is preferably used for fruits and vegetables production than grain crops. Therefore, the next productivity increase in arid and semiarid environments is expected to come largely from crop resilience to the stresses imposed by climatic variability.

Pearl millet (*Pennisetum glaucum* (L.) R. Br. syn. *Cenchrus americanus* L.) is one of the climate resilient crops having C_4 photosynthetic pathway which is very efficient in energy production (50% higher photosynthesis efficiency than C_3 crops) in hot and dry climate (Wang et al. 2012) and a widely grown cereal crop for food in SSA and the Indian subcontinent since time immemorial. Pearl millet is believed to be domesticated in West Africa about 4000–5000 years ago, probably in the southern margins of the present day Sahara Desert (Manning et al. 2011; Burgarella et al. 2018). Subsequently spread to eastern and southern Africa and introduced to India pearl millet became one of the staple crops. Today, its main center of diversity appears in the Sahel between Senegal and central Sudan (Brunken et al. 1977). Its high photosynthetic efficiency and rapid biomass production potential in harsh climatic conditions make the crop compatible for low soil moisture and soil fertility, and high temperature.

Pearl millet is thought to have originated in the Sahel region from Senegal to central Sudan (Oumar et al. 2008; Burgarella et al. 2018), and it is largely cultivated in West Africa and Asia. The archaeobotanical evidences indicate pearl millet domestication dating back at about 4500 BP (Manning et al. 2011), support the Sahara and Sahel hypothesis of origin and the broad distribution and cultivation in SSA (D'Andrea and Casey 2002). The wild progenitor of cultivated pearl millet has been identified as *Pennisetum glaucum* ssp. *monodii*, is native to the Sahel zone (Brunken 1977). It has widely been suggested that domestication of pearl millet is the result of multiple events (Marchais and Pernes 1985) leading to various morphological, agronomic and physiological modifications for use by the farmers and consumers. Differential selection pressures over centuries modified the crop to suit the needs of growers and consumers.

Pearl millet accounts for more than half of the total millet production in the world (FAOSTAT 2013). It is grown in the most dissolute environment where other major cereals, such as wheat (*Triticum aestivum*), maize (*Zea mays*), and rice (*Oryza sativa*), are likely to fail or produce no economic yield. Today, pearl millet is grown on more than 30 million ha in West Africa and the Indian subcontinent and provides basic sustenance to more than 90 million people in agriculturally marginal areas. Pearl millet has high nutritional value in terms of high levels of energy, dietary fiber, proteins with a balanced amino acid profile, essential minerals, vitamins, and antioxidants (Jukanti et al. 2016). Nevertheless, genetic improvement of pearl millet lags far behind other cereals. This is mainly because of limited resource allocation both from the public and private sector.

Pearl millet is also grown in the United States, Australia, and Brazil predominantly for forage and as a cover crop. Pearl millet was introduced into the United States in early date, grown and consumed by the early settlers until 1875 (Oelke et al. 1990), but with the improvement of other crops, its role for human consumption diminished. It is currently grown in small acreage primarily as a summer grazing forage crop. Until recently, pearl millet grain is widely known as a bird feed. However, with the recognition of its nutritional value, pearl millet is gaining importance slowly. There is a slight increase in productivity of millet since the year 2000, mainly in India as a result of adoption of high yielding hybrids (Fig. 6.1). However, the area has been decreasing over years probably because of an expansion of improved adaptive sorghum varieties into millet areas.

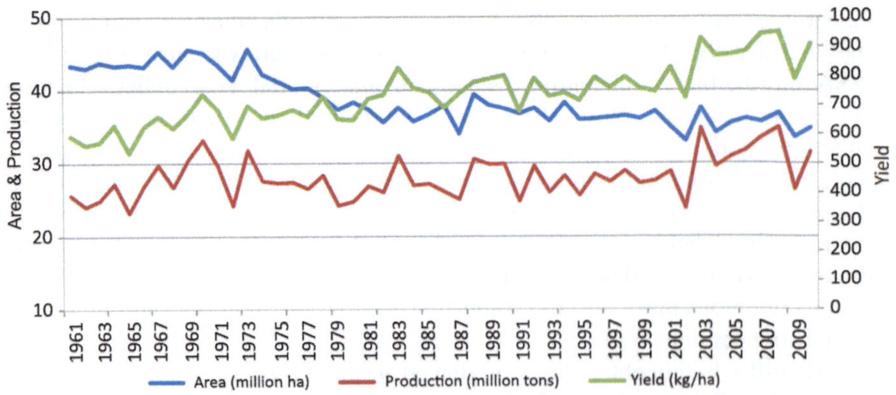

Fig. 6.1 Global trend in millets area, production, and productivity during 1961–2010 (FAO 2013)

Research on pearl millet improvement in the United States started back in 1936 (Burton and Powell 1968). Since then several breakthrough results have been recorded. The development of cytoplasmic male sterility (CMS) system for hybrid breeding, identification of dwarfing genes for reducing the plant height, development of early maturing photo-neutral genotypes, identification of resistance sources to major diseases such as downy mildew, rust and leaf spot, collection and preservation of diverse germplasm at ICRISAT, construction of comprehensive genetic linkage maps, detection of quantitative trait loci (QTL) for important traits, and the recent release of the genome sequence and re-sequencing based domestication are the important milestones.

To bring a quantum jump in the productivity, there are still big challenges to focus on, especially on improving the genetic yield potential, stress tolerance and seed characteristics for good stand establishment. Its importance as a subsistence crop in harsh environments and potential usefulness for large-scale production in the future, should have drawn much research attention as the yield harvested by farmers is still low. Disease pressure especially downy mildew is forcing most of the farmers in West Africa to stick to the traditional landraces rather than adopting genetically improved cultivars. Comprehensive research program geared towards addressing the priority production constraints is of paramount importance. This chapter portrays the development of milestones that have been made and the future focus on pearl millet improvement as a climate resilient crop in an abridged form.

6.1.1 Food, Nutrition, Energy and Environmental Security

Pearl millet is used to make a number of traditional foods and beverages across the Sahel of Africa and India. The unique characteristics of this cereal are that it can be as creamy as mashed potatoes or as fluffy as rice. In West Africa and the Indian subcontinent, pearl millet is regarded as one of the major sources of dietary energy and nutritional security. The crude protein content of pearl millet ranges from 9 to 18% with wide variation among genotypes and locations. Investigation into the nature of *Pennisetum* protein showed that only 6.8% of the total nitrogen is of the non-protein type (Swaminathan 1937). The embryo contains about 10% of the total protein, which is seven times as rich in protein as the endosperm on equal weight basis (Swaminathan et al. 1971). It also contains several essential minerals like phosphorus (P), iron (Fe), zinc (Zn), calcium (Ca), and magnesium (Mg). The essential amino acids and vitamins contributed to its therapeutic properties such as treatment of stomach ulcers by reducing excess acidity in the stomach after food intake. The high phosphorus content helps in bone growth and development. Because of its gluten-free nature, pearl millet is recommended as an alternative food for people with gluten allergy. The grain is also high in fiber content and the starch digests slowly. The slow digestion releases glucose into the blood stream at a slower rate than other cereals, and remarkably powerful for controlling diabetes (Kam et al. 2016).

Pearl millet is grown in the areas where it is too hot and too dry for other major cereals. Its production also involves minimum external input. The small-scale farmers in south Asia and SSA use minimum fertilizer and application of agro-chemicals in the form of herbicides, insecticides or fungicides is almost none. As a result it has significant economic and social significance with minimum impact on the environment. These undoubtedly ensure the sustainability of the production system.

6.1.2 Global Warming and Climate Change

Climate change (CC) is continuing as a major threat to sustainable agricultural production by embracing unpredictable extreme climatic events such as fluctuation in temperature and uneven rainfall patterns. This climate variability is expected to increase crop vulnerability in different agro-ecologies with drastic consequences on food security and economic growth. Pearl millet is a climate-smart crop with nutritious grains (Anuradha et al. 2017) and ideal for environments prone to drought and extreme heat, thereby reinforcing the fight against food insecurity in the arid and semiarid environments (Bailey et al. 1979; Buerkert et al. 2001; Jukanti et al. 2016).

A study on the effect of climate change on C_4 crop productivity in Africa and India indicated apparent temperature-driven yield reduction across the board with much more uncertainty in arid regions (Berg et al. 2013). Yield losses are induced by higher temperature leading to increased potential evapotranspiration, crop maintenance respiration and acceleration of the phenological cycle (Berg et al. 2013; Sultan et al. 2013) and impact pearl millet yield potential. Therefore, crop adaptation and improved agricultural practices would have a potential contribution in offsetting some of these negative impacts.

In the Sudanian and Sahelian savannas of West Africa, a study was conducted on the impact of climate change on pearl millet productivity (Sultan et al. 2013). This process-based crop model, calibrated and validated over multi-year field trials and surveys at eight contrasting sites in terms of climate and agricultural practices. Simulations under 35 future possible climate scenarios combining −20 to 20% precipitation anomalies and 0 to +6 °C temperature increases, indicated that most of the scenarios have a negative impact on yield than those recorded in the recent past (Fig. 6.2). This study implied up to 41% yield reduction could be incurred with 6 ° C increase in temperature and a 20% reduction in rainfall. The simulation study showed the photoperiod-sensitive pearl millet traditional cultivars seems more resilient to future climate conditions than improved cultivars with high genetic yield potential. Its fast growth rate as a result of high radiation use efficiency and large leaf area index result in high potential yield. The water saving, drought tolerance and climate change compliance of pearl millet make it a viable choice for the looming climate variability that will potentially affect crop production. In view of these circumstances, pearl millet cultivation is to be reclaimed by recognizing production options in context to anticipated climate change scenarios.

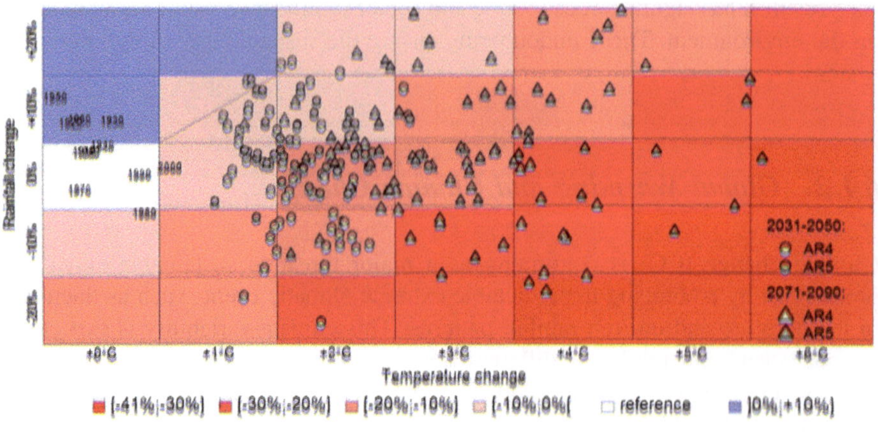

Fig. 6.2 Effects of rainfall and temperature changes on mean yield changes in pearl millet and sorghum relative to the 1961–90 baseline for 7 temperatures (*x*-axis) and 5 rainfall (*y*-axis) scenarios (Sultan et al. 2013). Results are shown as the average over the 35 stations across West Africa and the 6 cultivars of millet and sorghum. White triangles and circles are the projected anomalies computed by several CMIP3 (Climate Research Program's Coupled Model Inter-comparison Project phase 3 multi-model dataset) GCMs (Global Climate Model) and three Intergovernmental Panel on Climate Change (IPCC) assessment reports (AR) emission scenarios (B1, A1B, A2) for 2071–90 and 2031–50, respectively. Projections from CMIP5 GCMs and three Representative Concentration Pathways (RCPs) (4.5, 6.0 and 8.5) are represented by grey triangles and circles. Past observed climate anomalies from Climatic Research Unit (CRU) data are also projected by computing 10-year averages (e.g. '1940' is for 1941–50). All mean yield changes are significant at a 5% level except boxes with a diagonal line (Adopted from Sultan et al. 2013; Environ Res Lett 8:014040)

6.1.3 *Limitations of Traditional Breeding*

Successful improvement of pearl millet for genetic yield potential, tolerance to biotic and abiotic stresses, and grain quality characteristics would mend food security for millions of people depending on the crop as a staple grain. However, the development of high-yielding cultivars, drought tolerance, downy mildew resistance, and enhanced grain content is much below the demand. This is because of the lengthy conventional hybridization method that takes two fundamental steps that are lengthy in process. The first step involves generation of a breeding population that is highly variable for traits of interest. This is accomplished by identifying two or more parents having complementary traits and cross-pollinating the parents to initiate recombination. The second and longer fundamental step in conventional breeding involves raising several generations of segregating populations and selection of individual plants that combine useful traits of the parents among the population. These steps essentially follow the normal mating process with breeders' intervention with directed mating and selection of plants that suits human needs.

Marker-assisted breeding by and large has gained evident importance to address these long-standing challenges of plant breeding by enhancing the selection efficiency and cutting the lengthy phenotypic selection processes. Nevertheless, the application of molecular markers covers only very special aspects of plant breeding and will not replace the conventional breeding methods completely. Therefore, a systematic complementation of the conventional methods with the advanced genomic techniques is encouraged to lead to more efficient and expedited breeding procedures. The techniques such as 'speed breeding' and RGT (rapid generation turn over) in combination with molecular markers could help to reduce the overall time required in conventional breeding (Ochatt et al. 2002; Watson et al. 2018). The other game changer could be the use of DH (doubled haploid) lines to expedite the homozygosity by reducing the 6–8 generations to 2–3 generation thus saving 5–6 years.

6.2 Priority Target Traits

The relative importance of traits may vary across production conditions and the intended use of a crop. As pearl millet is grown for grain as well as forage use, it is commanding to target the traits exclusively for both purposes. In general, two evolutionary steps considerably modified plant phenotype during plant domestication process. The first step being the change in characteristic traits of domestic plants that have occurred under the conditions of primitive cultivation of wild populations (Hillman and Davis 1990). These groups of specific characters in cultivated plants are generally shared by all the components of the domestic gene pool of the same species and defined as "domestication syndrome" (Hammer 1984).

The second step, takes in the diversification of the domestic gene pool as a byproduct of local adaptations to new environments and to the needs and tastes of diverse mankind. The morphological and physiological diversity observed in the domestic gene pool, which largely exceeds what is usually observed in their wild counterparts, is the product of this evolutionary process (Lakis et al. 2012). For instance, population-based resequencing of cultivated barley accessions reveals that flowering time was the main target of this second step of the domestication process to adapt to new environmental conditions encountered during the expansion of the cultivation areas (Jones et al. 2008). Climatic variations dictate the selection for early or late maturing genotypes. The target traits that play important role in adaptation, yield performance, and stress tolerance in pearl millet are discussed hereunder.

6.2.1 Photoperiodism

Days to flowering is one of the most important plant characters that influence the cropping pattern in relation to the natural environment (Takei and Sakamoto 1987). Besides being an adaptive trait, flowering time and photoperiodic sensitivity are influencing yield and yield stability of pearl millet in arid and semi-arid regions of West and Central Africa (WCA). Pearl millet is generally a photoperiod sensitive crop and almost all landraces flower in short-day condition (Dave 1987). As photoperiodism affects the growth and development, it allows for flexible sowing dates by maintaining flowering and grain maturity at the end of the growing season. For instance, pearl millet varieties grown in the Sahel can mature in less than 90 days. In relation with the short duration of the rainy season in the Sahel, photoperiodism has an advantage in such hot and dry condition where the onset of rain varies from year to year.

In temperate regions, the minimum temperature during the light period of 12 h and less are mostly below optimum for pearl millet growth and development. This low suboptimal temperature for the crop can potentially extend the number of days from floral initiation to anthesis and slow down grain maturity (Burton 1965) which exposes the crop to likely frost in the autumn. Therefore, pearl millet breeders are intended to develop early maturing day-neutral varieties. So, the day-neutral cultivars can flower in the long-day condition of the summer by accumulating growing degree days and mature earlier before the onset of frost.

Selection in a flowering time pathway during domestication of pearl millet showed that genes for the trait underwent selection more frequently than expected (Clotault et al. 2012). Significant signatures of selection were found in six pearl millet flowering time genes and higher deviations from neutrality for circadian clock-associated genes, indicating that one category of genes of the flowering pathway were preferentially selected during pearl millet domestication. Crossing a late-maturing with an early maturing genotype revealed that photoperiodism was controlled by several genes with additive and minimal dominance effects (Burton 1965).

A study conducted on geo-referenced ICRISAT collection from a wide range of latitudes revealed that 45.6% of the accessions are photoperiod-sensitive (Upadhyaya et al. 2012). Generally, the late-maturing genotypes are short-day types, whereas the improved early-flowering ones are day-neutral (Burton 1951). Long-day delays the genetic tendency to flower by forcing the plant to wait for a specific signal (Dingkuhn et al. 2006). Latitude of origin of pearl millet genotypes affects photoperiod and day-length responses mainly in West Africa (Sanon et al. 2014). The critical photoperiod and temperature required to trigger flowering is cultivar-specific. As a result, some cultivars flower later than others due to difference in the level of photoperiodism.

Pearl millet varieties widely vary in growth cycle length (early: 45 days to late: 140 days) (Rao et al. 1985). The maturity duration of the cultivars grown in WCA can broadly be classified into three groups: (1) early types (flower in 45 to 70 days) are mostly facultative short-day and grown in the northern dry regions of the Sahel; (2) intermediate types (flower in 70 to 100 days) and (3) late landraces (flower in 100 to 140 days) are more abundant in the wetter southern regions and strictly sensitive to day length (absolute short-day plants). The world collection of pearl millet germplasm assembled at the ICRISAT gene bank, Patancheru, India, is from a wide range of latitudes and includes the typical temperature and photoperiod sensitive and insensitive accessions (Upadhyaya et al. 2012).

Selection during domestication left some regions of the genome depleted of diversity as compared to the wild genome, indicating potential locations for genes associated with domestication syndrome. Three flowering candidate genes, *PgHd3a*, *PgDwarf8*, and *PgPHYC* were cloned and studied for nucleotide diversity in the wild and domesticated pearl millet population (Sehgal et al. 2012). The sequence analysis revealed that the domesticated population has 84% nucleotide diversity found in the wild population which is attributed to gene flow between wild relatives and domesticated pearl millets. A positive selection was evidenced that *PgHd3a* and *PgDwarf8* were likely targeted by selection during domestication. Other genes that experienced the most unusual diversity loss, their putative functions are associated with regulation of response to hormones such as auxin and ethylene, regulation of the circadian clock and morphogenesis, along with transcription factors (Varshney et al. 2017).

6.2.2 Root System

The root system of vascular plants plays the key roles in acquiring water and mineral nutrients required for the survival as well as accumulation of metabolites for yield and nutritional quality. Moreover, the roots play a significant role in holding soil in place, carbon sequestration, reducing emissions of greenhouse gasses, and prevent the eutrophication of water bodies associated with the application of mineral fertilizers (White et al. 2013).

Sustainable crop production requires root systems optimized for growing conditions in the field. Many of the traits related to abiotic stress tolerance such as water and nutrient use efficiency, and yield performance are linked to the root properties. Deeper and profuse root systems could withstand drought effects by tapping extra water from the lower soil profile. Thus, root architecture have long been suggested mainly to improve crop adaptation and performance in water limited environments.

Drought tolerance in pearl millet is in large part related to the root system (McIntyre et al. 1995). Sustained water uptake ability by increasing total root length and maintenance of high leaf water status under soil drying conditions are the main factors for drought tolerance in pearl millet (Matsuura et al. 1996). Therefore, breeding pearl millet varieties with improved root traits promises to deliver benefits in water and nutrient acquisition in dry environments.

The hidden part of the plant, root system makes phenotyping very challenging. Recently, imaging technologies have become available that allow to illuminate the dynamics of root structure and function in the soil. Rhizotrons and micro-computed tomography (microCT) phenotyping approaches characterized early stage pearl millet root system development as a fast growing primary root and three distinct types of lateral roots formation on both primary roots and crown roots (Passot et al. 2016). Pearl millet inbred lines studied for the root system architecture exhibited significant variation in primary root length and lateral root density. It is presumed that primary root growth is associated with early stage drought stress tolerance. Search for genetic markers associated with primary root growth in a large panel of genetically fixed pearl millet inbred lines is a priority research area for drought tolerance. However, there is no genetic or genomic studies conducted on early root development traits to dissecting the genetic determinants controlling these key root phenotypes in pearl millet. Different phenotyping technologies that analyze processes at different spatial and temporal scales such as combined magnetic resonance imaging (MRI) and positron emission tomography (PET) may open up mechanistic understanding of root structure and function of the crop.

6.2.3 Cold Tolerance

Pearl millet is a sun loving plant and requires warmer temperatures (30–35 °C) for normal growth. Its sensitivity to cold weather limits its usefulness to only the hot temperature in the summer. Seed germination is also affected by soil temperature. Pearl millet seeds germinate only when the soil temperature is above 18 °C. Cold weather can stress plants and increase nitrate levels. Frost in the fall is also observed to stop the growth of pearl millet.

Nevertheless, the existence of genetic diversity for relative cold tolerance cannot be ruled out. Landraces from Yemen were found to be potential sources of variation for cold tolerance (Shivhare and Lata 2016). The existence of a complex gene regulatory network for stress tolerance was witnessed with the identification of about 2,494 differentially regulated transcripts in response to drought, salinity, and cold stress (Mishra et al. 2007).

6.2.4 Drought and Heat Tolerance

Drought inflicts an adverse effect on plant growth and development. The key plant biochemical processes such as photosynthesis and transpiration involve water molecules. Reduced water in the plant system also accelerates chloroplast damage (Stone 2001) and affects photosynthetic outputs as a result of reduced light interception and carbon reduction. Water is also required by plants for surface cooling from sun light radiation. About 99% of the water absorbed from the soil is usually evaporated from aerial parts such as- leaves, stems and flowers- into the atmosphere, in trade-off the plant parts remain in normal physiological status and intake CO_2.

Drought is the major constraint in pearl millet production as it is almost exclusively grown in the dry semiarid and arid environments. However, adaptive evolution and natural selection made pearl millet relatively the most drought and heat tolerant among other cereals. Morpho-physiological traits were developed through evolution, including both shoot and root characteristics naturally modified for efficient water absorption from the soil and conservation in the system help pearl millet adapt to the dry environment. Stomatal conductance, photosynthetic capacity, timing of phenological phases, starch availability during embryo development, stem reserve mobilization in drought stress, stay green, reduced leaf area, rooting depth and density, cuticular resistance and surface roughness, osmotic adjustment, membrane composition, antioxidative defense, and accumulation of stress-related proteins (Cattivelli et al. 2008) are some of the morpho-physiological traits of drought tolerance and relevant for yield performance. Drought often accompanying by increased air temperature results in reduced reproduction due to heat damage on the pollen grain (Bita and Gerats 2013), and increases sterility (Schoper et al. 1987; Jagadish et al. 2012).

Decreased CO_2 diffusion to the chloroplast through limited stomatal conductance affects the rate of photosynthesis (Pinheiro and Chaves 2011). Drought stress prompts plant system to accumulate metabolites such as proline in plant cell to serve as osmolyte, as a metal chelator, an antioxidative defense molecule, and a signaling molecule (Hayat et al. 2012). It also maintains osmotic balance, prevents electrolyte leakage by stabilizing membranes and proteins such as RUBISCO (Ribulose 1, 5-bisphosphate carboxylase) (Hayat et al. 2012). It also controls the reactive oxygen species (ROS) concentration in plants (Paleg et al. 1984; Hare et al. 1998) and maintains the integrity of mitochondrial electron transport complex II (Hamilton and Heckathorn 2001) to impart stress tolerance. However, molecular and biochemical evaluations conducted so far in pearl millet have not considered the effect of proline accumulation on drought tolerance. There are other numerous metabolic changes and enzymatic activities associated with water deficit in the plant system (Serba et al. 2017).

Traditional landraces from drier regions were good sources of genes for breeding drought tolerance (Kusaka et al. 2005; Yadav 2010). However, drought tolerance is a complex polygenic trait and the various morpho-physiological responses to drought are controlled by many genes and significantly influenced by the environment (Hu and Xiong 2014). Dynamic and complex cross-talk between different regulatory gene networks regulate physiological and morphological adaptation and plant stress responses (Saito and Matsuda 2010). Therefore, there are technical limitations prohibiting the classical breeding program to study plant responses to environmental stress. In that scenario, pre-breeding and genomics-assisted breeding will be very helpful for characterization of germplasm collection and introgression of desirable traits in adopted varieties (Varshney et al. 2018).

The mechanism how plants endure the effect of drought on their growth and development takes various forms. The ability of plants to grow and yield satisfactorily with limited moisture supply or under periodic soil water deficits is termed as 'drought resistance.' It is principally the constitutive plant traits of maintaining high plant water status (dehydration avoidance) that enables plants to survive in water limited environments (Blum 2005). This mechanism has virtuous relevance in terminal drought tolerance of pearl millet in which tolerant genotypes fill their grain and offset any drastic effect on their yield. Since phenotyping dehydration avoidance is difficult, not much study is devoted to understand the genetic bases of drought tolerance in pearl millet. However, a recent genome resequencing of a large number of genotypes shaded light on the genetic basis of drought tolerance in pearl millet. The heat and drought tolerance of pearl millet is associated with a stock of lipid biosynthesis genes for increased cuticular wax synthesis (Varshney et al. 2017). A detailed account of epicuticular wax in pearl millet is presented under the crop-specific traits below.

Root characteristics such as depth, density and architecture are also important traits in drought tolerance. Fine roots and root length density are some of the root traits known to contribute to productivity under drought stress (Comas et al. 2013). Genotypes with such root characteristics can exploit the water available deeper in the soil profile and better tolerate terminal drought. However, no information is available on the genomic regions or genes underlying such root characteristics in pearl millet. Further, any genetic variation in the germplasm for such traits and its relationship with terminal drought tolerance was not studied yet.

However, different researches conducted drought tolerance studies in pearl millet. Evaluation of hybrids against female and male inbred parents under sprinkler irrigation gradient highlighted the significant dry biomass reduction in low water level (Ibrahim et al. 1985). In this evaluation the performance of the hybrids was better than the parents for dry biomass, harvest index, and water use efficiency under severe drought stress.

High temperature stress can affect pearl millet plant at two stages; either at germination when soil temperatures can be very high or at flowering/reproductive stage. Field studies in the Sahel indicated that pearl millet seedlings are most vulnerable to high temperatures during the first 10 days of sowing, this was confirmed by field studies in the Indian Thar Desert (Yadav et al. 2012). Genetic differences in seedling survival under high soil surface temperatures were identified (Peacock et al. 1993). Pearl millet has good degree of tolerance to high temperatures of up to 42 °C during flowering. Hence, it has occupied considerable areas (about 1 m ha) in the hot and dry post-rainy season (locally referred to as summer) in the northern and western parts of India. Heat tolerance is required in summer season when air temperatures of >42 °C coincides with flowering period which leads to reproductive sterility and finally drastic reductions in grain yield. Large genetic variability for seed set at daily maximum air-temperature of $\geq$ 42 °C during flowering in pearl millet was reported (Gupta et al. 2015b). They indicated that seed set in pearl millet started declining when maximum air temperatures reaches 42 °C and decreased in curvilinear fashion to 20% till 46 °C. High temperature stress imposed for season long, or during reproductive stages of development for different durations (18 or 45 days) caused significant decrease in number of seeds per panicle, individual seed weight and seed yield per panicle (Djanaguiraman et al. 2017). Two periods (10–12 and 2–0 days before anthesis) were identified as most sensitive to short episodes (2 d) of high temperature stress, causing maximum decreases in pollen germination percentages and seed numbers per panicle.

6.2.5 *Salinity Tolerance*

In arid and semiarid regions, the extent of salt affected soils is steadily increasing mainly because of flooding irrigation system. Today, more than one billion hectares of land area are affected by soil salinity around the globe (Saade et al. 2016), and has become one of the major constraints on agricultural production.

Pearl millet is well-suited to grow in harsh conditions including saline soils. It is generally considered as an alternative crop for salt affected areas because of its fair tolerance to salinity (Ali et al. 2006). Research conducted on salt tolerance in semiarid areas found a significant interaction of NaCl concentration and pearl millet varieties (Yakubu et al. 2010). Though both plant growth parameters and nutrient contents significantly decreased with increasing soil salinization, they found a variety named Maiwa that was relatively tolerant to soil salinization as it expressed superior nutrient content and root and shoot growth when compared to the control. But the usefulness of these growth parameters for salinity response with respect to grain and forage yield needs to be further investigated.

6.2.6 Disease Resistance

Numerous fungal, bacterial, and viral pathogens that infect pearl millet were discovered decades ago (Ramakrishnan 1971). Downy mildew (*Sclerospora graminicola*), rust (*Puccinia substriata* var. *indica*), Pyricularia leaf spot or blast (*Pyricularia grisea*), and smut (*Moesziomyces penicillariae*) are the most important diseases causing major yield losses and quality reduction of pearl millet. High relative humidity (85–90%) and moderate temperature (20–30 °C) favor infection and disease development (Thakur et al. 2011). As a result, the severity of these diseases varies across different growing regions.

The first report of downy mildew (DM) pathogen on pearl millet was from India (Butler 1907). However, the disease remained sporadic and caused yield losses in poorly drained and low lying areas until the introduction of improved cultivars in 1960s (Singh 1995). DM is now a major biotic constraint in India and West Africa. The use of Tift 23A as a seed parent, increased the susceptibility of the first batch of hybrid cultivars in India. The wide spread of hybrids based on Tift $23A_1$ increased oospore inoculum and resulted in epidemics in 1971–72 (Singh 1995). DM incidence is estimated to cause pearl millet grain yield loss ranging from 10 to 80% (Gupta and Singh 1996) and gained more importance for resistance breeding.

Although several control measures—including seed sanitation, chemical control of the seed, soil, and air-borne inoculums—were suggested, a major emphasis has been placed on host-plant resistance for its effectiveness. Consequently, a great progress has been made in downy mildew resistance breeding in India. Studies in disease epidemiology and pathogen biology (Singh and Williams 1980) enabled development of reliable field and controlled environment screening techniques (Singh et al. 1997). All breeding material passes through the downy mildew-screening nursery and resistant varieties and hybrids have been bred since then. Pathogen isolation procedures, and severity rating scales, have been established for downy mildew and other major diseases such as rust, blast or leaf spot, ergot and smut at ICRISAT (Singh et al. 1997; Thakur et al. 2011). A good number of germplasm and breeding lines with high levels of stable resistance have been identified from West Africa (Singh 1990; Singh et al. 1997). These materials have strategically been used in breeding for resistance at ICRISAT and All India Coordinated Pearl Millet Improvement Program (AICPMIP) centers (Hash et al. 2006; Thakur et al. 2006). Development and commercial deployment of downy mildew resistant HHB 67, a popular hybrid being grown in North India, is the first successful story of marker-assisted breeding (MAB) in field crops in public domain in India (Hash et al. 2006; Thakur et al. 2006). However, the disease still remains the major constraint in West Africa. To offset the disease epidemics, farmers in the region are still growing landraces resistant to downy mildew but low in grain yield potential.

In addition to studying the biology of the pathogen, the *Sclerospora graminicola* genome was recently sequenced at 40x (Nayaka et al. 2017). A total of 299.9 Mb of the genome sequence was assembled and 65,404 genes predicted, of which 38,120 genes were annotated. This resource is important to study the pathogenicity, race dynamics, spread of the pathogen and as a powerful synergy to resistance breeding. It was also reported that endophytic *Trichoderma hamatum* UoM 13 isolated from pearl millet root suppressed downy mildew disease (Siddaiah et al. 2017).

Although sources of resistance are available, the mechanism of host resistance is still barely understood. An attempt was made to test enzyme lipoxygenase (LOX), known to play a role in disease resistance in many host-pathogen systems. LOX activity was tested in seeds of different genotypes of pearl millet with different level of susceptibility to downy mildew (Nagarathna et al. 1992). The LOX activity of seeds assay indicated a positive correlation between enzyme activity of genotypes and DM resistance in the field and thus can be used as a biochemical marker for downy mildew resistance. In a recent study to unravel inheritance, it was observed that resistance to DM is controlled by a single dominant gene in 834B and IP 18294-P1 and by two dominant genes in IP 18298-P1. A test for allelism inferred that a single dominant gene for resistance in 834B is nonallelic to that which governs resistance in IP 18294-1, whereas, one of the two dominant genes for DM resistance in IP 18298-P1 against the test isolates is allelic to the gene for DM resistance in 834B and a second gene is allelic to the resistance gene present in IP 18294-P1 (Raj et al. 2018).

Rust and leaf spot are the two important diseases of pearl millet in the United States (Wilson and Hanna 1992). Rust is known to cause up to 72% yield loss (Wilson et al. 1995). Rust disease control through other methods is more difficult as the pathogen is primarily disseminated by wind, and the spores can survive in the soil, on plant debris, volunteer pearl millet, and alternative hosts. The research for genetic resistance discovered a dominant rust resistance gene, Rr_1, in wild grass (*P. glaucum* ssp. *monodii*) from Senegal and introgressed into cultivated pearl millet via backcrossing (Hanna et al. 1985). This introgression only improved rust resistance provisionally because the gene was fast overcome by a shift in the virulence of the pathogen. Germplasm screening against single uredinal isolates of the rust pathogen discovered 10 new races with different level of prevalence (Tapsoba and Wilson 1995). Both race-specific and non-race-specific (horizontal) resistance is prevalent in the germplasm. Then the resistance breeding shifted to broadening the genetic basis of resistance by combining slow rusting genes with race-specific resistance in the development of improved forage pollinator lines (Wilson 2002). A major rust resistance QTL explaining 58% phenotypic variance was mapped on linkage group 1 (LG_1) using a F_7 recombinant inbred line population from the cross 81B-P6 x ICMP 451-P8 (Ambawat et al. 2016) and was found to confer a durable slow-rusting phenotype.

Pyricularia leaf spot (blast) is another common fungal disease affecting grain and forage pearl millet production in India and the United States. Unlike the rust pathogen, very little has been accomplished on pathogen characterization or management in the United States. Information on germplasm variability for host-plant resistance, precise phenotyping and screening techniques, and genetic mechanism of resistance are limited for this disease. Based on two resistant sources having diverse parentage, resistance to foliar blast in pearl millet is controlled by a single dominant gene (Gupta et al. 2011). Recently, six blast resistant lines having diverse parentage revealed the presence of single dominant gene governing resistance to two *Magnaporthe grisea* isolates, and were found allelic (Singh et al. 2018).

Variability from African germplasm has been exploited for disease resistance. However, fungal diseases, especially DM and smut, are still inflicting significant yield losses in major pearl millet-growing areas of West Africa. A priority breeding strategy that harness the germplasm available is highly needed to incorporate disease resistance with cultivar yield improvement to enhance acceptable to the growers.

6.2.7 Insect Resistance

There are about 100 species of insect pests attacking pearl millet as previously reviewed (Gahukar 1984; Nwanze and Harris 1992). Important insect pests in West Africa include stem borer (*Coniesta ignefusalis* Hampson), grain midge (*Geromyia penniseti* Felt), ear-head caterpillars (*Heliocheilus albipunctella* de Joannis), head beetle (*Rhinyptia infuscata* Burmeister), shoot flies (Athrigona spp), and leaf beetles (*Lema planifrons* Weise and *Chaetocnema tibialis* Illiger). Other general feeders such as armyworms (*Spodoptera exempta* Walker, *S. exigua* Hubner, *S. littoralis* Boisduval, *Mythimna loreyi* Duponchel), aphids (*Rhopalosiphum maidis*, Fitch (Aphididae: Homoptera)), grasshoppers (*Hieroglyphus nigrorepletus* (Bolivar), *H. banian* (Fab), *Colemania spheneroides* (Bolivar), *Chrotogonus spp*. (Acrididae: Orthoptera)), and locusts may cause complete defoilation and severe losses to the crop during prolonged drought early in the season. However, the occurrence of such pests are sporadic, and sometimes localized.

A recent study found that insects belonging to six orders and 11 families were found to herbivore on pearl millet in the southern United States (Obeng et al. 2015). Eastern leaf-footed stinkbug (*Leptoglossus phyllopus* (L.), Hemiptera: Coreidae), American bird grasshopper (*Schistocerca americana* Drury; Orthoptera: Acrididae) and the differential grasshopper (*Melanoplus differentialis* (Thomas: Orthoptera: Acrididae) were the most prevalent and dominant species.

Soil dwelling insects such as white grubs (*Holotrichia* spp, *Anomola* spp, (Melolonthidae: Coleoptera) and termites or white ants (*Odontotermes* spp, *Microtermes* spp, *Macrotermes* spp, (Termitidae: Isoptera) are attacking the root of pearl millet. Particularly, the grubs of *H. consanguinea* cut the roots causing wilting and death of plants in patches and is known to devastate the crop in large areas in central India.

Pearl millet is tolerant or essentially a poor host to sugarcane aphid (SCA) ((*Melanaphis sacchari* Zehntner, (Aphididae: Homoptera)). Field observations by producers, county extension agents, and millet breeders in several areas found little or no SCA in pearl millet field grown for forage. Controlled screening of pearl millet breeding lines along with a resistant and a susceptible grain sorghum (*Sorghum bicolor*) varieties at the K-State Agricultural Research Center-Hays

(ARCH) under controlled condition observed minimum damage (Serba and Michaud, unpublished). Representative seed parents, restorers and the germplasm were included in the study. The statistical analysis data found that there is significant difference among genotypes for aphid feeding damage but not among the different genotype categories (Fig. 6.3). However, none of the pearl millet genotypes affected to the level of susceptible sorghum, which implied pearl millet is a poor host for SCA. With the recent incidence of SCA on forage sorghum, pearl millet is recommended as an alternative to forage sorghums in the Central Great Plains of the United States (Trostle et al. 2015).

Inherent plant traits and its interaction with the external environmental factors affect plants tolerance to herbivory. Genetic tolerance with built-in compensation abilities of plants, climatic factors, and cultural practices are important factors in raising healthy pearl millet and keeping pests in bay.

6.2.8 Grain Micronutrients (Fe and Zn)

In the past decades, almost entire pearl millet research efforts have been directed towards the development of high yielding lines or cultivars through various breeding methods. Nutritionally, pearl millet is not less than any cereal crop and is rich in certain nutrition qualities, including micronutrients. Pearl millet is a healthy

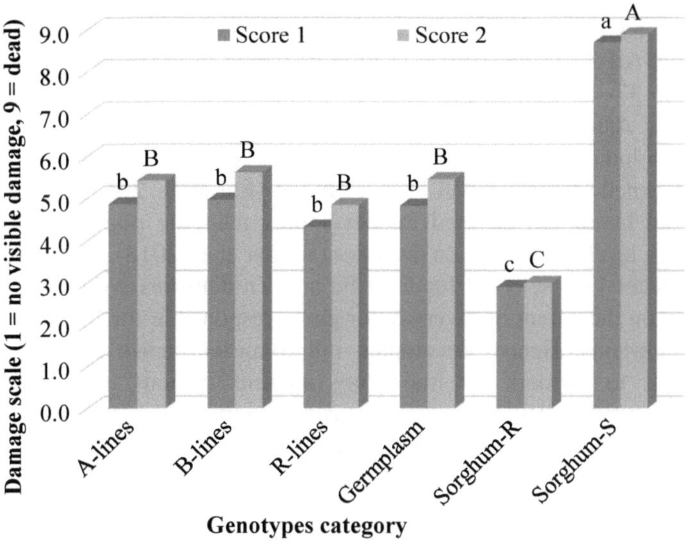

Fig. 6.3 Average SCA damage score on pearl millet parental lines and germplasm against resistant and susceptible sorghum using 1 to 9 damage scale (score 1 and 2 were made five and eight days after infestation, respectively). Bars labeled with the same later (upper or lowercase) were not significantly different (Serba and Michaud, unpublished)

and versatile grain worthy to add to anyone's diet. It has higher percentages of essential amino acid profile and also exhibit higher contents of Ca, Zn, Fe and Mg than sorghum (Rai and Virk 1999; Iren Leder 2004). Pearl millet is the cheapest source not only for energy and protein but also inexpensive source for Fe and Zn among all cereals and pulses (Rao et al. 2006). However, more research effort is needed to enhance the essential micronutrients in the grain beside yield per se. This will help to addresses the malnutrition of the consumers who mostly rely on pearl millet as a staple food.

Preliminary screening of several mainstream breeding populations at ICRISAT, including released open-pollinated varieties displayed substantial variability for Fe (30–75 mg kg^{-1}) and Zn (25–65 mg kg^{-1}) (Velu et al. 2007; Gupta et al. 2009; Rai et al. 2012). HarvestPlus-supported pearl millet biofortification research has shown much large genetic variability for grain Fe (31–125 mg kg^{-1}) and Zn (35–82 mg kg^{-1}) densities among advanced breeding lines, population progenies, hybrid parents. Also several lines and accessions with 90–100 mg kg^{-1} Fe and 70–80 mg kg^{-1} Zn densities were identified, indicating a good prospect of genetic enhancement for nutritional quality (Govindaraj et al. 2016). Considering these, both Fe and Zn become core traits in current ICRISAT pearl millet breeding program and demonstrated with dissemination of first surge of biofortified diverse hybrids parents (seed and restorer) with other preferred traits including diverse cytoplasm (Table 6.1).

Over the last four decades, ICRISAT-bred lines and populations have been extensively used by breeders, both in public and private sector to breed open-pollinated varieties (OPVs) and hybrids. All these available released OPVs and/or commercial hybrids were evaluated to examine the variability for Fe and Zn density (Fig. 6.4). The Fe density in hybrids varied from 46–56 mg kg^{-1} and Zn density from 37–44 mg kg^{-1}. In addition to high-Fe control ICTP 8203, very few hybrids viz., Ajeet 38 and 86M86 had high Fe and Zn densities. Systematic research also indicated Iniadi germplasm to be a valuable germplasm resource for Fe and Zn genetic improvement in pearl millet (Rai et al. 2013, 2015). The lines derived from Iniadi can be used for developing mapping populations to identify QTL for high levels of Fe and Zn densities (Kumar et al. 2018). Genomic studies of the Iniadi accessions selected for high Fe and Zn densities would provide useful information on the extent of diversity for genes responsible for high levels of these two micronutrients. Hence, developing micro-nutrient-dense seeds would be of great benefit to farmers by improving nutrient content of harvested grain. Biofortifying pearl millet is also the most cost effective and sustainable approach to enhance the levels of bioavailable micronutrients (Fe and Zn) and nutritional security (Vinoth and Ravindhran 2017).

Table 6.1 Biofortified diverse seed and restorer parents developed at ICRISAT (data are mean of 4 seasons)

Line	50% flowering time (days)	XRF Fe (mg kg^{-1})	XRF Zn (mg kg^{-1})	1000-grain weight (g)	CMS
Seed parents					
ICMA/B 1501	39	76	42	13.2	A4
ICMA/B 1502	43	92	50	13.6	A1
ICMA/B 1503	43	69	43	15.0	A4
ICMA/B 1504	47	97	55	15.5	A1
ICMA/B 1505	41	110	55	15.5	A1
ICMA/B 1506	45	96	53	9.9	A4
ICMA/B 1507	43	92	50	10.1	A4
ICMA/B 1508	53	73	44	15.0	A1
Restorer parents					
ICMR 1201	48	79	41	10.5	A1
ICMR 1202	50	89	47	14.2	A1
ICMR 1203	52	101	58	7.9	A4
ICMR 1301	55	91	52	12.7	A1
ICMR 1501	55	86	42	9.9	A1
ICMR 1502	51	110	62	12.4	A1
ICMR 1503	51	99	47	13.7	A4
ICMR 1504	57	96	51	8.8	A1
ICMR 1505	55	74	41	6.9	A4

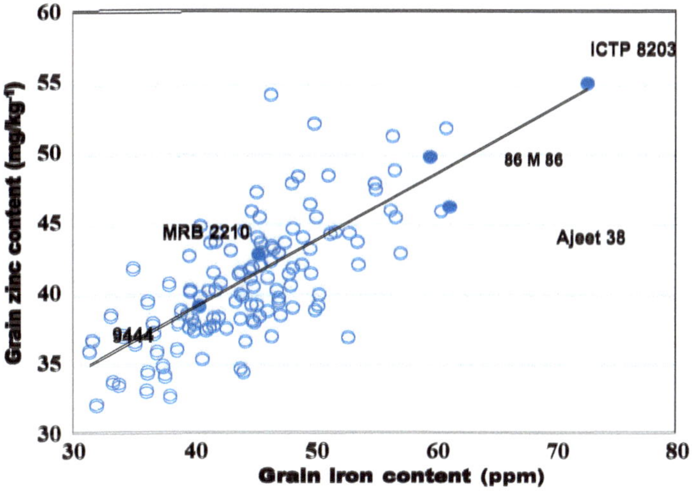

Fig. 6.4 Variability and relationship between grain Fe and Zn content in commercial hybrids of pearl millet

6.2.9 Nutrient Use Efficiency

Of the 16 essential nutrients needed for successful plant growth and development, nitrogen (N), phosphorus (P), and Potassium (K) are the three most important primary nutrients. Assuming soil moisture is conducive, nitrogen is the most important limiting factor in crop production. In fact, N is the important constituent of protein and protoplasm and its shortage leads to chlorosis (leaf yellowing) and slowdown the growth.

N ranks first among the applied inorganic fertilizers to maximize yield in agriculture. The global demand for N fertilizer for agricultural production, which already stands at approximately 110 million metric tons per year, is projected to increase to approximately 250 million metric tons by 2050 (www.fertilizer.org). Because nitrates are very mobile in the soil, a substantial amount (> 50% in some cases) of applied N is lost by leaching, run-off, and de-nitrification. In addition to an increase in production cost, in the long run these processes of N loss not only pollute the ground water and adversely affect soil structure but also have detrimental effects on the environment such as increase in nitric oxide, ozone, etc. Hence, developing crop varieties with improved efficiency for N absorption and utilization will help mitigate these problems to some extent (Frink et al. 1999; Good et al. 2004). In addition to its role in plant growth, N level (organic and inorganic forms) in the soil increases soil water-holding capacity and control wind-erosion (Bationo et al. 1993).

The prevalence of significant genotypic variation in biomass production and nitrogen use efficiency (NUE) were documented in pearl millet (Alagarswamy and Bidinger 1987). This study found that a genotype, Souna B from West Africa had NUE values 32% higher than the less efficient Indian genotype BJ 104, even though both had similar N uptake. In addition to genotype difference, N uptake and utilization in pearl millet was found to be influenced by the growing environment especially soil moisture (Maman et al. 2006).

Soil fertility is marginal and plant-available P is severely limited in the Sahelian west Africa (Lambers et al. 2015). Pearl millet is the cereal grown well to this adverse climatic condition and acid sandy soils. For economic production of pearl millet in such an environment assessment of genetic variability for P uptake, utilization in available germplasm, and identification of potential secondary selection traits are important endeavors. To overcome soil P-deficiency, exploitation of pearl millet genetic diversity is an economically worthwhile and environmentally feasible opportunity. Genetic variation was studied under low and high P level for uptake and utilization in both seedling and mature plants using 180 West African pearl millet inbred lines (Gemenet et al. 2015). Phosphorus utilization efficiency increased in low-P, but total P uptake was more important for grain production than P utilization under low-P conditions. Both seedling and mature plant morphological traits are potentially useful as secondary traits in selection of pearl millet for low-P adaptation. From this study, it was suggested that pearl millet breeding for low-P tolerance needs to be integrated with other system-oriented research, such as nutrient cycling and modest applications of locally available rock phosphate.

In general, pearl millet germplasm from West Africa unveils significant genetic variation for P-uptake and utilization efficiency, and grain yield under P-limited soils indicating the possibility of breeding P-efficient cultivars (Gemenet et al. 2015). To accelerate the breeding process for P uptake and utilization, mapping the genomic regions through marker-trait association and application of marker-assisted selection (MAS) are vital.

6.2.10 Water Use Efficiency

In water-limited environments, plant productivity is determined jointly by the amount of water available and the water use or evapotranspiration efficiency (Emendack et al. 2011). Yield and water use efficiency (WUE) of pearl millet vary across genotypes (Emendack et al. 2011) and moisture regimes (Sivakumar and Salaam 1999). Improvement in pearl millet WUE is likely surge yield by increasing the number of productive tillers and uniform maturity of the tillers with the main culm. Research conducted on soil nutrient status and WUE under water stressed condition found that, WUE is influenced by available P level in the soil, as cumulative transpiration increased with P level (Payne et al. 1992). In the arid environment of Niger, year effect on pearl millet yield was observed which is primarily because of variations in the amount and distribution of rainfall in relation to the potential demand for water (Sivakumar and Salaam 1999). Increase in soluble soil phosphate due to application of fertilizer increased WUE and yield. The beneficial effect of fertilizers could be attributed to the rapid early growth of leaves, which can contribute to reduction of soil evaporative losses and increased WUE.

WUE is also dependent on root architecture, as nodal and primary roots have distinct responses to soil moisture level (Rostamza et al. 2013). Association of greater nodal root length in pearl millet with increased shoot biomass is attributable to efficient water uptake and WUE. Six times longer nodal roots, mainly from 8-fold increase in branch root length, were observed with 12% soil water content than dry treatments. Enhanced plastic response to moisture around the nodal roots in pearl millet is attributed to faster growth and progression through ontogeny for earlier nodal root branch length and partitioning to nodal root length from primary roots, independent of shoot size. Genotypes with deeper nodal roots can be selected in a breeding program to shape root architecture. Enhanced response to soil moisture and rapid rate of plant development are important traits that favor nodal roots and WUE without any cost to shoot growth.

Reports show that WUE is more influenced by edaphic than climatic condition as significant reduction by waterlogging than drought observed (Zegada-Lizarazu and Iijima 2005). The drought resistance of pearl millet is explained by higher WUE but not by increased water uptake efficiency in deep soil layers as compared to barnyard millet, another drought-resistant millet species. However, there was not any study on the genetic factor underlying WUE variations in pearl millet. As WUE has a leveraging effect on drought tolerance, it necessitates a new frontier in genomic study.

6.2.11 Carbon Sequestration

Soil organic carbon (SOC) accumulation largely depends on vegetation cover. The pearl millet production system is characterized by a little, if any, recycling of organic matter. The crop is grown as a dual purpose crop in the arid and semi-arid areas where the stock is used as a fodder for animals or as a fuel for cooking. A study conducted on the effect of soil organic matter amendment on grain yield in rain-fed production systems in the semiarid tropics of India reported that for every Mg ha^{-1} increase in soil organic carbon stock in the root zone, pearl millet grain yield increases by 170 kg ha^{-1} (Srinivasarao et al. 2014). The grain yield response from pearl millet with the soil organic carbon amendment was the highest when compared to other crops such as groundnut (*Arachis hypogaea*), finger millet (*Eleusine coracana*), sorghum, soybean (*Glycine max*), castor (*Ricinus communis*), cluster bean (*Cyamopsis tetragonoloba*) and rice.

Comparison of five millet species [Japanese millet (*Echinochloa esculenta*), pearl millet, foxtail millet (*Setaria italica*), browntop millet (*Urochloa ramosa*), and proso millet (*Panicum miliaceum*)] for conservation use in the United States recommended pearl millet for fast-growing as a cover crop that can increase organic matter, scavenge residual nutrients, create large amounts of surface mulch, reduce compaction, and reduce root-lesion nematodes (USDA 2014).

6.2.12 Genome Plasticity

The adaptation of plants to a dynamic environment is defined by the controlled expression of genes both temporally in response to a change in the environment and regularly for normal growth and development. To survive in a changing environment and reproduce, genome plasticity acquired through inheritance and mutation enable a plant to compete for the resources it needs to grow. Changes in nucleotide sequence that occurs as a result of several forces, chemical and genetic, can alter the genetic content of a plant and genomic plasticity. Transposons, insertion sequence elements, DNA repetitions, introns, and DNA rearrangement are the reserve in the genome for changes in gene expression in response to external factors (Bennett 2004).

Early characterization of pearl millet nuclear genome with respect to its size, buoyant density, sequence organization, and association kinetics estimated the genome size at 0.22 pg $nucleus^{-1}$, GC content at 44.9%, and the melting temperature to be 49.7 °C (Wimpee and Rawson 1979). In other cereals, such as hexaploid wheat, new variation is rapidly generated because of the dynamic nature of wheat genomes through gene deletions and insertions of repetitive elements into coding and regulatory gene regions (Dubcovsky and Dvorak 2007). Pearl millet has an estimated genome size of 1.79 Gb (Varshney et al. 2017). The percentage of repetitive DNA in pearl millet is estimated at 80% (Wimpee and Rawson 1979; Paterson et al. 2009; Bennetzen et al. 2012), which is close to that of maize. It is presumed that this high level of repetitive DNA is a reflection for a high level of genome plasticity in pearl millet.

Remarkable intraspecific differences in gene expression were documented in pearl millet. Given the absence of detailed studies in pearl millet genome, it is difficult to present a detailed account of the genome organization. Yet it is not sufficiently studied how diversity bottlenecks are compensated. Although the wild relatives are growing in the same environment with the cultivated pearl millet, especially in Sahelian West Africa, the extent and significance of natural gene introgression between species has not been studied well. However, the transfer of CMS from the compatible primary gene pool to pearl millet warrants the gene flow from the wild relatives that enhance genome plasticity of the crop.

6.2.13 *Other Important Traits*

Pearl millet exhibits some characteristics that are relatively specific to the crop or shared among related species. Some of these specific traits are related to leaf area to extend the duration of active photosynthesis, leaf anatomical structure related to forage digestibility (such as brown mi-rib trait), accumulation of cuticular waxes on the above ground surface to minimize water loss in dry conditions, and purple pigmentation for ornamental value. These key traits are of interest in the improvement of pearl millet for specific usage.

6.2.13.1 Stay Green

Stay green is a characteristic of delayed leaf senescence. Some pearl millet genotypes extend the duration of active photosynthesis by delaying leaf senescence for a longer period of time than others. The relationship between carbon fixation capacity and spending over the life of a leaf alters the timing of senescence initiation and progression. In stay green variants the deconstruction of the photosynthetic apparatus during leaf senescence is partially or completely prohibited (Thomas and Howarth 2000). A complex signaling network that involves mutations activating cytokinin signaling or those suppressing ethylene, abscisic acid, brassinosteroid, and strigolactone signal transduction often result in stay-green phenotype (Kusaba et al. 2013). Also, impairment in the enzymatic steps responsible for chlorophyll breakdown leads to stay-green phenotypes.

A study conducted using some inbred lines found that a SNP in putative acetyl CoA carboxylase gene showed robust association with grain yield, harvest index, and panicle yield under irrigated and non-irrigated conditions and an InDel in putative chlorophyll a/b binding protein gene was significantly associated with both stay-green and grain yield under drought stress and the later may serve as a functional marker for selecting high yielding genotypes with 'stay green' phenotype (Sehgal et al. 2015). As increase in leaf area, daily duration of photosynthesis or leaf area duration is significantly associated with yield increases achieved in grain crops (Richards 2000), stay green characteristic is important for both yield increase and stress tolerance. A plant that maintains the integrity of its chlorophyll despite the soil water status during reproductive stage can continue photosynthesis and can fill the grain and yield a reasonable grain in drought stress. The semi-dwarf pearl millet inbred lines developed in the United States for hybrid breeding are stay-green. However, a detailed study on locating the genes/genome regions involved is not conducted.

Five scenarios were proposed for change in chlorophyll content and photosynthetic capacity (Kusaba et al. 2013) during the leaf senescence process in stay green leaves (Fig. 6.5). Disabled chlorophyll pigment degradation during the later stage is associated with stay green. Stay green characteristic also has an essential effect for the feed value of the fodder after grain harvested. Therefore, the trait has important especially in dual purpose pearl millet variety development.

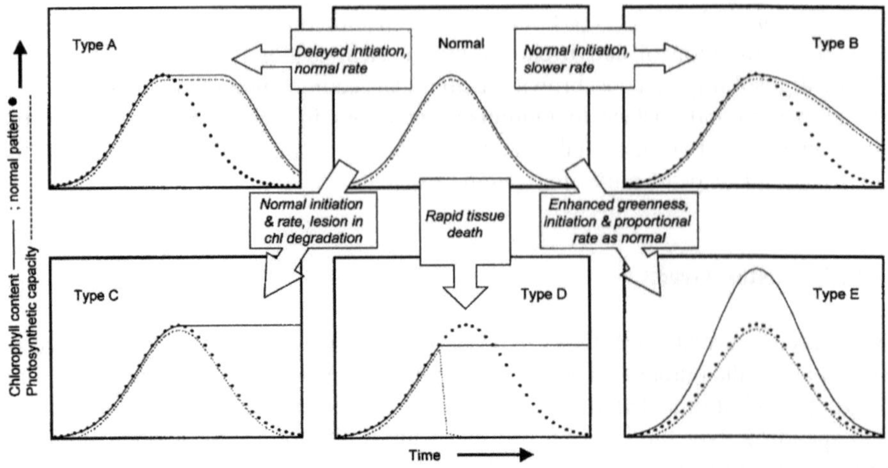

Fig. 6.5 Five ways to stay-green. Curves show chlorophyll content and photosynthetic capacity (arbitrary scale) for a representative leaf, whole plant or canopy. Type A stay-greens lose pigment and function at the normal rate after a delay in the start of senescence. In Type B, senescence is initiated on schedule, but subsequently proceeds more slowly. Type C stay-greens undergo functional senescence on a normal time-scale, but a lesion in pigment breakdown means they retain chlorophyll indefinitely. Type Ds are stay-green because they are dead. In Type E behavior, the photosynthetic capacity of an intensely green genotype may follow the normal ontogenetic pattern, but comparison of absolute pigment contents identifies it as a stay-green (Adopted with permission from Thomas and Howarth 2000; J. Exp. Bot. 51: 329–337)

6.2.13.2 Brown Mid-Rib (BMR)

Pearl millet is extensively used as a forage for livestock and improvement in digestibility and palatability may enhance feed value. High lignin content is the primary forage digestibility restricting factor by ruminants and reducing the shoot lignin content is the most effective way to increase the digestibility.

A brown coloration of the leaf mid veins (BMR) is associated with reduced lignin content and altered lignin composition of the midrib of maize, sorghum, sorghum x sudangrass, sudangrass, and pearl millet (Sattler et al. 2010). Therefore, brown midrib offers greater digestibility and palatability compared to conventional forage, thereby yielding greater returns in the form of weight gain in beef cattle or milk per ton of forage feed. Two sources of brown midrib lines have been reported in pearl millet (Gupta 1995). Pbmr developed at Purdue University through chemical mutagenesis (Cherney et al. 1988) and SDML 89107 developed by SADC/ICRISAT through germplasm selection from Zimbabwe (Gupta et al. 1993) were the two independent lines. Later, Tift-91 BMR was identified and bulked at the University of Georgia. In vitro dry matter digestibility (IVDMD) of pearl millet BMR was higher than its normal counterpart (Gupta 1995). Lower lignin content (23%) and 4% higher IVDMD was ($p \leq 0.01$) was reported in BMR forage lines (Cherney et al. 1990). BMR allelic inheritance study in pearl millet reported that BMR is controlled by homozygous recessive allele at a locus. The F_1s of the crosses between the two BMR and elite normal midrib inbred lines were found entirely normal, F_2 segregated 3:1 (normal/bmr) and testcross population segregated 1:1 confirm the trait under a major gene control (Gupta 1995).

6.2.13.3 Cuticular Wax

Abiotic stress arises from exposure to climatic extremes such as drought, heat, cold, and frost. Plants have evolved to exist in conditions which are unideal for maintenance of normal physiology and may be at the limit for survival. The above ground surfaces of terrestrial plants are covered with cuticular wax (Jetter et al. 2007). Epicuticular waxes are complex mixtures of hydrophobic molecules form the outermost layer of the hydrophobic cuticle, which is composed of cutin polyester membrane, and provides the last barrier to prevent uncontrolled water loss (Kosma and Jenks 2007). Cuticular wax plays a significant role in plant abiotic and biotic stress tolerance and has been implicated in defense mechanisms against excessive ultraviolet radiation, high temperature, bacterial and fungal pathogens, insects, high salinity, and low temperature tolerance (Xue et al. 2017). Cuticular waxes consist of homologous series of very-long-chain fatty acids, alcohols, aldehydes, alkanes, esters, and cyclic organic compounds (Cameron et al. 2006). Primary plant surfaces are coated with hydrophobic cuticular waxes to minimize non-stomatal water loss. Wax compositions differ greatly between plant species, organs, tissues, and developmental stages (Guo et al. 2017).

A comparison study between sorghum, maize, and pearl millet leaves showed no glossiness even on the pearl millet twelfth leaf (Traore 1985). The glossy character was correlated with a reduction or absence of observable wax deposits on the leaf surfaces and higher cuticular water loss than non-glossy leaves. A recent genomic study unfolded a substantial enrichment for wax biosynthesis genes (Varshney et al. 2017). Therefore, it is believed that cuticular wax deposition in the leaves that minimizes water loss and reflect radiation probably contributed to heat and drought tolerance in pearl millet. It was also demonstrated that changes in cuticle gene expression in drought exposed plants (Suh et al. 2005) provide evidence for alteration of cuticle traits for drought tolerance.

6.2.13.4 Purple Foliage

Another notable trait in pearl millet is purple foliage. It is known to be controlled by three plant pigments: anthocyanidins-cyanidin, delphinidin, and pelargonidin (Raju et al. 1985). A genetic study revealed that purple color is determined by three alleles at a single locus: Rp_2 (red), Rp_1 (purple), and *rp* (green) (Hanna and Burton 1992). Red is dominant over purple and normal green, whereas purple is dominant over the normal green ($Rp_2 > Rp_1 > rp$). Red plant can be distinguished at 5 days, whereas, purple can be distinguishable two weeks after emergence. The Rp_1-Rp_2 locus was independent of the trichomeless (*tr*), yellow (yn_1), female sterile (*fs*), and light green (lgn_1) loci but linked to the dwarf (d_2) locus (about 28% recombination). The D_2/d_2 plant height locus and the *P/p* foliage color locus are linked and mapped to pearl millet linkage group 4 (Azhaguvel et al. 2003). A purple foliage pearl millet was identified in a population of 500 plants derived from a combined mutagenic treatment with a 20 Kr dose of gamma rays and 0.1% aqueous solution of ethyl methane sulphonate (EMS) (Varalakshmi et al. 2012).

6.2.13.5 Bristle Panicle

Bristliness in pearl millet is another important trait. The inflorescence consists of a central rachis covered with short hairs and bears fascicles on rachillae. Each fascicle contains spikelets surrounded by a wall of bristles (i.e., involucre). The extent of prolongation of the fascicle axis limits the length of bristles. Long bristles have an advantage in deterring birds feeding on the grain. A comparison of varieties with long and short bristles observed that long bristle showed reduced damage by blister beetle (*Psalydolytta fusca* Olivier) in the Sahel (Gahukar 1988, 1991). Bristled panicle is a dominant trait and hence easy to work in hybrids, when one of the parents is bristled.

6.2.13.6 Dwarfing Genes

Plant height is a vital component for carbon gain strategy (Moles et al. 2009) as it determines a plant's ability to compete for light (Falster and Westoby 2003). On the other hand, plant height has agronomic importance as plants with short stature entail improved lodging resistance. Furthermore, cultivars with short plant height are preferred in modern agriculture for input responsiveness and mechanized harvesting. Dwarfing genes have been successfully utilized in developing short statured cultivars of cereals such as wheat, rice, barley, sorghum, and pearl millet. Semi-dwarf grain cultivars respond better to high levels of nitrogen application than tall plants because of their reduced lodging vulnerability.

Five different sources dwarf plants, named D_1–D_5, were first discovered in 1960s (Burton and Fortson 1966). Through crossing with the tall counterparts, the dwarfness in source lines were found to be controlled largely by one or two recessive genes. When transferred to near-isogenic backgrounds, dwarfness from source lines D_1 and D_2 is controlled by single independently segregating recessive genes, designated as d_1 and d_2. The d_2 dwarfing gene is known to reduce plant height by 50% through a reduction in the length of all stem internodes except the peduncle, leading to a higher proportion of leaves (Rai and Hanna 1990). The d_2 also has several pleiotropic effects. In addition to increased leaf percentage, improved nutritional quality of the stem fraction of pearl millet forage was attributed to d_2 (Burton et al. 1969). Subsequently, d_2 has been deployed widely in commercial cultivars grown in India, the United States, and Australia. Comparison of tall and dwarf near-isogenic F_1 hybrids found dwarf varieties have virtually no impact on the production that might restrict the release of dwarf pearl millet cultivars (Bidinger and Raju 1990).

After the discovery of dwarfing genes and inheritance studies, the research was focused on knowing the genomic location of the genes, the biochemical function of the candidate genes controlling the trait. Identifying the actual gene is particularly important to develop molecular markers that can be used to screen for the presence of the gene long before the effects of the gene can be visually observed. Using restriction fragment length polymorphism (RFLP) markers in an F_2 mapping population developed from a IP19283 x Tift 238D1 cross, the d_1 and d_2 plant height loci were mapped as morphological traits in pearl millet linkage group 1 (LG1) and linkage group 4 (LG4), respectively (Azhaguvel et al. 2003). QTLs for plant height were detected in parallel on respective linkage groups.

Genetic mapping using large population and haplotype analysis of three tall and three dwarf inbred lines delineated the d_2 region in the genome (Parvathaneni et al. 2013). Comparative analysis defines the region in sorghum genome 410 kb with 40 annotated genes. One of the sorghum genes annotated within this region is $ABCB_1$, which encodes a P-glycoprotein involved in auxin transport. This gene had previously been shown to underlie the economically important dw_3 dwarf mutation in sorghum. The co-segregation of $ABCB_1$ with the d_2 phenotype, its differential expression in the tall inbred ICMP 451 and the dwarf inbred Tift 23DB, and the similar phenotype of stacked lower internodes in the sorghum dw_3 and pearl millet d_2 mutants suggest that $ABCB_1$ is a likely candidate for d_2. ATP binding cassette subfamily B member 1 (*Abcb*$_1$) gene that plays a role in polar auxin transport, was identified as the likely candidate underlying d_2 in pearl millet (Parvathaneni et al. 2013). The *ABCB*$_1$ gene encodes a P-glycoprotein and underlies dwarfing traits in maize (*br*$_2$), sorghum (*dw*$_3$), and pearl millet (d_2) displayed considerable variation in intron composition (Parvathaneni et al. 2017). Physiological analysis confirmed that the gene affects the downward transport of auxin. If this gene is on, the auxin flows freely, and millet will grow to its full height, about 3 m. If it is off, the millet plant may only grow 1–1.5 m in height.

In addition to the five dwarfing genes, 13 dwarf plants were identified from world collection (Appa Rao et al. 1986). F_2 from the cross between three (IP 8056, IP 8210 and IP 8214) and tall inbred showed continuous variation for plant height suggesting that dwarfness was controlled by more than one gene. Only two of the 10 crosses with either d_1 (Tift 238) or d_2 (Tift 23 DB) dwarfs, produced tall F_1 hybrids and they segregated for height in F_2 indicating that these two new dwarfs were non-allelic to d_1 and d_2. Reciprocal crosses of these two dwarfs produced tall F_1 hybrids and showed a dihybrid segregation of 9:3:4 in F_2 indicating that the dwarfing genes of these two parents were non-allelic to each other. These non-allelic dwarfs were recognized as new sources and assigned the gene symbols d_3 (IP 10401), and d_4 (IP 10402).

6.3 Genetic Resources of Climate-Smart Genes

The genus *Pennisetum* consists of approximately 140 highly diverse species (Brunken 1977). The genus is a heterogeneous assemblage of species with four different basic chromosome numbers (n = 5, 7, 8 and 9) (Jauhar 1981); ploidy levels ranging from diploid to octoploid, both sexual and apomictic reproductive behaviors, and annual, biennial or perennial life cycles (Martel et al. 1997). Centered on morphological differences the genus is divided into five sections, namely *Gymnothrix* (P. Beauv.) Benth. & Hook. f., *Brevivalvula* Doll, *Heterostachya* Stapf & C. E. Hubb., *Eupennisetum* Stapf, and *Penicillaria* (Willd.) Benth. & Hook. f. (= *Pennisetum*) (Stapf and Hubbard 1934).

6.3.1 Gene pool

Genetic diversity studies in *Pennisetum* recognized three gene pools and delineated as primary, secondary, and tertiary gene pools. These genepools were first proposed based on genetic relationship among the species (Harlan and De-Wet 1971). Later, these gene pools were identified on the basis of crossability and cross fertility of the wild species with the domesticated diploid cultivated *P. glaucum* and gene transfer complexity between genepools. Some of the sections classified under secondary and tertiary genepools have different life cycle, mode of reproduction, and/or basic chromosome numbers (Table 6.2).

Table 6.2 Life cycle, reproductive behavior, and chromosome numbers of few species of the genus Pennisetum classified under primary, secondary, and tertiary genepools

Genepool/section/ species	Life cycle	Mode of reproduction	Chromosome		References
			(2*n*)	x	
Primary genepool					
Section Pennisetum					Jauhar (1981), Martel et al. (1996), Jauhar and Hanna (1998), Robert et al. (2011)
P. glaucum ssp. *glaucum*	a	Sx	14	7	
P. glaucum ssp. *Monodii*					
Ecotype 1. *P. violaceum*	a	Sx	14	7	
Ecotype 2. *P. mollissimum*	a	Sx	14	7	
Secondary genepool					
Section Pennisetum					Jauhar (1981), Martel et al. (2004), Robert et al. (2011)
P. purpureum	p	Sx, Ap	28	7	
Section Heterostachya					
P. squamulatum	p	Ap	54, 54	7, 9	
Tertiary genepool					
Section Brevivalvula					Jauhar (1981), Martel et al. (1997), Jauhar and Hanna (1998), Martel et al. (2004), Robert et al. (2011)
P. pedicellatum	a	Ap	36, 54	9	
P. polystachion	a/p	Sx, Ap	18, 36, 54	9	
P. hordeoides	a	Ap	36, 54	9	
P. subangustum	a	Sx, Ap	18, 36, 45	9	
P. setosum	a	Sx, Ap	54	9	
Section Gymnothrix					
P. mezianum	p	Ap	32	8	
P. hohenacken	p	Sx, Ap	18	9	
P. alopecuroides	p	Sx	18	9	
P. ramosum	a, b	Sx, Ap	10	5	

Life cycle: a = annual, b = biannual, p = perennial; Reproduction: Sx = Sexual, Ap = apomictic

6.3.1.1 Primary Gene Pool

The primary gene pool includes all forms of cultivated, weedy, and wild diploids ($2n = 2x = 14$). Pearl millet is a diploid with seven pairs of homologous chromosomes ($2n = 2x = 14$), have annual growth habit. Harlan and De-Wet (1971) classified that the cultivated *P. glaucum* and the wild relative *P. glaucum* ssp. *monodii* are the two members of the primary genepool. *P. violaceum* and *P. mollissimum* are currently considered as the two ecotypes of subspecies *monodii,* and are the wild diploids ($2n = 2x = 14$) also belonging to the primary gene pool.

Cytogenetic investigation of these wild relatives along with the cultivated form of this gene pool have shown high similarity between their karyotypes (Khalfallah et al. 1993). And in situ hybridization analysis confirmed similar localization of rDNA among members of the primary gene pool (Martel et al. 1996) that substantiated their conserved high level of genome similarity with the cultivated pearl millet.

On the other hand, *P. schweinfurthii* Pilger is another diploid species that have $2n = 2x = 14$ chromosomes and annual in growth habit (Hanna and Dujardin 1986). Its chromosomes were also reported to be similar in size as those of pearl millet, but non-homologous. Inter-specific hybrids between *P. schweinfurthii* and pearl millet were morphologically intermediate to both species. They were also found as male sterile and partially female sterile. In Sahelian West Africa, the intermediate weedy form, *P. stenostachyum* is widely found.

Enzyme polymorphism study in the *Pennisetum* gene pool reported variations in leaf esterases (EST), 6-phosphogluconate dehydrogenase (PGD), shikimate dehydrogenase (SKDH), leucine aminopeptidase (AMP), phosphoglucomutase (PGM) and malate dehydrogenase (MDH) genes (Lagudah and Hanna 1989). In the primary gene pool, two loci, *Est-1* and *Est-2*, were identified to controlling leaf esterases.

6.3.1.2 Secondary Gene Pool

The secondary pool solely consisted of the tetraploid *P. purpureum* (Shum.) ($2n = 4x = 28$). *P. purpureum* is a fast-growing perennial grass native to the African grasslands. Pearl millet (*P. glaucum*) with AA genome and *P. purpureum* (Napier grass or elephant grass) with A'A'BB genomes are the two economically important species of the genus *Pennisetum*. It is widely grown in tropical and subtropical regions of the world as one of the most important tropical forage crops. The genetic proximity between *P. glaucum* and *P. purpureum* enables triploid hybrids ($2n = 3x = 21$) that yield higher quality forage from both species. Conventional cytogenetic techniques revealed the presence of homeology between the genomes A and A', and with the genome B (dos Reis et al. 2014). Genomic in situ hybridization (GISH) confirmed the homeology between the genomes A of pearl millet and A'B of Napier grass, and showed that there are differences in the distribution and proportion of homologous regions after hybridization (dos Reis et al. 2014).

These days, the apomictic octaploid *P. squamulatum* ($2n = 8x = 56$) is considered as the member of secondary gene pool (Pattanashetti et al. 2016). The perennial *P. squamulatum* is stress tolerant and can be utilized for pearl millet or Napier grass improvement. Genome similarity, crossability, genetic relatedness, and cytology of hybrids supported the placement of *P. squamulatum* in secondary genepool (Kaushal et al. 2008).

6.3.1.3 Tertiary Gene Pool

The broad tertiary gene pool comprises *Pennisetum* species of various ploidy levels and growth habits (Dujardin and Hanna 1989) that are distantly related to primary and secondary genepools. This comprises many among the approximately 140 diverse species of the genus. Gene flow from the tertiary genepool to pearl millet is limited because of strong reproductive barrier with the primary and secondary genepools. The species have either annual or perennial growth habit and have economic importance as a forage, ornamental, or landscaping purposes.

The whole section Brevivalvula belongs to the tertiary genepool. It consists of six morphological taxa: *P. atrichum* Stapf & Hubb., *P. hordeoides* (Lam.) Steud., *P. pedicellatum* Trin., *P. polystachion* (L.) Schult., *P. setaseum* (Swartz) L. Rich., and *P. subangustum* (Schum.) Stapf & Hubb., which together form a polyploid and agamic complex. Four euploid ($x = 9$) and twelve aneuploid chromosome levels have been reported so far. The polyploids are apomictic, while the diploid populations of *P. polystachion* and *P. subangustum* are considered sexual.

6.3.2 Races

Pearl millet is categorized under the section *Pennisetum*. Based on grain shape which follows geographic pattern, the world collection of cultivated pearl millet were classified into four races (*typhoides, nigritarium, globosum,* and *leonis*) (Brunken et al. 1977). Pearl millet domestication has produced many landraces displaying a broad diversity in environmental adaptation, cycle length, morphological traits, genetic yield potential, stresses tolerance, and grain quality characteristics.

Race *typhoides* is mainly cultivated in India and characterized by obovate caryopses that are obtuse and terete in cross section. Inflorescences are mostly cylindrical in shape. Morphologically it is the most variable among the four races and is also most widely distributed. It occurs in entire Africa. It is the only basic race found outside Africa and the predominant race grown in India.

Race *nigritarum* caryopsis is angular in cross-section with three and six facets per grain. Inflorescences are candle-like. The apex of the grain is usually truncate and often tinged purple. The mature grain is generally longer and protrudes beyond the floral bracts. This race is generally found in western Sudan to northern Nigeria (Brunken et al. 1977). Race *nigritarum* is dominantly grown in eastern Sahel.

Race *globosum* is dominantly grown in western Sahel and has spherical shaped caryopsis with each of its dimensions being approximately equal. Depth of the grain always exceeds 2.4 mm. The grain is otherwise terete and obtuse. Inflorescences are candle shaped and often exceed 1 m in length. It is the most common race in central Nigeria, Niger, Ghana, Togo, and Benin (Bono 1973).

Race *leonis* is dominant in coastal regions of West Africa and characterized by an acute, oblanceolate, and terete caryopsis. The most distinct character of the leonis grain is its acute apex, which is terminated by the remnants of the stylar base.

At maturity, approximately one-third of the grain protrudes beyond the floral bracts. Inflorescence shape is candle-like. It is specific to Sierra Leone but also grown in Senegal and Mauritania (Brunken et al. 1977).

6.4 Genetic Diversity

Genetic diversity is a reflection of a variety of genes in a given species which are important for survival of natural selection, for tailoring adaptation to a changing environment, and conservation of desired traits. Genetic diversity occurs as an outcome of mutations, recombination, genetic drift, migration, and selection. Natural and human selection processes over thousands of years have generated diverse cultivars of pearl millet adapted to different biophysical conditions, suited to various production systems and socio-economic conditions, and well-matched various consumer preferences (Brunken 1977).

6.4.1 Phenotypic Diversity

Different phenotypic traits such as flowering time, panicle length, grain and stover characteristics, grain nutritional composition, and tolerance to biotic and abiotic stresses in cultivated pearl millet (Bhattacharjee et al. 2007; Amadou et al. 2013) have been used to study the genetic diversity at different times. A study conducted in Nigeria on 25 pearl millet genotypes collected for diverse morphological variation showed that farmers' husbandry practice resulted in the isolation of ideotypes, making landrace names tradeoff for genetic diversity (Danjuma and Mohammed 2014). The results suggested that artificial selection had greater influence in shaping the population than environmental factors. ICRISAT pearl millet breeding program at India has developed diverse range of seed (A-/B- pairs) and restorer parents (R-lines) utilizing diverse sources of germplasm, and characterized 99 A-/B- pairs and 114 restorer parents based on 26 morphological traits (Rai et al. 2009).

Besides per se performance and geographic origin, selection of parents based on genetic diversity yields significant yield improvement. A very recent study into the effect of genetic distance of inbred parents on heterosis of the hybrid reported positive correlation between phenotypic distance of parents and better-parent heterosis for grain yield (Gupta et al. 2018). But the correlation was not strong enough to be used as a major selection criterion for parental selection for heterosis breeding to improve grain and stover yield.

The wide range of climatic conditions in the center of diversity and farmers preferences and utilization habits created landraces with local adaptation that maintained broad genetic variability. The prominent early-maturing and productive landrace from West Africa, Iniadi, contributed desirable traits towards genetic improvement of pearl millet (Andrews and Kumar 1996). The traits contributed by

Iniadi included adaptation, productivity and grain nutritional quality. However, the genetic control and heritability of the nutritional composition of the grain need further investigation for effective bio-fortification with essential micronutrients.

A study attempted to quantify the degree of diversity in 122 commercial hybrids of pearl millet cultivated in India and to understand the relationship among various phenotypic and quality traits, showed large variation for flowering time (42–58 days), tillering (1.1–4.4 panicles/plant), individual grain size (7.6–17.3 mg), plant height (185–268 cm), panicle length (20–33 cm) and grain yield (3.5–7.2 tons ha^{-1}). Hybrids resulted in 7 distinct clusters, which highlighted the successful efforts of Indian national program of pearl millet improvement toward genetic diversification of hybrids (Yadav et al. 2016).

6.4.2 Genotypic Diversity

Analysis of molecular variation within and between Indian landraces (Bhattacharjee et al. 2002) and among inbred lines derived from WCA landraces (Stich et al. 2010) revealed more than two-fold variation between than within landraces. Population structure analysis among the later, categorized the landraces into five subgroups depicting the diversity of West African germplasm.

Genotypic diversity investigated on 213 parental lines (99 seed and 114 restorer parents) of pearl millet indicated significant diversity (Nepolean et al. 2012). Analysis of 379 hybrid parents developed at ICRISAT (current 166 parents and 213 previously developed hybrid parents) carried out using SSRs detected 12.7 alleles per locus. Also, the seed and restorer parents were clearly separated from each other, indicating existence of two diverse and broad-based pools in hybrid parents of pearl millet (Gupta et al. 2015a). Restorers (R-lines) were found more diverse than seed parents (B-lines), as higher average gene diversity was detected among R-lines (0.70) than B-lines (0.56).

Genetic diversity analysis in pearl millet inbred germplasm association panel (PMiGAP) representing cultivated germplasm from Africa and Asia, elite improved open-pollinated cultivars, hybrid parental inbred lines and inbred mapping population parents showed an average gene diversity of 0.54 and six subpopulations within PMiGAP (Sehgal et al. 2015). The PMiGAP panel along with several other lines were also studied for genome wide diversity (Varshney et al. 2017).

6.4.3 Relationship with Other Cultivated Species and Wild Relatives

Pennisetum sect. Penicillaria is a morphologically diverse taxon native to tropical Africa. Following morphological and biosystematic evidence the section is divided into two biological species. *Pennisetum purpureum* Schmach. is a tetraploid, perennial species of the wet tropics. *Pennisetum americanum* (L.) K. Schum., a native of the arid and semi-arid tropics, is annual and diploid and is segmented into three sub-species each having different adaptations to domestication. Sub-species monodii, a wild taxon from the Sahel of West Africa, is identified as the progenitor of pearl millet. Subspecies stenostachyum is morphologically intermediate between subsp. americanum and monodii and includes the mimetic weeds often found in association with pearl millet. A genetic diversity and species relationship study among eight wild species of the genus *Pennisetum* including pearl millet using RFLP markers revealed wide variation among the species (George et al. 2005).

6.4.4 Relationship with Geographical Distribution

There are many landraces (varieties named by farmers) grown in different ecological niches by small-scale farmers, but the role of sociological factors in the evolutionary dynamics of the crop is not well understood. A study assessed the connection between ethno-linguistic diversity and extent genetic diversity in pearl millet grown in the western side of the lake Chad in Central Africa revealed the existence of a genetic structure concomitant with ethno-linguistic differences (Naino Jika et al. 2017). On the other hand, a high seed and pollen-mediated gene flow among different pearl millet growing regions of Sudan was inferred as molecular variation of pearl millet accessions within the regions was much higher than among the regions (Bashir et al. 2015). As there are no linguistic barriers among the regions, the theory of more seed exchange between villages of the same language group contributed for the variation. This outcome suggests that gene flow is limited between landraces grown by different ethno-linguistic groups probably because of language barriers that limit seed exchange among the farmers' communities. Irrespective of its outcrossing nature, the lack of genetic exchange among the landraces might have also been contributed by physical delineations along the linguistic differences.

This type of strong concurrence between ethno-linguistic boundaries and genetic discontinuities was also observed for cassava (Delêtre et al. 2011) signifying the influence of sociological factors in genetic exchange. Similarly, population genetic structure was also coincided with the main language families for sorghum landraces in Africa (Westengen et al. 2014). A genetic diversity study on maize landraces grown by different ethno-linguistic groups in southern Mexico is emphasizing the role of farmers' local selection in influencing genetic structure rather than environmental factors (Perales et al. 2005).

On the other hand, geographic distance from the center of diversity was associated with high genetic divergence suggesting a major effect of isolation by distance on divergence (Hu et al. 2015). Moreover, adaptation to a new environment can bring novel variations as a result of ecological opportunity and competition

(Andrew et al. 2010). But, the genetic basis of environmental adaptation in pearl millet has not been well understood.

6.4.5 *Extent of Genetic Diversity*

Understanding the extent of genetic variability available and bio-geography of genetic resources have paramount importance for optimal improvement and conservation of pearl millet. Moreover, it is an indispensable raw material for success in breeding new cultivars for improved yield, quality, and stress tolerance. Hence, valid estimation and thorough understanding of the array of genetic diversity present in pearl millet would have practical application in breeding for different objectives. Assessment of genetic variability permits to identify contrasting parental materials to enhance hybrid vigor and yield stability in variable climates (Haussmann et al. 2012).

To broaden the extent of genetic variability available to breeders, ICRISAT was engaged in germplasm collection and characterization. By the year 2007, ICRISAT collected more than 20,844 cultivated pearl millet accessions and 750 wild relatives through 76 collection missions in 28 countries (Upadhyaya et al. 2007). Most of these collections were made from the center of diversity, WCA. To facilitate identification of potential parents for a genetic improvement program, a core collection comprising of 2094 accessions ($\sim$10% of the entire collection) were identified using data on 11 quantitative traits (Bhattacharjee et al. 2007). Then, a mini-core collection comprising 238 accessions (1.1% of the collection) was established following proportional sampling strategy, using the same 11 qualitative and 8 quantitative traits data on the accessions (Upadhyaya et al. 2011). This stratified information about the ICRISAT collection warrants a persistent availability of germplasm for further improvement of pearl millet mainly in India.

6.5 Association Studies

Dissecting marker–trait relationship using allelic variations accumulated through historical recombination events in natural populations is encouraged for QTL mapping and candidate gene identification. While conventional linkage mapping using bi-parental populations has identified a number of important quantitative traits in pearl millet, the resolution provided by two parent-derived mapping populations is severely limited. Some efforts of association mapping made in pearl millet created valuable information about genetic variability and linkage disequilibrium (LD). It is believed that a good understanding and identification of underlying genes, alleles or QTLs for stress responsiveness and adaptation traits may facilitate breeding for increased climate-resilience (Kole et al. 2015).

6.5.1 Extent of Linkage Disequilibrium

Genes that control phenotypic traits can be identified with high resolution using association studies based on LD. However, LD patterns across the pearl millet genome have not been systematically studied yet. Periodic studies targeting a certain genome region or genes have been conducted. From the available information, a phenomenon of rapid decay of inter-loci and intra-locus LD (Hu et al. 2015; Varshney et al. 2017) is observed.

6.5.2 Target Gene-Based LD Studies

Several target gene or genome region based assessment of LD was conducted in pearl millet. Assessment of the extent of LD among 1,575 pairs of loci mapped on LG_2 in a total of 250 inbred lines of PMiGAP found that a total of 441 (28%) of the marker pairs were in LD ($P < 0.01$) (Sehgal et al. 2015). But, when LD was calculated separately within each of the six sub-populations formed by population structure analysis, the frequency of pairs of loci with significant LD ($P < 0.05$) was reduced by more than half.

The extent of LD in *PgD8*, a gene in gibberellic acid (GA) pathway and conditioning photoperiod in pearl millet was found lower than that of the homologous maize D8 (Li et al. 2018). Likewise, the effect of selection on LD in the vicinity of Phytochrome C gene (*PHYC*) was assessed with a panel of 90 pearl millet inbred lines using 75 markers in 100-kb region identified the best candidate markers on the PHYC gene (Saïdou et al. 2014). However, in genomic regions containing polymorphisms for genes that have been targeted for selection, it has been speculated that LD may extend over a relatively long distance (Saïdou et al. 2009).

Attempt of mining favorable alleles for grain Fe and Zn content through association mapping in pearl millet identified three SSR markers consistently associated with elevated Fe and Zn with more than 11% coefficient of determination (r^2-value) (Anuradha et al. 2017). As Fe plays a role in the production of oxygen-carrying proteins hemoglobin and myoglobin its deficiency creates health and developmental problems in the developing world. Lack of sufficient Fe in nutrition can lead to different forms of anemia.

6.5.3 Genome-Wide LD Studies

A characterization of two pearl millet diversity panels using genotyping by sequencing (GBS) of *PstI-MspI* reduced representation libraries identified 83,875 single nucleotide polymorphisms (SNPs) (Hu et al. 2015). In this study much faster LD decay in Senegalese landraces compared to global accessions. Rapid LDD of

0.5 kb in B- and R- lines (48 bp) as well as in PMiGAP lines (84–444 bp) was also observed using genome-wide SNPs detected through resequencing (Varshney et al. 2017). This indicated that LD in pearl millet is similar with that of maize, which is also an allogamous species.

In the model species foxtail millet nonlinear regression showed rapid decline of LD with distance and rapid LD decays to half the initial value within ca. 1.2 kb (He et al. 2015). Similarly in maize, LD decayed rapidly with distance between sites within loci, but there was substantial variation among genes (Remington et al. 2001). In this study, predicted r^2 values declined to less than 0.1 within 2,000 bp in two-third of the loci.

6.5.4 *Potential of Association Studies for Germplasm Enhancement*

Diversity in plant genetic resources (PGR) provides opportunity for plant breeders to develop new and improved cultivars with desirable traits such as genetic yield potential, biotic stress tolerance, environmental adaptation, and quality attributes. Genetic diversity within and between plant populations is usually assessed using morphological, biochemical (allozyme), and molecular marker (DNA) analysis.

In this postgenomic era, abundant single nucleotide polymorphism (SNPs) and the affordability genotyping stimulated the application of genomic-assisted breeding and efficient utilization of genetic diversity for crop improvement. Association mapping, also known as linkage disequilibrium (LD) mapping, utilizes ancestral recombination events in germplasm collections or natural populations to make marker-trait associations attributed to the strength of linkage disequilibrium between markers and functional polymorphisms across a set of diverse germplasm (Zhu et al. 2008). It also offers an alternative means of allele mining by enabling to survey a much larger and diverse genepool having thousands of recombination events, lending high mapping resolution, and potential to identify the causal polymorphism within a gene and/or QTL.

The existence of tremendous phenotypic variability is documented in pearl millet germplasm for many agronomic traits (Bhattacharjee et al. 2007). In spite of the extent of genetic variability available, only a small fraction of the gene pools are used for developing pearl millet varieties and hybrids. This limited utilization of the available germplasm is probably because of lack of access to the resource by the breeders, limited preservation or lack of systematic documentation of the genetic diversity. In pearl millet improvement, utilization of wild relatives as donors of specific traits such as apomixes (Hanna 1987), or resistance to pests and diseases (Wilson et al. 2004) is also scarce. Further, most of the allele mining for agronomically important traits including biotic and abiotic stress resistance has been conducted using bi-parental mapping populations (Yadav et al. 2002, 2004; Bidinger et al. 2007; Sehgal et al. 2012; Vengadessan et al. 2013).

For a suburb germplasm conservation and genetic improvement of pearl millet, understanding the genomic diversity, population structure of the germplasm, and association mapping are the important areas to be assessed. The recent release of genome sequence is expected to facilitate association mapping and genome selection in pearl millet. GWAS across 288 testcross progenies of PMiGAP was also conducted for 20 grain yield and stover yield component trait and identified 1054 strongly significant marker–trait association for 15 of the traits (Varshney et al. 2017). A study of genomic diversity, population structure, and linkage disequilibrium conducted on 398 accessions from different geographic regions provided a noteworthy insight (Serba et al. 2019). Data analysis using a total of 82,112 genome-wide SNPs discovered through GBS revealed hierarchical genetic structure of six subgroups that mostly overlap with the geographic origins (west Africa, east Africa, Southern Africa, the Middle East, India, and lines developed in USA). The result also confirmed that germplasm from west Africa rooted the dendrogram of the phylogeny analysis with much diversity and greater LD decay, indicating a long history of recombination among landraces. Assessment of genetic differentiation between population subgroups and genome-wide patterns of nucleotide variation within each subpopulation indicated that the Indian subpopulation is less differentiated from all other subpopulations ($F_{ST} = 0.006$). This is probably attributed to the strong breeding program led by ICRISAT that utilized diverse germplasm. On the contrary, the Middle East subpopulation was relatively highly differentiated from all others (average $F_{ST} = 0.072$), followed by the inbred lines from US breeding programs ($F_{ST} = 0.060$). The later indicates the limited germplasm utilization by the US breeding program. It is also discovered that the lowest average nucleotide diversity ($\pi = 4.23 \times 10^{-4}$) was found in the inbred lines from the United States as compared to the average genome-wide nucleotide diversity for the whole population ($\pi = 5.0 \times 10^{-4}$).

6.6 Molecular Mapping of Genes and QTLs

The application of genomics has become an indispensable component of breeding programs and proven to be useful for identifying novel genes for traits of agronomic importance and stress tolerance. Marker assisted-selection (MAS) has a great practical importance to facilitate gene introgression into desirable genetic backgrounds and crop improvement by shortening the lengthy phenotypic evaluation and by increasing selection accuracy.

Genomic research for pearl millet is however lagging behind the major cereals such as rice, maize, wheat, and sorghum. It is a poor man's crop and the public and private investment towards the improvement of the crop is limited. As a result, addressing the genetic yield barriers with the help of next-generation sequencing (NGS) technology and use of available germplasm is inadequate. Nevertheless, with the continuous efforts of ICRISAT and few national programs especially India, UK, some basic genomic information has become available. Most recently, pearl millet genome sequence has been released for public use (Varshney et al. 2017) as a reference for further development of genomics-assisted breeding. This genome sequence, provides a genetic blueprint of the species and apparently facilitates the development of genomic tools that would expedite the breeding process and improve selection gains through the application of marker-assisted breeding.

6.6.1 *Brief History of Mapping*

Pearl millet is one of the orphan crops with minimum public investment in the research and development of the crop. Compared to other cereals such as rice, sorghum, maize, wheat, and barley, research on the development and application of molecular markers is limited in pearl millet. However, development of molecular markers for the improvement of pearl millet as a potential crop for the hot and dry environments was started in 1991 (Gale et al. 2005). As a result, the first genetic linkage map was constructed using RFLP earlier than other orphan crops (Liu et al. 1994). Linkage groups corresponding with the seven haploid chromosome number of the species were successfully formed and this RFLP based map was used as the basis for subsequent pearl millet marker-based studies. Further marker development and mapping efforts utilized different marker systems and mapping populations.

The first pearl millet genetic linkage map constructed using RFLP markers served as a foundation for subsequent mapping using polymerase chain reaction (PCR) based markers and currently using the high throughput SNP markers. Then, significant efforts have been made in developing other molecular markers, mapping the genome, trait-specific QTL and genes in pearl millet. Development of genetic linkage maps using RFLP, simple sequence repeats (SSR), and single nucleotide polymorphism (SNPs), identification of QTL for drought tolerance, downy mildew resistance, grain quality traits, and genes for plant height are among the molecular resources have become available for pearl millet breeding.

A genetic linkage map from four different crosses have been integrated to develop a consensus map of 353 RFLP and 65 SSR markers where extreme localization of recombination toward the chromosome ends that affect transfer of genes controlling important agronomic traits from donor to elite pearl millet germplasm (Qi et al. 2004). To provide improved genome coverage, pearl millet genetic linkage map was integrated with diversity arrays technology (DArT) and SSR markers following PstI/BanII complexity reduction digestion (Supriya et al. 2011). A total of 321 loci (258 DArTs and 63 SSRs) that spanned 1148 cM with an average adjacent-marker interval length of 3.7 cM was constructed.

PCR-based linkage map was constructed on the basis of a recombinant inbred line (RILs) population resulting from a cross between Tift 23DB and PI 536400 (Pedraza-Garcia et al. 2010). The objective of this mapping was to obtain more evenly distributed markers throughout the linkage map and improve genome coverage of the gaps in earlier maps characterized by high concentration of markers in the centromere region than the distal regions of the linkage groups. An average marker density of per 9.2 cM was achieved with 196 PCR-based DNA markers (66 sequence-related amplified polymorphisms (SRAPs), 63 random amplified polymorphic DNA (RAPDs), 27 inter-simple sequence repeats (ISSRs), 31 pearl millet, 6 sorghum, and 3 maize SSRs) involving 152 recombinant inbred lines (RILs) but mapped into nine linkage groups.

6.6.2 Evolution of Marker Types: RFLP to SNPs

After the construction of the first linkage map using RFLP, development and application of other marker systems continued. Construction of a bacterial artificial chromosome (BAC) library using nuclear DNA of pearl millet contains a total of 159,100 clones with an average insert size of 90 kb and corresponding to 5.8 haploid genome equivalents (Allouis et al. 2001) was used as a resource for the isolation of SSR sequences. PCR-based screening of 4.7 haploid genome equivalents using five sequence-tagged site (STS) and six SSR markers identified an average of 5.4 positive superpools.

With the development of PCR, a program at ICRISAT has developed 100 SSR markers and mapped 60 of them (Gale et al. 2005). That facilitated the integration of the pearl millet map in the grass consensus map and subsequently the establishment of the plant genome database MilletGenes. This database provided genome related data (maps, markers, DNA sequences, and images) on pearl millet, finger millet (*Eleusine coracana*), foxtail millet and tef (*Eragrostis tef*) (http://jicbio.bbsrc.ac.uk/cereals, Accessed on 18 August, 2018). Probe information, end-sequences of RFLP probes, RFLP and STS polymorphism data, autoradiograph and gel images, segregation data, genetic maps, QTL data, and morphological data were used to be deposited in this database.

SNP marker development was first conducted using rice genome as a reference (Bertin et al. 2005). Using pearl millet expressed sequence tags (ESTs) and annotated rice genomic sequences, 299 homologues of single-copy rice genes with precise prediction of the intron positions were identified as single-strand conformational polymorphism (SSCP)-SNPs. Analysis of the fragments amplified by PCR primers designed to amplify approximately 500-bp genomic fragments on SSCP gels revealed considerable polymorphism. Further sequence analysis of the fragments over a panel of eight inbred genotypes estimated about one SNP or InDel (insertion-deletion) every 59 bp in the introns, but considerably fewer in the exons.

With the application of NGS, a large number of SNPs markers were developed and applied in mapping and QTL identification (Sehgal et al. 2012; Moumouni et al. 2015; Punnuri et al. 2016), population genomics and genome-wide association mapping (Hu et al. 2015; Sehgal et al. 2015). Application of GBS is proved to be a quick and low cost marker development and genotyping platform. With the current availability of the genome sequence, this approach has a great potential in characterizing the genome and mapping of important traits.

6.6.3 Mapping Populations

Genetic linkage maps are necessary for applied genetics and marker-assisted breeding of pearl millet. To this end, significant efforts have been made in constructing genetic linkage maps, improving the genome coverage, marker density, and integrating the map for better resolution. The first linkage map based on RFLP marker used an inter-varietal F_2 population (Liu et al. 1994). The first integrated map was constructed using four F_2 populations developed from LGD x ICMP 85410, 81B x ICMP 451, ICMB 841 x 863B, and PT 732B x P1449-2 crosses (Qi et al. 2004). The F_2 population of the ICMB 841-P3 x 863B-P2 cross was also used to integrate a newly developed SSR markers in previous maps (Senthilvel et al. 2008).

Subsequently, RILs were used for genetic linkage mapping of pearl millet genome. RILs developed from Tift23DB x PI536400 cross (Pedraza-Garcia et al. 2010), H 77/833-2 x PRLT 2/89-33 (Supriya et al. 2011), H77/833-2 x PRLT 2/89-33 (Sehgal et al. 2012) crosses were used for genetic linkage mapping. Four RILs populations developed from ICMB 841-P3 x 863B-P2, H 77/833-2 x PRLT 2/89-33, 81B-P6 x ICMP 451-P8, and PT 732B-P2 x P1449-2-P1 were used to construct a consensus linkage map (Rajaram et al. 2013). An F_2 population of a cross between wild pearl millet (116_11-(PS202-14)-121) and a cultivated pearl millet (SOSAT-IBL-197) was also used to map high density map (Moumouni et al. 2015). To circumvent the problems of segregation distortion and masking of minor-effect alleles/QTLs, a set of chromosome segment substitution lines (CSSLs) for all the seven LGs was developed by introgression of overlapping chromosome segments from 863B line into the genetic background of elite ICMB 841 line (Kumari et al. 2014). These are valuable genetic stocks for minor-effect QTL detection, fine mapping, and trait mechanism studies, especially for complex traits in pearl millet.

The second-generation mapping populations such as nested-association mapping (NAM) and multi-parent advanced generation inter-cross (MAGIC) populations are derived from multiple elite breeding lines with a combination of useful traits. These populations are useful for precise QTL mapping and for use in cultivar development. However, their development in pearl millet is not yet emphasized.

6.6.4 Genetic Linkage Maps

Obviously, the marker system and speed of genotyping evolved over time. Much improvement has been made in genome coverage, marker density, and resolution since the first map of pearl millet (Table 6.3). The coverage is more than double and the density and distribution of markers dramatically increased. For comparison, the first RFLP based (Fig. 6.6a) and the recent haplotype based (Fig. 6.6b) maps are presented below.

Table 6.3 List of genetic linkage maps of pearl millet constructed using different populations and marker systems

Population and cross	Marker system	Loci mapped	Total length (cM)	Marker density (cM)	References
F_2 LGD x ICMP 85410	RFLP	181	303	2	Liu et al. (1994)
RILs (152) Tift23DB x PI536400	RAPD, SRAP, ISSR, SSR	196	1796	9.2	Pedraza-Garcia et al. (2010)
F_2 (four) LGD x ICMP 85410, 81B x ICMP 451, ICMB 841 x 863B, PT 732B x P1449-2	RFLP, SSR	242[a]	473	2	Qi et al. (2004)
F2 (149) I CMB 841-P3 x 863B-P2	SSR	27[b]			Senthilvel et al. (2008)
RILs (140) H 77/833-2 x PRLT 2/89-33	DArT, SSR	321	1148	3.7	Supriya et al. (2011)
RILs (88) H77/833-2 x PRLT 2/89-33	EST-SSR, SNP, CISP	133	815.3	6.1	Sehgal et al. (2012)
RILs (four) ICMB 841-P3 x 863B-P2, H 77/833-2 x PRLT 2/89-33, 81B-P6 x ICMP 451-P8, PT 732B-P2 x P1449-2-P1	EST-SSR, gSSR, STS	174	899	5.2	Rajaram et al. (2013)
F_2 (93) 116_11-(PS202-14)-121 x SOSAT-IBL-197	SNP	314[c]	640	2.1	Moumouni et al. (2015)
RILs (150) Tift 99DB x Tift 454	SNP	16,650	716.7	2.1	Punnuri et al. (2016)

[a]The information presented is for the consensus map; [b]Newly developed SSR markers integrated in previous maps (Yadav et al. 2004; Bidinger et al. 2007); [c]Non-redundant haplotypes

6.6.5 Mapped QTLs

Prompted by the straight forward diploid genetics and high levels of polymorphism, several QTL controlling important traits in pearl millet have been detected. As drought tolerance is the most important trait of the crop, mapping QTL for terminal drought tolerance (Yadav et al. 2002) has provided a more-targeted approach to

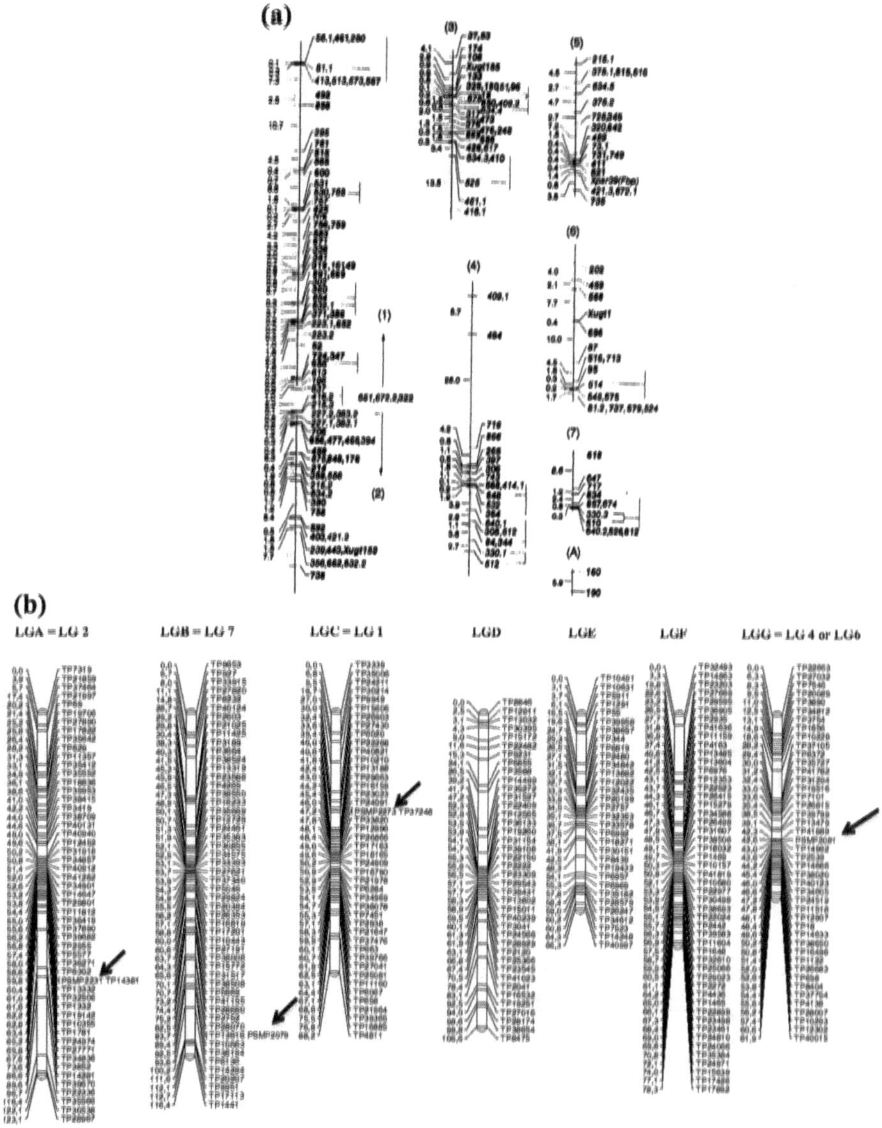

Fig. 6.6 The first genetic linkage map of pearl millet constructed using RFLP markers in F_2 population of LCD-I-B-I0 x ICMP 85410 cross (Liu et al. 1994) (**a**), Haplotype-based genetic map of pearl millet developed using SNP markers developed by genotyping-by-sequencing (GBS) (Moumouni et al. 2015) (**b**) (Adopted with permission from **a** Liu et al. 1994, Theoretical and Applied Genetics 89(4):481–7, and **b** Moumouni et al. 2015, Molecular Breeding 35:5)

improving the drought tolerance and yield in water-limited environments (Howarth and Yadav 2002).

Consequently, some efforts have been made in genome mapping and detection of QTL for specific traits (Table 6.4). With the aim of identifying genes underlying drought tolerance (DT) QTL, facilitate understanding of molecular mechanisms of drought tolerance, and accelerate genetic improvement of pearl millet through MAS, a genetic linkage map based on gene-based markers was constructed (Sehgal et al. 2012). In considerably short period of time, hundreds of pearl millet molecular markers were developed (Liu et al. 1994; Qi et al. 2001; Allouis et al. 2001) and genetic linkage maps were produced (Liu et al. 1994; Devos et al. 2000). These maps became the platform for the detection of QTL for downy mildew resistance (Jones et al. 1995; Jones et al. 2002).

6.7 Marker-Assisted Breeding

Valuable genetic improvements have been made to several important traits of pearl millet through conventional breeding. It is presumed that considerable increase in efficiency can be achieved through the application of MAS. Thus far numerous QTL for important traits such as yield and yield components, disease resistance, and drought tolerance have been mapped.

QTL mapping for yield and yield components (Poncet et al. 2000; Yadav et al. 2003), drought tolerance (Yadav et al. 2002), downy mildew (Jones et al. 1995, 2002; Yadav et al. 1999) and rust resistance have been conducted. A major QTL mapped for terminal drought tolerance on linkage group 2 (LG_2) (Yadav et al. 2002) was validated through introgression of the QTL region from the donor parent genome to a hybrid parent (Serraj et al. 2005). Similarly, introgression of a downy mildew resistance QTL into an elite parental lines through RFLP-based marker-assisted backcrossing was successfully done (Thakur et al. 2008). Recently there are QTL mapped for different traits using high throughput marker system. This QTL based on high resolution maps and the associated markers are readily available for MAS. It was suggested that development of second-generation mapping population and study of market-trait association studies using high throughput markers will advance the breeding process (Serba and Yadav 2016).

6.7.1 Germplasm Characterization and Distinctness, Uniformity, and Stability (DUS)

It is evident that the extent of genetic variability is the determining factor for plant breeding success. A vast genetic variability is available in pearl millet. Natural and human selection processes imposed since its domestication, resulted in the

Table 6.4 Major and minor effect QTLs and genes detected for various traits in pearl millet

Trait	QTL locus/markers	LG	LOD value	PVE/ Effect	References
Grain yield	Xpsm592-Xpsm356	2	6.91	/−17.27	Yadav et al. (2003, 2004)
	Xpsm464-Xpsm716	4	2.32	/−10.62	
	Xpsm588-Xpsm514	6	2.21	/11.26	
	Xpsm322-Xpsm2050	2	6.07	33.2/	
	Xpsmp2064-Xpsm345	5	2.39	16.9/	
Stover yield	Xpsm295-Xpsm416.3	4	4.12	/−21.86	Yadav et al. (2003), Poncet et al. (2000)
	Xpsm87.1-Xpsm514	6	4.96	/23.84	
	Xpsm322-Xpsm2050	2	2.82	30.0/	
	Xpsmp2064-Xpsm345	5	4.36	22.7/	
	Xpsm514-Xwg110	6	2.86	15.5/	
	Xpsmp2074-Xpsmp2027	7	6.83	40.8/	
Biomass yield	Xpsm592-Xpsm443	2	8.22	/−49.79	Yadav et al. (2003)
	Xpsm716-Xpsm265	4	3.84	/−35.03	
	Xpsm87.1-Xpsm514	6	3.62	/37.72	
Harvest index	Xpsm618-Xpsm717	6	2.20	/−0.76	Yadav et al. (2003, 2004), Kannan et al. (2014)
	Xpsmp2248_166	6	4.33	–	
	Xpsm322-Xpsm2050	2	5.35	–	
Plant height	D_1/d_1	1	–	–	Azhaguvel et al. (2003), Kannan et al. (2014), Kumar et al. (2017), Poncet et al. (2000)
	D_2/d_2	4	–	–	
	Xpsmp2085_175	4	–	–	
	Xpsmp2224_157	7	–	–	
	Pgpb6112-pgpb9106	1	6.4	30.2/2.9	
	Pgpb9498-Xipes017	1	8.54	32.3/−3.0	
	Xipes203-Xpsmp2273	1	8.8	33.0/5.9	
	Pgpb12094-pgpb10685	2	6.67	28.7/2.1	
	Xpsmp322-Xipes181	2	6.45	25.5/2.7	
	Xipes162-Xipes163	2	5.7	22.9/3.6	
	pgpb7379-Xipes2227	3	5.91	23.6/1.8	
	Pbpb6901-pgpb8757	3	9.83	39.5/−3.1	
	Xpsmp2085-Xipes225	4	8.87	34.1/6.1	
	Xipes207-Xicmp3058	6	5.78	23.2/−0.4	
	Xipes153-Xpsmp2040	7	9.7	35.8/−6.0	
	Pbpb8626-Xipes205	7	6.42	25.4/−2.9	
	Xipes198-Xipes082	7	11.13	39.8/−6.4	
	Hmax6	6	–	25.3/	
	hmax7	7	2.73	16.1/	
Panicle length	los1	1	2.45	13.2/1.27	Poncet et al. (2000), Kumar et al. (2017)
	los2	2	7.37	35.9/−2.13	
	pgpb9498-Xipes17	1	6.93	27.1/−1.4	
	Xipes226-Xicmp3032	1	7.27	28.2/0.7	
	Xipes4-Xipes229	1	6.42	25.4/0.4	
	Xipes7-Xpsmp2088	2	7.34	28.4/1.7	
	pgpb9647-Xicmp3027	5	6.08	24.4/−0.4	
	Xipes200-Xicmp3002	6B	8.61	32.5/−0.4	
	pgpb10687-pgpb10299	6B	5.94	24.6/0.9	
	pgpb12322-pgpb8782	6B	5.95	26.0/1.1	
	pbpb9915-Xipes206	7	6.23	24.7/−1.0	
	pbpb9819-Xpsmp2074	7	6.65	26.2/0.2	
	Xipes198-Xipes82	7	7.84	30.1/−1.0	

(continued)

Table 6.4 (continued)

Trait	QTL locus/markers	LG	LOD value	PVE/ Effect	References
Panicle number	*Xpsm858-Xpsm565*	1	2.65	/0.68	Yadav et al. (2003)
	Xpsm592-Xpsm443	2	8.29	/−1.87	
	Xpsm95-Xpsm575	6	2.34	/−1.07	
	Xpsm618-Xpsm717	7	4.04	/−1.31	
Flowering time	*Xpsm59-Xpsm443*	2	6.79	/−26.98	Yadav et al. (2003), Kannan et al. (2014), Kumar et al. (2017), Poncet et al. (2000), Punnuri et al. (2016)
	Xpsm416.3-Xpsm196.2	4	3.92	/−0.75	
	Xpsm87.1-Xpsm95	6	9.41	/1.26	
	Xpsmp2248_162	6	5.78	/0.30	
	Pgpb6981-Xipes226	1	8.42	32.9/	
	Xipes236-Xpsmp2059	2	6.46	25.2/	
	Pgpb11647- Xipes166	3	5.85	23.4/	
	Pgpb7379-Xpsmp2227	3	14.68	48.8/	
	Pgpb9967-Xpsmp11527	4	6.05	24.1/	
	Pgpb10505-Xipes230	5	10.05	36.8/	
	head5	5	10.75	59.0/10.93	
	head7	7	3.68	24.2/−5.52	
	S1_1423-S1_3590	1	2.61	3.03/1.8	
	S2_1896-S2_2803	2	4.86	6.00/−2.0	
	S5_0012-S5_1669	5	2.38	4.75/1.5	
	S7_0244-S7_2067	7	2.48	0.49/1.3	
Panicle threshing percentage	Xpsmp2085_175	4	2.4		Kannan et al. (2014)
1000 grain weight	Xipes200-Xicmp3002	6	7.36	28.5/0.8	Kannan et al. (2014)
Iron (Fe)	Xpsmp2214-Xipes142	3	4.68	20.5/8.5	Kumar et al. (2016, 2017)
	Xpsmp2214-Xipes142	3	4.68	20.5/8.5	
	Xipes017-pgpb 12900	1	6.22	9.0/4.0	
	pgpb10531-pgpb9130	1	25.36	31.9/9.7	
	Xipes188-pgpb6069	3	6.59	9.5/0.4	
	pgpb9502-pgpb6039	4B	6.87	10.4/−0.6	
	pgpb11956-pgpb9273	7	7.25	12.5/−1.9	
	pgpb8427-pgpb13221	7	8.58	12.2/5.3	
	pgpb11938-pgpb8987	7	8.83	12.5/4.9	
	pgpb6825-Xipes195	7	9.70	14.0/0.1	
	pgpb8445-pgpb11206	A	7.67	12.4/4.0	
	pgpb10660-pgpb8626	D	7.00	11.6/1.2	
	pgpb10727-Xipes179	E	9.36	14.3/3.1	
Zinc (Zn)	Xpsmp2214-Xipes142	3	9.66	32.3/8.5	Kumar et al. (2016), 2017)
	pgpb10531-pgpb9130	1	23.93	30.4/6.7	
	pgpb10397-pgpb10394	1	6.50	9.4/1.7	
	pgpb9502-pgpb6039	4B	7.33	11.1/−0.6	
	pgpb10483-pgpb11463	4B	6.68	11.6/−2.2	
	pgpb13229-pgpb12681	5	8.17	11.6/2.7	
	Xipes198-pgpb8427	7	7.16	10.2/2.7	
	pgpb12329-pgpb9721	7	7.58	10.9/2.8	
	pgpb8779-pgpb12691	H	6.68	11.6/2.1	

(continued)

Table 6.4 (continued)

Trait	QTL locus/markers	LG	LOD value	PVE/ Effect	References
Downy mildew	*M413-M93* *M543-M380* *M37-M248* *M390-M318*	1 2 3 5	27.4 7.2 5.0 5.4		Jones et al. (2002)
Rust	*Rr_1* *Rust Res(ICMP83506)*	3 4		– –	Morgan et al. (1998)
Leafspot	S2_7773-S2_8331 S3_0019-S3_4763 S5_2145-S5_4145 S7_0738-S7_3864	2 3 5 7	2.18 2.25 4.56 3.01	1.78/−0.6 1.82/−0.5 4.83/0.9 5.05/0.9	Punnuri et al. (2016)
Drought tolerance*	M214-M443 M716-M416 M87.1-M514	2 4 6			Yadav et al. (1999)

PVE = Phenotypic value explained (%)

development of diverse cultivars adapted to different environments, suited to various production systems, and aligned with different consumer preferences (Brunken 1977). There are wide variations in various morphological traits, phenological, yield, and quality traits often corresponding to adaptation zones especially in the center of diversity.

To ensure the availability of these genetic variations for breeding programs, collection, conservation, characterization, evaluation, and documentation of the germplasm are very important pre-breeding activities. A concerted effort has been made mainly by ICRISAT to collect from different geographic areas and also consolidate collections done by different centers and different groups around the world (Upadhyaya et al. 2007; Yadav et al. 2017). Accordingly, ICRISAT genebank collected and conserved a total of 22,288 pearl millet germplasm. In addition to ICRISAT, Institute of Research for Development (IRD, France) has 3,968 accessions collected from 16 countries, Canadian Genetic Resources Program (Saskatoon, Canada) has 3,821 accessions of cultivated *P. glaucum* and related species, and the US Germplasm Resource Information Network (GRIN) collected and preserved more than 1283 accessions (Serba et al. 2017; Yadav et al. 2017).

6.8 BAC Library

With the objective to develop resources from which SSR markers can be developed for pearl millet, a bacterial artificial chromosome (BAC) library was constructed (Allouis et al. 2001). Through a novel way to developing new markers, 25 SSR markers that can anchor individual BACs to the genetic maps were developed from 40 BAC pools, comprising a total of 384 clones (Qi et al. 2001).

6.9 Genomics-Assisted Breeding

6.9.1 Genome Sequence

A draft whole genome sequence of 1.79 Gb has been assembled using whole genome shotgun (WGS) and BAC library sequencing of a reference genotype Tift 23D2B1-P1-P5 (Varshney et al. 2017). This genome resource is expected to bring a paradigm shift to pearl millet molecular breeding by providing a foundation for accelerating gene mapping and characterization. The unparalleled heat and drought tolerance has been attributed to the enrichment of wax biosynthesis genes discovered through the genome sequence. In addition to the whole genome sequence, 963 pearl millet inbred lines and 31 wild accessions preserved at ICRISAT were re-sequenced to generate data that give insight into the population structure, genetic diversity, and domestication of the crop. The genome sequence and the resequencing data is expected to facilitate future marker-trait association studies, to define heterotic pools, and predict hybrid performance.

Tift23D2B1, an important ancestral genotype of many seed parents of pearl millet hybrids in use for forage and grain as well as dual-purpose hybrids was used for the genome sequence. Tift23D2B1 was developed at the Coastal Plain Experiment Station (Tifton, Georgia, USA) by introducing d_2 dwarfing gene into the genetic background of elite seed parent maintainer line Tift 23B1. The genome features such as repetitive DNA, gene density, and SNPs identified using resequencing of a pearl millet inbred germplasm association panel (PMiGAP) is visualized (Fig. 6.7). A total of 38,579 gene models with mean coding sequence length of 1,014.71 bp were annotated. Higher gene density was observed in the telomeric region than the peri-centromeric region of the pseudomolecules (exact centromere location unknown). Conversely, repetitive DNA density was higher in peri-centromeric region than telomeric region. The average GC content of pearl millet (47.9%) is higher than that of foxtail millet (46.1%), sorghum (44.5%), barley (*Hordeum vulgare*, 44.4%), and rice (43.5%). These resources are important preludes for the application of genomic-assisted breeding to improve this climate-resilient crop for agronomic traits and yield potential.

The assembled pearl millet genome accounts for more than 77% of the repetitive elements. This percentage of its repetitive DNA is consistent with the previous estimate of 80% repetitive DNA of the crop (Paterson et al. 2009; Bennetzen et al. 2012) which is close to the proportion of repetitive DNA found in the 2.3-Gb maize genome (nearly 85%) (Schnable 2009), and considerably higher than 730-Mb of sorghum (~61%) (Paterson et al. 2009), ~400-Mb of foxtail millet (~46%) (Bennetzen et al. 2012) or 466-Mb of rice (~42%) (Yu 2002) genomes. More than 50% of its nuclear genome repetitive DNA is classified as long-terminal repeat retrotransposons (Varshney et al. 2017).

In pearl millet, there were several independent domestication events that resulted in four different races (Brunken et al. 1977). Enriching the genome sequence through resequencing of the populations used in the QTL mapping and in entirety

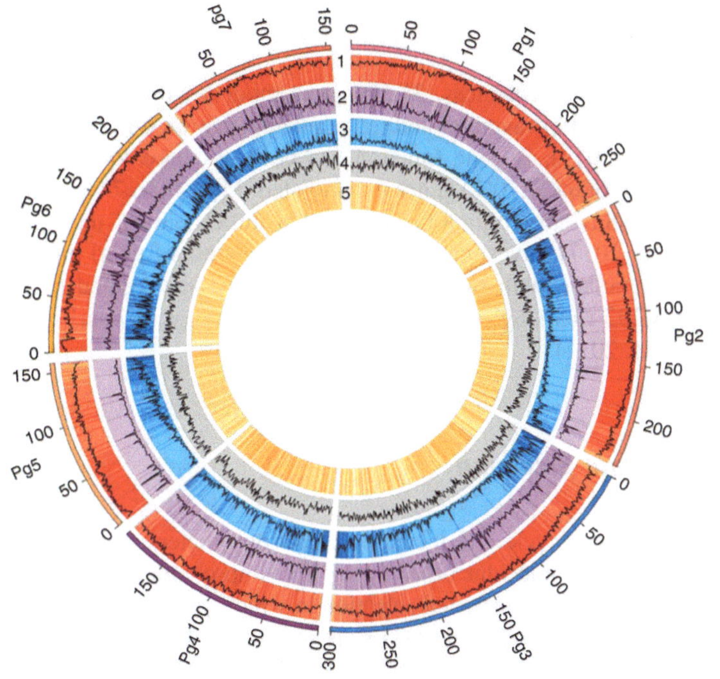

Fig. 6.7 Genome features in 1-Mb intervals across the seven pseudomolecules (Varshney et al. 2017). Units on the circumference are megabase values of pseudomolecules. (1) Repeat density, (2) tandem repeat density, (3) gene density, (4) GC content and (5) SNPs identified by resequencing PMiGAP lines in 1-Mb bins. The genome assembly furnished an average GC content of 47.9% and contained 38,579 gene models with mean coding sequence length of 1,014.71 bp (Adopted from Varshney et al. 2017, *Nature Biotechnology* 35:969–976)

of independently sequencing a genotype from other race thus becomes highly desirable. To conduct inclusive characterization of the genetic factors responsible for the seed morphological variations used as a basis for the classification, integrating the known races into the genome sequence would broaden the application of the resource.

With the assumption that origin of domestication corresponds with the greatest genetic diversity (Vavilov 1992), domestication origin of pearl millet is recently inferred based on whole genome sequence of a large number of wild forms and cultivated landraces (Burgarella et al. 2018) representative of the geographical diversity of pearl millet (Varshney et al. 2017). The result indicated Western Sahara as the original center of domestication of pearl millet. It was also used to predict the onset of cultivated of pearl millet in Africa back to 4,900 years ago. The genome sequence established wild-to-crop gene flow that increased cultivated genetic diversity leading to diversity hotspots in western and eastern Sahel and adaptive introgression of 15 genomic regions. The result of this study reconciled the genetic and archaeological data available for the crop.

6.9.2 Gene Annotation

Based on the genome sequences, 38,579 gene models with an average transcript size of 3,945 bp and an average coding sequence size of 687 bp were estimated (Varshney et al. 2017). A gene function has been assigned to 27, 893 (72.30%) genes, leaving 10,686 (27.70%) genes unannotated. More than 74% of the predicted pearl millet proteins have orthologues in foxtail millet, one of the related species (Varshney et al. 2017). Investigation of the completeness of pearl millet genes was studied in comparison with rice genes. The comparison revealed that more than 90% of the 4,202 rice single-copy genes have homology in pearl millet genome (Varshney et al. 2017).

6.9.3 Genomic Selection

Accelerated breeding cycle and increased selection efficiency are the two reasons behind the drive for molecular breeding in crop improvement. Genomic selection (GS) is a system of marker-assisted selection in which genetic markers covering the whole genome are used so that all QTLs are in linkage disequilibrium with at least one marker (Goddard and Hayes 2007). It is advocated to accelerate the breeding cycle and facilitate efficient selection of superior genotypes (Crossa et al. 2017). It is an advanced molecular breeding approach in which high throughput SNPs in the whole genome are indiscriminately used in the genomic-enabled prediction of the breeding value of candidate genotypes for selection. This accelerated genetic gain approach integrated with a precise phenotyping is promising to identify superior alleles in a germplasm. Therefore, it has become a favored genomics-assisted breeding approach that can be integrated in germplasm enhancement and efficient development of climate resilient cultivars (Varshney et al. 2018).

Markers developed from a resequencing data were used to carry out GS to predict grain yield for test crosses in favorable, early stress, late stress, and across environments (Varshney et al. 2017). High prediction accuracy, measured as the Pearson correlation coefficient between the predicted and observed values, standardized with the square root of the heritability (h = 0.78), amounting to 0.6 was observed for the performance across environments. As a modeling study in wheat indicates GS with this level of prediction accuracy could substantially improve selection gain per year (Longin et al. 2015).

Another GS study for its application in the hybrid breeding observed that phenotypic data from inbred parents can improve genomic prediction in pearl millet hybrids (Liang et al. 2018). In this study, data from hybrids gave high prediction accuracies for 1000-grain weight (0.73–0.74), days to flowering (0.87–0.89), and plant height (0.72–0.73), followed by grain yield (0.48–0.51). When BLUPs were used to control for the effects of heterosis, the inclusion of inbred phenotypic datasets moderately improved genomic predictions of the hybrid genomic estimated breeding values. But for traits with little to no heterosis, no changes in prediction accuracy were observed between hybrid only and hybrid/inbred data.

6.10 Defining Heterotic Gene Pools

Enhancement of grain and stover yield is the most important breeding objective. Lately there has been a lot of emphasis on the development of high yielding single-cross hybrids. Categorization of hybrid parents into different heterotic groups is a basic prerequisite for maximum exploitation of heterosis in single-cross hybrids. In pearl millet, development of diverse sets of seed (A-/B-) and pollen/restorer (R-lines) inbred parents from genetically distant gene pools are a starting point for the development of a hybrid breeding program. In this direction, a first attempt was made by ICRISAT and partners to identify heterotic groups for grain yield using EST and genomic SSR markers (Ramya et al. 2018). The study revealed that largely the maintainer and restorer lines form two broad gene pools, which can be broken down into sub-groups for maximization of heterosis for grain yield. On the basis of SSR genotyping data, the B- lines were differentiated into 10 sub-clusters (B1 through B10), and R- lines into 11 sub-clusters (R1 through R11). Based on per se performance, high specific combining ability (SCA) effects and standard heterosis, a total of seven heterotic groups were identified. These were B10R5, B3R5, B3R6, B4UD, B5R11, B2R4, and B9R9 (Fig. 6.8).

6.11 Role of Bioinformatics

To conduct an integrated genomics research on pearl millet, the utilization of a common communication tool among different centers is crucial. Having a common tool facilitates efficient use of resources that are generated at different centers, such as genome information and sharing of research results. In this regard, bioinformatics plays a vital role not only in boiling down the vast NGS data and generating useful information but also communication. The availability of genome, comparative, gene expression and protein databases provide useful information about the crop.

To empower pearl millet crop improvement genome sequencing and assembly was conducted recently. To unravel population structure, genetic diversity, evolution and domestication history, 994 pearl millet genotypes were re-sequenced. As a result, a genome database (https://www.ncbi.nlm.nih.gov/id=4543) was created.

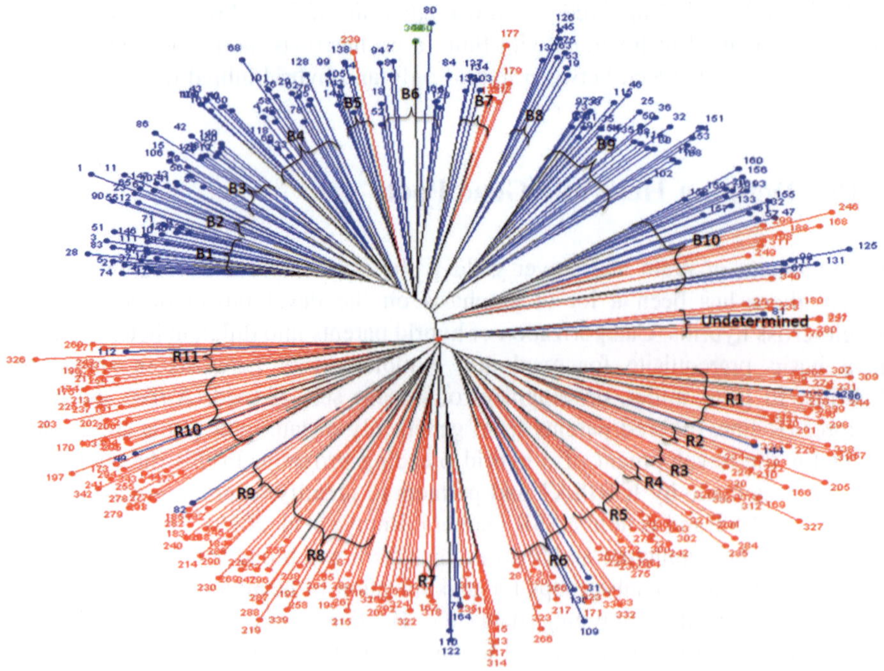

Fig. 6.8 Unweighted neighbor joining tree based on the simple matching dissimilarity matrix of 88 SSR marker data for 347 hybrid parental lines. Lines in blue are B-lines, line in green is Tift 23D2B1-P1-P5 (B-line, and the world reference genotype), lines in red are R-lines; B1 to B10 are the 10 sub-clusters of the B-lines; while R1–R11 are the 11 sub-clusters of the R-lines (Adopted from Ramya et al. (2018). Front. Plant Sci. 8:1934. https://doi.org/10.3389/fpls.2017.01934

6.12 Future Perspectives

Today, fewer crop species are feeding the world than several decades ago. This scenario raises a serious concern about the resilience of the global food system in the wake of global climate change. With lack of on-farm species diversity and dietary intake, more and more people are now exposed to harvest and health failures than ever. Many people are affected by gluten allergy. Gluten free, protein and micronutrient rich crops that can withstand the climatic variability would have a great chance of coming to the main stream production in certain parts of the world where the consequence of climate change would be impactful.

6.12.1 Potential for Expansion of Productivity

Among crops with high potential of production expansion, pearl millet is expected to come to the forefront. This is attributed to its exceptional heat and drought tolerance and high nutritional value. Productivity has been low; however, some improvements have been made in India. Improved hybrids have been widely adopted by Indian farmers with the result that the crop productivity has gone up from 305 kg ha^{-1} during 1951–1955 to 998 kg ha^{-1} during 2008–2012, registering a 227% improvement which assumes greater significance is being given despite more than 90% of pearl millet is grown as rainfed and often on marginal lands (Yadav and Rai 2013). Launching hybrid breeding in major pearl millet growing countries in west Africa have a glimmer of hope to bring about a paradigm shift in productivity in the region.

6.12.2 Potential for Expansion into Nontraditional Areas

A small number of crop species that are being produced in large scale account for overall human diet. These crops have become the key in global food security. With the looming climate change, the current production system based on a limited number of crop species is vulnerable to the effect of climate variability. To avert such anticipated effect of the climate change, diversity in crop species is an ultimate solution.

Pearl millet is a drought resilient crop with heat and salinity tolerance makes it a potential for expansion in dryland as well as agriculturally favorable areas of the world. However, yield advantage of other crops like sorghum is impeding its popularity among the commercial producers in many areas. With the application of genomics, genetic yield potential, biotic and abiotic stresses tolerance, and nutritional quality improvements is expected to progress in the near future. With the expected improvement and sizable production areas of major crops challenged by climatic changes, the prospect of pearl millet to be considered as a major crop is high.

6.12.3 Potential for Nutritional Enhancement

Germplasm and breeding lines have much higher nutritional level than farmers growing OPVs or hybrids. So far, millet breeders are focused on productivity of gain like other crops. The genetic enchantment of essential nutrients of new cultivars has to be achieved with no penalty on grain yield productivity. Micronutrient screening is highly expensive to deal with larger germplasm and breeding lines when resources are limited. With availability of X-ray fluorescent (XRF) tool, cost of biofortification breeding will decrease over time, and micronutrient content built into the gene pool will not affect future breeding for yield traits. Apparently micronutrient traits are not affected by genetic erosion (just like d_2 gene revolution) and involve little maintenance breeding after the genes are transferred to elite lines. Therefore, to cope with climate change and nutritional importance, diverting

25–30% of breeding investment as part of mainstreaming these traits at NARS and seed companies would be highly helpful in delivering nutritionally-dense pearl millet cultivars to farmers in India and SSA in near future.

Acknowledgements Funding for the senior author was provided by the United States Agency for International Development under Cooperative Agreement No. AID-OAA-A-13-00047 with the Kansas State University Sorghum and Millet Innovation Lab (SMIL). The contents are solely the responsibility of the authors and do not necessarily reflect the views of USAID or others. This is contribution number 19-043-B from the Kansas Agricultural Experiment Station.

References

Alagarswamy G, Bidinger FR (1987) Genotypic variation in biomass production and nitrogen use efficiency in pearl millet [*Pennisetum americanum* (L.) Leeke]. In: Gabelman WH, Loughman BC (eds) Genetic aspects of plant mineral nutrition: proceedings of the second international symposium on genetic aspects of plant mineral nutrition, University of Wisconsin, Madison, June 16–20, 1985. Springer Netherlands, Dordrecht, pp 281–286

Ali GM, Murtaza N, Collins JC, McNeilly T (2006) Study of salt tolerance parameters in pearl millet *Pennisetum americanum*. J Cent Eur Agri 7:365–376

Allouis S, Qi X, Lindup S, Gale MD, Devos KM (2001) Construction of a BAC library of pearl millet, *Pennisetum glaucum*. Theor Appl Genet 102:1200–1205

Amadou I, Gounga ME, Le GW (2013) Millets: nutritional composition, some health benefits and processing—a review. Emir J Food Agri 25:501–508. https://doi.org/10.9755/ejfa.v25i7.12045

Ambawat S, Senthilvel S, Hash CT, Nepolean T, Rajaram V, Eshwar K, Sharma R, Thakur RP, Rao VP, Yadav RC, Srivastava RK (2016) QTL mapping of pearl millet rust resistance using an integrated DArT- and SSR-based linkage map. Euphytica 209:461–476

Andrew RL, Wallis IR, Harwood CE, Foley WJ (2010) Genetic and environmental contributions to variation and population divergence in a broad-spectrum foliar defence of *Eucalyptus tricarpa*. Ann Bot 105:707–717

Andrews D, Kumar KA (1996) Use of the west African pearl millet iniadi in cultivar development. Plant Genet Resour Newsl 105:15–22

Anuradha N, Satyavathi CT, Bharadwaj C, Nepolean T, Sankar SM, Singh SP, Meena MC, Singhal T, Srivastava RK (2017) Deciphering genomic regions for high grain iron and zinc content using association mapping in pearl millet. Front Plant Sci 8:412

Appa Rao S, Mengesha MH, Reddy CR (1986) New sources of dwarfing genes in pearl millet (*Pennisetum americanum*). Theor Appl Genet 73:170–174

Azhaguvel P, Hash CT, Rangasamy P, Sharma A (2003) Mapping the *d1* and *d2* dwarfing genes and the purple foliage color locus *P* in pearl millet. J Hered 94:155–159

Bailey AV, Piccolo B, Sumrell G, Burton GW (1979) Amino acid profiles, chemical scores, and mineral contents of some pearl millet inbred lines. J Agri Food Chem 27:1421–1423

Bashir EMA, Ali AM, Ali AM, Mohamed ETI, Melchinger AE, Parzies HK, Haussmann BIG (2015) Genetic diversity of Sudanese pearl millet (*Pennisetum glaucum* (L.) R. Br.) landraces as revealed by SSR markers, and relationship between genetic and agro-morphological diversity. Genet Resour Crop Evol 62:579–591

Bationo A, Christianson CB, Klaij MC (1993) The effect of crop residue and fertilizer use on pearl millet yields in Niger. Fertil Res 34:251–258

Bennett PM (2004) Genome plasticity. In: Woodford N, Johnson AP (eds) Genomics, proteomics, and clinical bacteriology: methods and reviews. Humana Press, Totowa, NJ, pp 71–113

Bennetzen JL, Schmutz J, Wang H, Percifield R, Hawkins J, Pontaroli AC, Estep M, Feng L, Vaughn JN, Grimwood J, Jenkins J, Barry K, Lindquist E, Hellsten U, Deshpande S, Wang X, Wu X, Mitros T, Triplett J, Yang X, Ye C-Y, Mauro-Herrera M, Wang L, Li P, Sharma M, Sharma R, Ronald PC, Panaud O, Kellogg EA, Brutnell TP, Doust AN, Tuskan GA, Rokhsar D, Devos KM (2012) Reference genome sequence of the model plant Setaria. Nat Biotechnol 30:555–561

Berg A, de Noblet-Ducoudré N, Sultan B, Lengaigne M, Guimberteau M (2013) Projections of climate change impacts on potential C4 crop productivity over tropical regions. Agri For Meteorol 170:89–102

Bertin I, Zhu JH, Gale MD (2005) SSCP-SNP in pearl millet-a new marker system for comparative genetics. Theor Appl Genet 110:1467–1472

Bhattacharjee R, Bramel P, Hash C, Kolesnikova-Allen M, Khairwal I (2002) Assessment of genetic diversity within and between pearl millet landraces. Theor Appl Genet 105:666–673

Bhattacharjee R, Khairwal IS, Bramel P, Reddy KN (2007) Establishment of a pearl millet [*Pennisetum glaucum* (L.) R. Br.] core collection based on geographical distribution and quantitative traits. Euphytica 155:35–45

Bidinger FR, Raju DS (1990) Effects of the *d2* dwarfing gene in pearl millet. Theor Appl Genet 79:521–524

Bidinger FR, Nepolean T, Hash CT, Yadav RS, Howarth CJ (2007) Quantitative trait loci for grain yield in pearl millet under variable postflowering moisture conditions. Crop Sci 47:969–980

Bita CE, Gerats T (2013) Plant tolerance to high temperature in a changing environment: scientific fundamentals and production of heat stress-tolerant crops. Front Plant Sci 4:1–18

Blum A (2005) Drought resistance, water-use efficiency, and yield potential—are they compatible, dissonant, or mutually exclusive? Aust J Agri Res 56:1159–1168

Bono M (1973) Contribution à la morpho-systématique des Pennisetum annuels cultivés pour leur grain en Afrique Occidentale francophone. L'Agronomie Trop Série 3. Agron Générale Etudes Sci 28:229–356

Brunken JN (1977) A systematic study of *Pennisetum* sect. *Pennisetum* (Gramineae). Am J Bot 64:161–176

Brunken J, de Wet JMJ, Harlan JR (1977) The morphology and domestication of pearl millet. Econ Bot 31:163–174

Buerkert A, Moser M, Kumar AK, Fürst P, Becker K (2001) Variation in grain quality of pearl millet from Sahelian West Africa. Field Crops Res 69:1–11

Burgarella C, Cubry P, Kane NA, Varshney RK, Mariac C, Liu X, Shi C, Thudi M, Couderc M, Xu X, Chitikineni A, Scarcelli N, Barnaud A, Rhoné B, Dupuy C, François O, Berthouly-Salazar C, Vigouroux Y (2018) A western Sahara centre of domestication inferred from pearl millet genomes. Nat Ecol Evol 2:1377–1380

Burton GW (1951) Quantitative inheritance in pearl millet (*Pennisetum glaucum*). Agron J 43:409–417

Burton GW (1965) Photoperiodism in pearl millet, *Pennisetum typhoides*. Crop Sci 5:333–335

Burton GW, Fortson JC (1966) Inheritance and utilization of five dwarfs in pearl millet (*Pennisetum typhoides*) breeding. Crop Sci 6:69–72

Burton GW, Powell JB (1968) Pearl millet breeding and cytogenetics. Adv Agron 20:49–89

Burton GW, Monson WG, Johnson JC, Lowrey RS, Chapman HD, Marchant WH (1969) Effect of the *d2* dwarf gene on the forage yield and quality of pearl millet 1. Agron J 61:607–612

Butler EJ (1907) Some diseases of cereals caused by *Sclerospora graminicola*. Memiors of the Department of Agriculture in India. Bot Ser 2:1–24

Cameron KD, Teece MA, Smart LB (2006) Increased accumulation of cuticular wax and expression of lipid transfer protein in response to periodic drying events in leaves of tree tobacco. Plant Physiol 140:176–183

Cattivelli L, Rizza F, Badeck FW, Mazzucotelli E, Mastrangelo AM, Francia E, Marè C, Tondelli A, Stanca AM (2008) Drought tolerance improvement in crop plants: an integrated view from breeding to genomics. Field Crops Res 105:1–14

Cherney JH, Axtell JD, Hassen MM, Anliker KS (1988) Forage quality characterization of a chemically induced brown-midrib mutant in pearl millet. Crop Sci 28:783–787

Cherney DJ, Patterson JA, Johnson KD (1990) Digestibility and feeding value of pearl millet as influenced by the brown-midrib, low-lignin trait. J Anim Sci 68:4345–4351

Clotault J, Thuillet AC, Buiron M, De Mita S, Couderc M, Haussmann BIGG, Mariac C, Vigouroux Y (2012) Evolutionary history of pearl millet [*Pennisetum glaucum* (L.) R. Br.] and selection on flowering genes since its domestication. Mol Biol Evol 29:1199–1212

Comas LH, Becker SR, Cruz VMV, Byrne PF, Dierig DA (2013) Root traits contributing to plant productivity under drought. Front Plant Sci 4:442

Crossa J, Pérez-Rodríguez P, Cuevas J, Montesinos-López O, Jarquín D, de los Campos G, Burgueño J, González-Camacho JM, Pérez-Elizalde S, Beyene Y, Dreisigacker S, Singh R, Zhang X, Gowda M, Roorkiwal M, Rutkoski J, Varshney RK (2017) Genomic selection in plant breeding: methods, models, and perspectives. Trends Plant Sci 22:961–975

D'Andrea AC, Casey J (2002) Pearl millet and Kintampo subsistence. Afr Arch Rev 19:147–173

Danjuma MN, Mohammed S (2014) Genetic diversity of pearl millet (*Pennisetum typhoides*) cultivars in semi-arid northern Nigeria. J Nat Sci Res 4:34–42

Dave HR (1987) Pearl millet hybrids. In: Witcombe JR, Beckerman SR (eds) Proceedings of the international pearl millet workshop. ICRISAT, Patancheru, AP 502324, India, pp 121–126

Delêtre M, McKey DB, Hodkinson TR (2011) Marriage exchanges, seed exchanges, and the dynamics of manioc diversity. Proc Natl Acad Sci USA 108:18249–18254

Devos KM, Pittaway TS, Reynolds A, Gale MD (2000) Comparative mapping reveals a complex relationship between the pearl millet genome and those of foxtail millet and rice. Theor Appl Genet 100:190–198

Dingkuhn M, Singh BB, Clerget B, Chantereau J, Sultan B (2006) Past, present and future criteria to breed crops for water-limited environments in West Africa. Agri Water Manage 80:241–261

Djanaguiraman M, Perumal R, Ciampitti IA, Gupta SK, Prasad PVV (2017) Quantifying pearl millet response to high temperature stress: thresholds, sensitive stages, genetic variability and relative sensitivity of pollen and pistil. Plant, Cell Environ 41:993–1007

dos Reis GB, Mesquita AT, Torres GA, Andrade-Vieira LF, Vander Pereira A, Davide LC (2014) Genomic homeology between *Pennisetum purpureum* and *Pennisetum glaucum* (Poaceae). Comp Cytogenet 8:199–209

Dubcovsky J, Dvorak J (2007) Genome plasticity a key factor in the success of polyploid wheat under domestication. Science 316:1862–1866

Dujardin M, Hanna WW (1989) Crossability of pearl millet with wild *Pennisetum* species. Crop Sci 29:77–80

Emendack Y, Herzog H, Götz K-P, Malinowski D (2011) Mid-Season water stress on yield and water use of millet (*Panicum miliaceum*) and sorghum (*Sorghum bicolor* L. Moench)

Falster DS, Westoby M (2003) Plant height and evolutionary games. Trends Ecol Evol 18:337–343

FAO (2013) Food and Agriculture Organization of the United Nations, Rome, Italy

FAOSTAT (2013) FAO Statistics YearBook 2013, Food and Agriculture Organization of the United Nations, Rome, Italy

Frink CR, Waggoner PE, Ausubel JH (1999) Nitrogen fertilizer: retrospect and prospect. Proc Nat Acad Sci 96(4):1175–1180

Gahukar RT (1984) Insect pests of pearl millet in West Africa: a review. Trop Pest Manage 30:142–147

Gahukar RT (1988) Problems and perspectives of pest management in the Sahel: a case study of pearl millet. Trop Pest Manage 34:35–38

Gahukar RT (1991) Pest status and control of blister beetles in West Africa. Trop Pest Manage 37:415–420

Gale MD, Devos KM, Zhu JH, Allouis S, Couchman MS, Liu H, Pittaway TS, Qi XQ, Kolesnikova-Allen M, Hash CT (2005) New molecular marker technologies for pearl millet improvement. SAT eJournal 1:1–7

Gemenet DC, Hash CT, Sanogo MD, Sy O, Zangre RG, Leiser WL, Haussmann BIG (2015) Phosphorus uptake and utilization efficiency in West African pearl millet inbred lines. Field Crops Res 171:54–66
George S, Prashanth SR, Parida A (2005) Diversity and species relationship in pearl millet (*Pennisetum typhoides*) and related species. J Plant Biochem Biotechnol 14:141–147
Goddard ME, Hayes BJ (2007) Genomic selection. J Anim Breed Genet 124:323–330
Good AG, Shrawat AK, Muench DG (2004) Can less yield more? Is reducing nutrient input into the environment compatible with maintaining crop production?. Trends Plant Sci 9(12):597–605
Govindaraj M, Rai KN, Kanatti A, Velu G, Shivade H (2016) Breeding high-iron pearl millet cultivars: present status and future prospects. In: 2nd international conference on global food security. October 11–14, 2015, Ithaca, NY, USA
Guo Y, Busta L, Jetter R (2017) Cuticular wax coverage and composition differ among organs of *Taraxacum officinale*. Plant Physiol Biochem 115:372–379
Gupta SC (1995) Inheritance and allelic study of brown midrib trait in pearl millet. J Hered 86:301–303
Gupta GK, Singh D (1996) Studies on the influence of downy mildew infection on yield and yield-contributing plant characters of pearl millet in India. Intl J Pest Manag 42:89–93
Gupta SC, Monyo ES, Rao SA (1993) Registration of SDML 89107 brown midrib pearl millet germplasm. Crop Sci 33:882
Gupta SK, Velu G, Rai KN, Sumalini K (2009) Association of grain iron and zinc content with grain yield and other traits in pearl millet. Crop Improv 36:4–7
Gupta SK, Sharma R, Rai KN, Thakur RP (2011) Inheritance of foliar blast resistance in pearl millet (*Pennisetum glaucum*). Plant Breed 131:217–219
Gupta SK, Nepolean T, Sankar SM (2015a) Patterns of molecular diversity in current and previously developed hybrid parents of pearl millet [Pennisetum glaucum (L.) R. Br.]. Amer J Plant Sci 6:1697–1712
Gupta SK, Rai KN, Singh P, Ameta VL, Gupta SK, Jayalekha AK, Mahala RS, Pareek S, Swami ML, Verma YS (2015b) Seed set variability under high temperatures during flowering period in pearl millet (*Pennisetum glaucum* L. (R.) Br.). Field Crops Res 171:41–53
Gupta SK, Nepolean T, Shaikh CG, Rai K, Hash CT, Das RR, Rathore A (2018) Phenotypic and molecular diversity-based prediction of heterosis in pearl millet (*Pennisetum glaucum* L. (R.) Br.). Crop J 6:271–281
Hamilton EW, Heckathorn SA (2001) Mitochondrial adaptations to NaCl. Complex I is protected by anti-oxidants and small heat shock proteins, whereas complex II is protected by proline and betaine. Plant Physiol 126:1266–1274
Hammer K (1984) Das domestikationssyndrom. Die Kult 32:11–34
Hanna WW (1987) Utilization of wild relatives of pearl millet. In: Witcombe JR, Beckman SR (eds) Proceedings of the international pearl millet workshop, ICRISAT, Patancheru, AP, India, pp 33–42
Hanna WW, Burton GW (1992) Genetics of red and purple plant color in pearl millet. J Hered 83:386–388
Hanna WW, Dujardin M (1986) Cytogenetics of Pennisetum schweinfurthii Pilger and its hybrids with pearl millet. Crop Sci 26:449–453
Hanna WW, Wells HD, Burton GW (1985) Dominant gene for rust resistance in pearl millet. J Hered 76:134
Hare PD, Cress WA, Van Staden J (1998) Dissecting the roles of osmolyte accumulation during stress. Plant Cell Environ 21:535–553
Harlan JR, De-Wet JMJ (1971) Toward a rational classification of cultivated plants. Taxon 20:509–517
Hash CT, Thakur RP, Rao VP, Bhaskar RAG (2006) Evidence for enhanced resistance to diverse isolates of pearl millet downy mildew through gene pyramiding. Intl Sorghum Millets Newsl 47:134–138

Haussmann BIG, Fred Rattunde H, Weltzien-Rattunde E, Traoré PSC, vom Brocke K, Parzies HK (2012) Breeding strategies for adaptation of pearl millet and sorghum to climate variability and change in West Africa. J Agron Crop Sci 198:327–339

Hayat S, Hayat Q, Alyemeni MN, Wani AS, Pichtel J, Ahmad A (2012) Role of proline under changing environments: a review. Plant Signal Behav 7:1456–1466

He S, Yang Y, Morrell PL, Yi T (2015) Nucleotide sequence diversity and linkage disequilibrium of four nuclear loci in foxtail millet (*Setaria italica*). PLoS ONE 10:e0137088

Hillman GC, Davis MS (1990) Domestication rates in wild-type wheats and barley under primitive cultivation. Biol J Linn Soc 39:39–78

Howarth CJ, Yadav RS (2002) Successful marker assisted selection for drought tolerance and disease resistance in pearl millet. Iger Innov 18–21

Hu H, Xiong L (2014) Genetic engineering and breeding of drought-resistant crops. Annu Rev Plant Biol 65:715–741

Hu Z, Mbacké B, Perumal R, Guèye MC, Sy O, Bouchet S, Prasad PVV, Morris GP (2015) Population genomics of pearl millet (*Pennisetum glaucum* (L.) R. Br.): comparative analysis of global accessions and Senegalese landraces. BMC Genomics 16:1048

Ibrahim YM, Marcarian V, Dobrenz AK (1985) Evaluation of drought tolerance in pearl millet (*Pennisetum americanum* (L.) Leeke) under a sprinkler irrigation gradient. Field Crops Res 11:233–240

Jagadish SVK, Septiningsih EM, Kohli A, Thomson MJ, Ye C, Redona E, Kumar A, Gregorio GB, Wassmann R, Ismail AM, Singh RK (2012) Genetic advances in adapting rice to a rapidly changing climate. J Agron Crop Sci 198:360–373

Jauhar PP (1981) Cytogenetics and breeding of pearl millet and related species. Alan R Liss, New York, USA

Jauhar PP, Hanna WW (1998) Cytogenetics and genetics of pearl millet. Adv Agron 64:1–26

Jetter R, Kunst L, Samuels AL (2007) Composition of plant cuticular waxes. Ann Plant Rev 23:145–181

Jones ES, Liu CJ, Gale MD, Hash CT, Witcombe JR (1995) Mapping quantitative trait loci for downy mildew resistance in pearl millet. Theor Appl Genet 91:448–456

Jones ES, Breese WA, Liu CJ, Singh SD, Shaw DS, Witcombe JR, Jonesa ES, Breeseb WA, Liuc CJ, Singhd SD, Shawb DS, Witcombe JR (2002) Mapping quantitative trait loci for resistance to downy mildew in pearl millet: Field and glasshouse screens detect the same QTL. Crop Sci 42:1316–1323

Jones H, Leigh FJ, Mackay I, Bower MA, Smith LMJ, Charles MP, Jones G, Jones MK, Brown TA, Powell W (2008) Population-based resequencing reveals that the flowering time adaptation of cultivated barley originated east of the Fertile Crescent. Mol Biol Evol 25:2211–2219

Jukanti AK, Gowda CLL, Rai KN, Manga VK, Bhatt RK (2016) Crops that feed the world 11. Pearl Millet (*Pennisetum glaucum* L.): an important source of food security, nutrition and health in the arid and semi-arid tropics. Food Secur 8:307–329

Kam J, Puranik S, Yadav R, Manwaring HR, Pierre S, Srivastava RK, Yadav RS (2016) Dietary interventions for type 2 diabetes: how millet comes to help. Front Plant Sci 7:1454

Kannan B, Senapathy S, Bhasker Raj AG, Chandra S, Muthiah A, Dhanapal AP, Hash CT (2014) Association analysis of SSR markers with phenology, grain, and stover-yield related traits in pearl millet (*Pennisetum glaucum* (L.) R. Br.). Sci World J 2014:562327

Kaushal P, Khare A, Nath Zadoo S, Roy A, Malaviya D, Agrawal A, Ali Siddiqui S, Nath Choubey R (2008) Sequential reduction of *Pennisetum squamulatum* genome complement in *P. glaucum* (2n = 28) × *P. squamulatum* (2n = 56) hybrids and their progenies revealed its octoploid status. Cytologia 73(2):151–158

Khalfallah N, Sarr A, Siljak-Yakovlev S (1993) Karyological study of some cultivated and wild stocks of pearl millet from Africa (*Pennisetum typhoides* Stapf et Hubb. and *P. violaceum* (Lam.) L. Rich.). Caryologia 46:127–138

Kole C, Muthamilarasan M, Henry RJ, Edwards D, Sharma R, Abberton M, Batley J, Bentley A, Blakeney M, Bryant J, Cai H, Cakir M, Cseke LJ, Cockram J, de Oliveira AC, De Pace C, Dempewolf H, Ellison S, Gepts P, Greenland A, Hall A, Hori K, Hughes S, Humphreys MW, Iorizzo M, Ismail AM, Marshall A, Mayes S, Nguyen HT, Ogbonnaya FC, Ortiz R, Paterson AH, Simon PW, Tohme J, Tuberosa R, Valliyodan B, Varshney RK, Wullschleger SD, Yano M, Prasad M, Kole C, Muthamilarasan M, Henry RJ, Edwards D, Sharma R, Abberton M, Batley J, Bentley A, Blakeney M, Bryant J, Cai H, Cakir M, Cseke LJ, De Oliveira AC, De Pace C, Dempewolf H, Ellison S, Gepts P, Hall A, Hori K, Howe GT, Hughes S, Humphreys MW, Iorizzo M, Abdelbagi M, Marshall A, Mayes S, Nguyen HT, Ogbonnaya FC, Ortiz R, Paterson AH (2015) Application of genomics-assisted breeding for generation of climate resilient crops: progress and prospects. Front Plant Sci 6:563

Kosma DK, Jenks MA (2007) Eco-physiological and molecular-genetic determinants of plant cuticle function in drought and salt stress tolerance. In: Jenks MA, Hasegawa PM, Jain SM (eds) Advances in molecular breeding toward drought and salt tolerant crops. Springer, Netherlands, Dordrecht, pp 91–120

Kumar S, Hash CT, Thirunavukkarasu N, Singh G, Rajaram V, Rathore A, Senapathy S, Mahendrakar MD, Yadav RS, Srivastava RK (2016) Mapping quantitative trait loci controlling high iron and zinc content in self and open pollinated grains of pearl millet [*Pennisetum glaucum* (l.) R. Br.]. Front Plant Sci 7:1636

Kumar S, Hash CT, Nepolean T, Satyavathi CT, Singh G, Mahendrakar MD, Yadav RS, Srivastava RK (2017) Mapping QTLs controlling flowering time and important agronomic traits in pearl millet. Front Plant Sci 8:1731

Kumar S, Hash CT, Nepolean T, Mahendrakar MD, Satyavathi CT, Singh G, Rathore A, Yadav RS, Gupta R, Srivastava RK (2018) Mapping grain iron and zinc content quantitative trait loci in an iniadi-derived immortal population of pearl millet. Genes 9:1–17

Kumari BR, Kolesnikova-Allen MA, Tom Hash C, Senthilvel S, Nepolean T, Kavi Kishor PB, Riera-Lizarazu O, Witcombe JR, Srivastava RK (2014) Development of a set of chromosome segment substitution lines in pearl millet. Crop Sci 54:2175

Kusaba M, Tanaka A, Tanaka R (2013) Stay-green plants: what do they tell us about the molecular mechanism of leaf senescence. Photosynth Res 117:221–234

Kusaka M, Lalusin AG, Fujimura T (2005) The maintenance of growth and turgor in pearl millet (*Pennisetum glaucum* [L.] Leeke) cultivars with different root structures and osmo-regulation under drought stress. Plant Sci 168:1–14

Lagudah ES, Hanna WW (1989) Species relationship in the *Pennisetum* gene pool: enzyme polymorphism. Theor Appl Genet 78:801–808

Lakis G, Navascués M, Rekima S, Simon M, Remigereau M-SS, Leveugle M, Takvorian N, Lamy F, Depaulis F, Robert T (2012) Evolution of neutral and flowering genes along pearl millet (*Pennisetum glaucum*) domestication. PLoS ONE 7:e36642

Lambers H, Martinoia E, Renton M (2015) Plant adaptations to severely phosphorus-impoverished soils. Curr Opin Plant Biol 25:23–31

Leder Iren (2004) Sorghum and millets, in cultivated plants, primarily as food sources. In: Füleky G (ed) Encyclopedia of life support systems (EOLSS), developed under the auspices of the UNESCO, vol 1. Eolss Publishers, Oxford, UK, p 66

Li Y, Bhosale S, Haussmann BI, Stich B, Melchinger AE, Parzies HK (2018) Genetic diversity and linkage disequilibrium of two homologous genes to maize D8: sorghum *SbD8* and pearl millet *PgD8*. J Plant Breed Crop Sci 2(5):117–128

Liang Z, Gupta SK, Yeh C-T, Zhang Y, Ngu DW, Kumar R, Patil HT, Mungra KD, Yadav DV, Rathore A, Srivastava RK, Gupta R, Yang J, Varshney RK, Schnable PS, Schnable JC (2018) Phenotypic data from inbred parents can improve genomic prediction in pearl millet hybrids. Genes Genomes|Genet 8:2513–252

Liu CJ, Witcombe JR, Pittaway TS, Nash M, Hash CT, Busso CS, Gale MD (1994) An RFLP-based genetic map of pearl millet (*Pennisetum glaucum*). Theor Appl Genet 89:481–487

Longin CF, Mi X, Wurschum T (2015) Genomic selection in wheat: optimum allocation of test resources and comparison of breeding strategies for line and hybrid breeding. Theor Appl Genet 128:1297–1306

Maman N, Mason SC, Lyon DJ (2006) Nitrogen rate influence on pearl millet yield, nitrogen uptake, and nitrogen use efficiency in Nebraska. Commun Soil Sci Plant Anal 37:127–141

Manning K, Pelling R, Higham T, Schwenniger JL, Fuller DQ (2011) 4500-Year old domesticated pearl millet (*Pennisetum glaucum*) from the Tilemsi Valley, Mali: New insights into an alternative cereal domestication pathway. J Arch Sci 38:312–322

Marchais L, Pernes J (1985) Genetic divergence between wild and cultivated pearl millets (*Pennisetum typhoides*). I. Male sterility. Z Pflanzenzüchtg 95:103–112

Martel E, Ricroch A, Sarr A (1996) Assessment of genome organization among diploid species (2n = 2x = 14) belonging to primary and tertiary gene pools of pearl millet using fluorescent *in situ* hybridization with rDNA probes. Genome 39:680–687

Martel E, De Nay D, Siljak-Yakovlev S, Brown S, Sarr A (1997) Genome size variation and basic chromosome number in Pearl millet and fourteen related *Pennisetum* species. J Hered 88:139–143

Martel E, Poncet V, Lamy F, Siljak-Yakovlev S, Lejeune B, Sarr A (2004) Chromosome evolution of *Pennisetum* species (Poaceae): implications of its phylogeny. Plant Syst Evol 249:139–149

Matsuura A, Inanaga S, Sugimoto Y (1996) Mechanism of interspecific differences among four gramineous crops in growth response to soil drying. Jpn J Crop Sci 65:352–360

McIntyre BD, Riha SJ, Flower DJ (1995) Water uptake by pearl millet in a semiarid environment. Field Crops Res 43:67–76

Mishra RN, Reddy PS, Nair S, Markandeya G, Reddy AR, Sopory SK, Reddy MK (2007) Isolation and characterization of expressed sequence tags (ESTs) from subtracted cDNA libraries of *Pennisetum glaucum* seedlings. Plant Mol Biol 64:713–732

Moles AT, Warton DI, Warman L, Swenson NG, Laffan SW, Zanne AE, Pitman A, Hemmings FA, Leishman MR (2009) Global patterns in plant height. J Ecol 97:923–932

Morgan RN, Wilson JP, Hanna WW, Ozias-Akins P (1998) Molecular markers for rust and pyricularia leaf spot disease resistance in pearl millet. Theor Appl Genet 96:413–420

Moumouni KH, Kountche BA, Jean M, Hash CT, Vigouroux Y, Haussmann BIG, Belzile F (2015) Construction of a genetic map for pearl millet, *Pennisetum glaucum* (L.) R. Br., using a genotyping-by-sequencing (GBS) approach. Mol Breed 35:1–10

Nagarathna KC, Shetty SA, Bhat SG, Shetty HS (1992) The possible involvement of lipoxygenase in downy mildew resistance in pearl millet. J Exp Bot 43:1283–1287

Naino Jika AK, Dussert Y, Raimond C, Garine E, Luxereau A, Takvorian N, Djermakoye RS, Adam T, Robert T (2017) Unexpected pattern of pearl millet genetic diversity among ethno-linguistic groups in the Lake Chad Basin. Heredity 118:491–502

Nayaka SC, Shetty HS, Satyavathi CT, Yadav RS, Kishor PBK, Nagaraju M, Anoop TA, Kumar MM, Kuriakose B, Chakravartty N, Katta AVSKM, Lachagari VBR, Singh OV, Sahu PP, Puranik S, Kaushal P, Srivastava RK (2017) Draft genome sequence of *Sclerospora graminicola*, the pearl millet downy mildew pathogen. Biotechnol Rep 16:18–20

Nepolean T, Gupta SK, Dwivedi SL, Bhattacharjee R, Rai KN, Hash CT (2012) Genetic diversity in maintainer and restorer lines of pearl millet. Crop Sci 52:2555–2563

Nwanze KF, Harris KM (1992) Insect pests of pearl millet in West Africa. Rev Agri Entomol 80:1133–1155

Obeng E, Cebert E, Ward R, Nyochembeng LM, Mays DA, Singh HP, Singh BP (2015) Insect incidence and damage on pearl millet (*Pennisetum glaucum*) under various nitrogen regimes in Alabama. Florida Entomol 98:74–79

Ochatt S, Sangwan R, Marget P, Yves Placide A, Rancillac M, Perney P (2002) New approaches towards the shortening of generation cycles for faster breeding of protein legumes. Plant Breed 121:436–440

Oelke EA, Oplinger ES, Putnam DH, Durgan BR, Doll JD, Undersander DJ (1990) Alternative field crops manual: millets. University of Wisconsin Cooperative or Extension Service, Department of Agronomy, Madison and Center for Alternative Plant and Animal Products, University of Minnesota, St. Paul, MN

Oumar I, Mariac C, Pham J-LL, Vigouroux Y (2008) Phylogeny and origin of pearl millet (*Pennisetum glaucum* [L.] R. Br) as revealed by microsatellite loci. Theor Appl Genet 117:489–497

Paleg LG, Stewart GR, Bradbeer JW (1984) Proline and glycine betaine influence protein solvation. Plant Physiol 75:974–978. https://doi.org/10.1104/pp.75.4.974

Parvathaneni RK, Jakkula V, Padi FK, Faure S, Nagarajappa N, Pontaroli AC, Wu X, Bennetzen JL, Devos KM (2013) Fine-mapping and identification of a candidate gene underlying the *d2* dwarfing phenotype in pearl millet, *Cenchrus americanus* (L.) Morrone. Genes Genomes Genet 3:563–572

Parvathaneni RK, DeLeo VL, Spiekerman JJ, Chakraborty D, Devos KM (2017) Parallel loss of introns in the *ABCB1* gene in angiosperms. BMC Evol Biol 17:238

Passot S, Gnacko F, Moukouanga D, Lucas M, Guyomarc'h S, Ortega BM, Atkinson JA, Belko MN, Bennett MJ, Gantet P, Wells DM, Guédon Y, Vigouroux Y, Verdeil J-L, Muller B, Laplaze L (2016) Characterization of pearl millet root architecture and anatomy reveals three types of lateral roots. Front Plant Sci 7:1–11

Paterson AH, Bowers JE, Bruggmann R, Dubchak I, Grimwood J, Gundlach H, Haberer G, Hellsten U, Mitros T, Poliakov A, Schmutz J, Spannagl M, Tang H, Wang X, Wicker T, Bharti AK, Chapman J, Feltus FA, Gowik U, Grigoriev I V, Lyons E, Maher C a, Martis M, Narechania A, Otillar RP, Penning BW, Salamov AA, Wang Y, Zhang L, Carpita NC, Freeling M, Gingle AR, Hash CT, Keller B, Klein P, Kresovich S, McCann MC, Ming R, Peterson DG, Mehboob-ur-Rahman, Ware D, Westhoff P, Mayer KFX, Messing J, Rokhsar DS (2009) The *Sorghum bicolor* genome and the diversification of grasses. Nature 457:551–556

Pattanashetti SK, Upadhyaya HD, Dwivedi SL, Vetriventhan M, Reddy KN (2016) Pearl millet. In: Singh M, Upadhyaya HD (eds) Genetic and genomic resources for grain cereals improvement. Academic Press, San Diego, CA, pp 253–289

Payne WA, Malcolm DC, Hossner LR, Lascao RJ, Onken AB, Wendt CW (1992) Soil phosphorus availability and pearl millet water-use efficiency. Crop Sci 32:1010–1015

Peacock JM, Soman P, Jayachandran R, Rani AU, Howarth CJ, Thomas A (1993) Effects of high soil surface temperature on seedling survival in pearl millet. Exp Agri 29:215–225

Pedraza-Garcia F, Specht JE, Dweikat I (2010) A new PCR-based linkage map in pearl millet. Crop Sci 50:1754–1760

Perales HR, Benz BF, Brush SB (2005) Maize diversity and ethnolinguistic diversity in Chiapas, Mexico. Proc Natl Acad Sci USA 102:949–954. https://doi.org/10.1073/pnas.0408701102

Pinheiro C, Chaves MM (2011) Photosynthesis and drought: can we make metabolic connections from available data? J Exp Bot 62:869–882

Poncet V, Lamy F, Devos KM, Gale MD, Sarr A, Robert T (2000) Genetic control of domestication traits in pearl millet (*Pennisetum glaucum* L., Poaceae). Theor Appl Genet 100:147–159

Punnuri SM, Wallace JG, Knoll JE, Hyma KE, Mitchell SE, Buckler ES, Varshney RK, Singh BP (2016) Development of a high-density linkage map and tagging leaf spot resistance in pearl millet using genotyping-by-sequencing markers. Plant Genome 9:1–13

Qi X, Lindup S, Pittaway TS, Allouis S, Gale MD, Devos KM (2001) Development of simple sequence repeat markers from bacterial artificial chromosomes without subcloning. Biotechniques 31:355–362

Qi X, Pittaway TS, Lindup S, Liu H, Waterman E, Padi FK, Hash CT, Zhu J, Gale MD, Devos KM (2004) An integrated genetic map and a new set of simple sequence repeat markers for pearl millet, Pennisetum glaucum. Theor Appl Genet 109:1485–1493

Rai KN, Hanna WW (1990) Morphological characteristics of tall and dwarf pearl millet isolines. Crop Sci 30:23–25

Rai KN, Virk DS (1999) Breeding methods. In: Khairwal IS, Rai KN, Andrews DJ, Harinarayana G (eds) Pearl millet breeding. Oxford & IBH Publishing, New Delhi, India, pp 185–211

Rai KN, Gupta SK, Bhattacharjee R, Kulkarni V, Singh AK, Rao AS (2009) Morphological characteristics of ICRISAT-bred pearl millet hybrid seed parents. J SAT Agri Res 7:1–169

Rai KN, Govindaraj M, Rao AS (2012) Genetic enhancement of grain iron and zinc content in pearl millet. Qual Assur Saf Crop Foods 4:119–125

Rai KN, Yadav OP, Rajpurohit BS, Patil HT, Govindaraj M, Khairwal IS, Rao AS, Shivade H, Pawar VY (2013) Breeding pearl millet cultivars for high iron density with zinc density as an associated trait. J SAT Agri Res 11:1–7

Rai KN, Velu G, Govindaraj M, Upadhyaya HD, Rao AS, Shivade H, Reddy KN, Rai KN, Velu G, Govindaraj M, Upadhyaya HD, Rao AS, Shivade H, Reddy KN (2015) Iniadi pearl millet germplasm as a valuable genetic resource for high grain iron and zinc densities. Plant Genet Resour 13:75–82. https://doi.org/10.1017/S1479262114000665

Raj C, Sharma R, Pushpavathi B, Gupta SK, Radhika K (2018) Inheritance and allelic relationship among downy mildew resistance genes in pearl millet. Plant Dis 102:1136–1140

Rajaram V, Nepolean T, Senthilvel S, Varshney RK, Vadez V, Srivastava RK, Shah TM, Supriya A, Kumar S, Kumari BR, Bhanuprakash A, Narasu ML, Riera-Lizarazu O, Hash CT (2013) Pearl millet [Pennisetum glaucum (L.) R. Br.] consensus linkage map constructed using four RIL mapping populations and newly developed EST-SSRs. BMC Genomics 14:159

Raju NSN, Rao YS, Rao MVS, Manga V (1985) Anthocyanidins of purple and sub-red pigmentation in pearl millet (*Pennisetum americanum*). Indian J Bot 8:185–186

Ramakrishnan TS (1971) Diseases of millets. Indian Council of Agriculture Research, New Delhi, India

Ramya AR, Ahamed ML, Satyavathi CT, Rathore A, Katiyar P, Raj AGB, Kumar S, Gupta R, Mahendrakar MD, Yadav RS, Srivastava RK (2018) Towards defining heterotic gene pools in pearl millet [*Pennisetum glaucum* (L.) R. Br.]. Front Plant Sci 8:1934

Rao SA, Mengesha MH, Sharma D (1985) Collection and evaluation of pearl millet (*Pennisetum americanum*) germplasm from Ghana. Econ Bot 39:25–38

Rao PP, Birthal PS, Reddy BVS, Rai KN, Ramesh S (2006) Diagnostics of sorghum and pearl millet grains-based nutrition in India. Intl Sorghum Millets Newsl 47:93–96

Remington DL, Thornsberry JM, Matsuoka Y, Wilson LM, Whitt SR, Doebley J, Kresovich S, Goodman MM, Buckler ES (2001) Structure of linkage disequilibrium and phenotypic associations in the maize genome. Proc Natl Acad Sci USA 98:11479–11484

Richards RA (2000) Selectable traits to increase crop photosynthesis and yield of grain crops. J Exp Bot 51:447–458

Robert T, Khalfallah N, Martel E, Lamy F, Poncet V, Allinne C, Remigereau M-S, Rekima S, Leveugle M, Lakis G, Siljak-Yakovlev S, Sarr A (2011) Pennisetum. In: Kole C (ed) Wild crop relatives: genomic and breeding resources, millets and grasses. Springer, Berlin, Heidelberg, pp 217–255

Rostamza M, Richards RA, Watt M (2013) Response of millet and sorghum to a varying water supply around the primary and nodal roots. Ann Bot 112:439–446

Saade S, Maurer A, Shahid M, Oakey H, Schmöckel SM, Negrão S, Pillen K, Tester M (2016) Yield-related salinity tolerance traits identified in a nested association mapping (NAM) population of wild barley. Sci Rep 6:32586

Safriel U, Adeel Z, Niemeijer D, Puigdefabregas J, White R, Lal R, Winslow M, Ziedler J, Prince S, Archer E, King C, Shapiro B, Wessels K, Nielsen T, Portnov B, Reshef I, Thonell J, Lachman E, Mcnab D (2005) Dryland systems. Washington DC, USA

Saïdou AA, Mariac C, Luong V, Pham JL, Bezançon G, Vigouroux Y (2009) Association studies identify natural variation at PHYC linked to flowering time and morphological variation in pearl millet. Genetics 182:899–910

Saïdou A-A, Clotault J, Couderc M, Mariac C, Devos KM, Thuillet A-C, Amoukou IA, Vigouroux Y (2014) Association mapping, patterns of linkage disequilibrium and selection in the vicinity of the Phytochrome C gene in pearl millet. Theor Appl Genet 127:19–32

Saito K, Matsuda F (2010) Metabolomics for functional genomics, systems biology, and biotechnology. Annu Rev Plant Biol 61:463–489

Sanon M, Hoogenboom G, Traoré SB, Sarr B, Garcia Y, Garcia A, Somé L, Roncoli C (2014) Photoperiod sensitivity of local millet and sorghum varieties in West Africa. NJAS—Wageningen J Life Sci 68:29–39

Sattler SE, Funnell-Harris DL, Pedersen JF (2010) Brown midrib mutations and their importance to the utilization of maize, sorghum, and pearl millet lignocellulosic tissues. Plant Sci 178:229–238

Schnable PS (2009) The B73 maize genome: complexity, diversity, and dynamics. Science 326:1112–1115

Schoper JB, Lambert RJ, Vasilas BL, Westgate ME (1987) Plant factors controlling seed set in maize: the influence of silk, pollen, and ear-leaf water status and tassel heat treatment at pollination. Plant Physiol 83:121–125

Sehgal D, Rajaram V, Armstead IP, Vadez V, Yadav YP, Hash CT, Yadav RS (2012) Integration of gene-based markers in a pearl millet genetic map for identification of candidate genes underlying drought tolerance quantitative trait loci. BMC Plant Biol 12:9

Sehgal D, Skot L, Singh R, Srivastava RK, Das SP, Taunk J, Sharma PC, Pal R, Raj B, Hash CT, Yadav RS (2015) Exploring potential of pearl millet germplasm association panel for association mapping of drought tolerance traits. PLoS ONE 10:e0122165

Senthilvel S, Jayashree B, Mahalakshmi V, Kumar PS, Nakka S, Nepolean T, Hash C (2008) Development and mapping of simple sequence repeat markers for pearl millet from data mining of expressed sequence tags. BMC Plant Biol 8:119

Serba DD, Yadav RS (2016) Genomic tools in pearl millet breeding for drought tolerance: status and prospects. Front Plant Sci 7:1724

Serba DD, Perumal R, Tesso TT, Min D (2017) Status of global pearl millet breeding programs and the way forward. Crop Sci 57:2891–2905

Serba DD, Muleta KT, St Amand P, Bernardo A, Bai G, Ramasamy P, Bashir E (2019) Genetic diversity, population structure, and linkage disequilibrium in pearl millet. The Plant Genome. https://doi.org/10.3835/plantgenome2018.11.0091

Serraj R, Tom Hash C, Rizvi SMH, Sharma A, Yadav RS, Bidinger FR (2005) Recent advances in marker-assisted selection for drought tolerance in pearl millet. Plant Prod Sci 8:334–337

Shivhare R, Lata C (2016) Exploration of genetic and genomic resources for abiotic and biotic stress tolerance in pearl millet. Front Plant Sci 7:2069

Siddaiah CN, Satyanarayana NR, Mudili V, Kumar Gupta V, Gurunathan S, Rangappa S, Huntrike SS, Srivastava RK (2017) Elicitation of resistance and associated defense responses in Trichoderma hamatum induced protection against pearl millet downy mildew pathogen. Sci Rep 7:43991

Singh SD (1990) Sources of resistance to downy mildew and rust in pearl millet. Plant Dis 74:871–874

Singh SD (1995) Downy mildew of pearl millet. Plant Dis 79:545–550

Singh SD, Williams RJ (1980) The role of sporangia in the epidemiology of pearl millet downy mildew. Phytopathology 70:1187–1190

Singh SD, Wilson JP, Navi SS, Talukdar BSS, Hess DE, Reddy KN (1997) Screening techniques and sources of resistance to downy mildew and rust in pearl millet. ICRISAT, 110 pp, Patancheru, India

Singh S, Sharma R, Pushpavathi B, Gupta SK, Durgarani CV, Raj C (2018) Inheritance and allelic relationship among gene(s) for blast resistance in pearl millet [*Pennisetum glaucum* (L.) R. Br.]. Plant Breed 137:573–584

Sivakumar MVK, Salaam SA (1999) Effect of year and fertilizer on water-use efficiency of pearl millet (*Pennisetum glaucum*) in Niger. J Agri Sci 132:139–148

Srinivasarao C, Lal R, Kundu S, Babu MBBP, Venkateswarlu B, Singh AK (2014) Soil carbon sequestration in rainfed production systems in the semiarid tropics of India. Sci Total Environ 487:587–603

Stapf O, Hubbard CE (1934) Pennisetum. In: Prain D (ed) Flora of tropical Africa, vol 9 part 6, Gramineae (Maydeae–Paniceae). Reeve, London, pp 954–1070

Stich B, Haussmann BI, Pasam R, Bhosale S, Hash CT, Melchinger AE, Parzies HK (2010) Patterns of molecular and phenotypic diversity in pearl millet [*Pennisetum glaucum* (L.) R. Br.] from West and Central Africa and their relation to geographical and environmental parameters. BMC Plant Biol 10:216

Stone P (2001) The effects of heat stress on cereal yield and quality. In: Basra AS (ed) Crop responses and adaptations to temperature stress. Food Products Press, Binghamton, NY, USA, pp 243–291

Suh MC, Samuels AL, Jetter R, Kunst L, Pollard M, Ohlrogge J, Beisson F (2005) Cuticular lipid composition, surface structure, and gene expression in Arabidopsis stem epidermis. Plant Physiol 139:1649 LP-1665

Sultan B, Roudier P, Quirion P, Alhassane A, Muller B, Dingkuhn M, Ciais P, Guimberteau M, Traore S, Baron C (2013) Assessing climate change impacts on sorghum and millet yields in the Sudanian and Sahelian savannas of West Africa. Environ Res Lett 8:014040

Supriya A, Senthilvel S, Nepolean T, Eshwar K, Rajaram V, Shaw R, Hash CT, Kilian A, Yadav RC, Narasu ML (2011) Development of a molecular linkage map of pearl millet integrating DArT and SSR markers. Theor Appl Genet 123:239–250

Swaminathan M (1937) The relative value of the proteins of certain foodstuffs in nutrition. Indian J Med Res 24:767–786

Swaminathan MS, Naik MS, Kaul AK, Austin A (1971) Choice of strategy for the genetic upgrading of protein properties in cereals, millets and pulses. Indian J Agril Res 41:393–406

Takei E, Sakamoto S (1987) Geographical variation of heading response to daylength in foxtail millet (*Setaria italica* P. BEAUV.). Jpn J Breed 37:150–158

Tapsoba H, Wilson JP (1995) Isolation of pathogenic races of *Puccinia substriata* var. *indica* with new sources of rust resistant pearl millet. In: Peare ID (ed) Proceedings of the first national grain pearl millet symposium. University of Georgia, Tifton, GA, USA, pp 57–60

Thakur RP, Shetty HS, Khairwal IS (2006) Pearl millet downy mildew research in India: progress and perspectives. J SAT Agri Res 2:1–6

Thakur RP, Rai KN, Khairwal IS, Mahala RS (2008) Strategy for downy mildew resistance breeding in pearl millet in India. SAT eJournal 6:1–11

Thakur RP, Sharma R, Rao VPP, Rajan S, Rao VPP (2011) Screening techniques for pearl millet diseases. Information Bulletin No 89. ICRISAT, Patancheru, India

Thomas H, Howarth CJ (2000) Five ways to stay green. J Exp Bot 51:329–337

Traore M (1985) Physiological and morphological mechanisms of drought resistance in sorghum and pearl millet. I. Effects of leaf treatment with abscisic acid. II. Seed pre-treatment with abscisic acid. III. Comparative shoot and root development. IV. Leaf surface morphol. University of Nebraska, Lincoln

Trostle C, Corriher-Olson V, Knutson A (2015) Hybrid pearl millet as an alternative to sugarcane aphid-susceptible sorghum family forages. Texas A&M AgriLife Extension Service

Upadhyaya HD, Reddy KN, Gowda CLL (2007) Pearl millet germplasm at ICRISAT genebank—status and impact. SAT eJournal 3:1–5

Upadhyaya HD, Yadav D, Reddy KN, Gowda CLL, Singh S (2011) Development of pearl millet minicore collection for enhanced utilization of germplasm. Crop Sci 51:217–223

Upadhyaya H, Reddy K, Ahmed MI, Dronavalli N, Laxmipathi Gowda C (2012) Latitudinal variation and distribution of photoperiod and temperature sensitivity for flowering in the world collection of pearl millet germplasm at ICRISAT genebank. Plant Genet Resour 10:59–69

USDA (2014) Comparison of five millet species for conservation use in the United States. USDA Plant Materials Technical Note No 2, 12 pp

Varalakshmi P, Mohan Dev Tavva SS, Arjuna Rao PV, Subba Rao MV, Hash CT (2012) Genetic architecture of purple pigmentation and tagging of some loci to SSR markers in pearl millet, *Pennisetum glaucum* (L.) R. Br. Genet Mol Biol 35:106–118

Varshney RK, Shi C, Thudi M, Mariac C, Wallace J, Qi P, Zhang H, Zhao Y, Wang X, Rathore A, Srivastava RK, Chitikineni A, Fan G, Bajaj P, Punnuri S, Gupta SK, Wang H, Jiang Y, Couderc M, Katta MAVSK, Paudel DR, Mungra KD, Chen W, Harris-Shultz KR, Garg V, Desai N, Doddamani D, Kane NA, Conner JA, Ghatak A, Chaturvedi P, Subramaniam S, Yadav OP, Berthouly-Salazar C, Hamidou F, Wang J, Liang X, Clotault J, Upadhyaya HD, Cubry P, Rhoné B, Gueye MC, Sunkar R, Dupuy C, Sparvoli F, Cheng S, Mahala RS, Singh B, Yadav RS, Lyons E, Datta SK, Tom Hash C, Devos KM, Buckler E, Bennetzen JL, Paterson AH, Ozias-Akins P, Grando S, Wang J, Mohapatra T, Weckwerth W, Reif JC, Liu X,

Vigouroux Y, Xu X (2017) Pearl millet genome sequence provides a resource to improve agronomic traits in arid environments. Nat Biotechnol 35:969–976
Varshney RK, Singh VK, Kumar A, Powell W, Sorrells ME (2018) Can genomics deliver climate-change ready crops? Curr Opin Plant Biol 1–7. https://doi.org/10.1016/j.pbi.2018.03.007
Vavilov NI (1992) Origin and geography of cultivated plants. Cambridge University Press, Cambridge
Velu G, Rai KN, Muralidharan V, Kulkarni VN, Longvah T, Raveendran TS (2007) Prospects of breeding biofortified pearl millet with high grain iron and zinc content. Plant Breed 126:182–185
Vengadessan V, Rai KN, Kannan Bapu JR, Hash CT, Bhattacharjee R, Senthilvel S, Vinayan MT, Nepolean T (2013) Construction of genetic linkage map and QTL analysis of sink-size traits in pearl millet (*Pennisetum glaucum*). ISRN Genet 2013:1–14
Vinoth A, Ravindhran R (2017) Biofortification in millets: a sustainable approach for nutritional security. Front Plant Sci 8:29
Wang C, Guo L, Li Y, Wang Z (2012) Systematic comparison of C3 and C4 plants based on metabolic network analysis. BMC Syst Biol 6:S9–S9
Watson A, Ghosh S, Williams MJ, Cuddy WS, Simmonds J, Rey M-D, Asyraf Md Hatta M, Hinchliffe A, Steed A, Reynolds D, Adamski NM, Breakspear A, Korolev A, Rayner T, Dixon LE, Riaz A, Martin W, Ryan M, Edwards D, Batley J, Raman H, Carter J, Rogers C, Domoney C, Moore G, Harwood W, Nicholson P, Dieters MJ, DeLacy IH, Zhou J, Uauy C, Boden SA, Park RF, Wulff BBH, Hickey LT (2018) Speed breeding is a powerful tool to accelerate crop research and breeding. Nat Plants 4:23–29
Westengen OT, Okongo MA, Onek L, Berg T, Upadhyaya H, Birkeland S, Kaur Khalsa SD, Ring KH, Stenseth NC, Brysting AK (2014) Ethnolinguistic structuring of sorghum genetic diversity in Africa and the role of local seed systems. Proc Natl Acad Sci USA 111:14100–14105
White PJ, George TS, Gregory PJ, Bengough AG, Hallett PD, McKenzie BM (2013) Matching roots to their environment. Ann Bot 112:207–222
Wilson JP (2002) Disease of pearl millet in the Americas. In: Leslie JF (ed) Sorghum and millets diseases. Iowa State Press, Ames, Iowa, pp 465–469
Wilson JP, Hanna WW (1992) Effects of gene and cytoplasm substitutions in pearl millet on leaf blight epidemics and infection by Pyricularia grisea. Phytopathology 82:839–842
Wilson JP, Hanna WW, Gascho G, Wilson DM (1995) Pearl millet grain yield loss from rust infection. In: Teare ID (ed) Proceedings of the first national grain pearl millet symposium. University of Georgia, Tifton, GA, USA, pp 54–56
Wilson JP, Hess DE, Hanna WW, Kumar KA, Gupta SC (2004) *Pennisetum glaucum* subsp. *monodii* accessions with Striga resistance in West Africa. Crop Protec 23:865–870
Wimpee CF, Rawson JRY (1979) Characterization of the nuclear genome of pearl millet. Biochim Biophys Acta—Nucl Acids Protein Synth 562:192–206
Xue D, Zhang X, Lu X, Chen G, Chen Z-H (2017) Molecular and evolutionary mechanisms of cuticular wax for plant drought tolerance. Front Plant Sci 8:621
Yadav OP (2010) Drought response of pearl millet landrace-based populations and their crosses with elite composites. Field Crops Res 118:51–56. https://doi.org/10.1016/j.fcr.2010.04.005
Yadav OP, Rai KN (2013) Genetic improvement of pearl millet in India. Agri Res 2:275–292
Yadav RS, Hash CT, Bidinger FR, Howarth CJ (1999) QTL analysis and marker-assisted breeding of traits associated with drought tolerance in pearl millet. In: Ito O, O'Toole J, Hardy B (eds) Genetic improvement of rice for water-limited environments. International Rice Research Institute, Manila, Philippines, pp 221–233
Yadav RS, Hash CT, Bidinger FR, Cavan GP, Howarth CJ (2002) Quantitative trait loci associated with traits determining grain and stover yield in pearl millet under terminal drought-stress conditions. Theor Appl Genet 104:67–83

Yadav R, Bidinger F, Hash C, Yadav Y, Yadav O, Bhatnagar S, Howarth C (2003) Mapping and characterisation of QTL × E interactions for traits determining grain and stover yield in pearl millet. Theor Appl Genet 106:512–520

Yadav RS, Hash CT, Bidinger FR, Devos KM, Howarth CJ (2004) Genomic regions associated with grain yield and aspects of post-flowering drought tolerance in pearl millet across stress environments and tester background. Euphytica 136:265–277

Yadav OP, Rai KN, Gupta SK (2012) Pearl millet: genetic improvement in tolerance to abiotic stresses. In: Improving crop productivity in sustainable agriculture, pp 261–288

Yadav O, Rai K, Yadav H, Rajpurohit B, Gupta S, Rathore A, Karjagi C (2016) Assessment of diversity in commercial hybrids of pearl millet in India. Indian J Plant Genet Resour 29:130

Yadav OP, Upadhyaya HD, Reddy KN, Jukanti AK, Pandey S, Tyagi RK (2017) Genetic resources of pearl millet: status and utilization. Indian J Plant Genet Resour 30:31–47

Yakubu H, Ngala AL, Dugje IY (2010) Screening of millet (Pennisetum glaucum L.) varieties for salt tolerance in semi-arid soil of Northern Nigeria. World J Agri Sci 6:374–380

Yu J (2002) A draft sequence of the rice genome (*Oryza sativa* L. ssp. *indica*). Science 296:79–92

Zegada-Lizarazu W, Iijima M (2005) Deep root water uptake ability and water use efficiency of pearl millet in comparison to other millet species. Plant Prod Sci 8:454–460

Zhu C, Gore M, Buckler ES, Yu J (2008) Status and prospects of association mapping in plants. Plant Genome 1:5

Chapter 7
Genomic Designing for Climate Smart Finger Millet

Damaris Achieng Odeny, Adnan Niazi, Kassahun Tesfaye, Dagnachew Lule, Solomon Wanyonyi and Josephine Sarah Kunguni

Abstract Finger millet is a nutritious cereal crop mainly grown in eastern Africa and southern parts of India. The crop has an incredible ability to adapt to adverse agro-ecological conditions, and is therefore a favorite of smallholder farmers in the tropics, especially women. Finger millet grain is gluten free and exceptionally rich in micronutrients including calcium, folic acid and iron. Despite its unique quality, there has been limited research investment in finger millet resulting in the lack of genetic and genomic resources for more efficient breeding. The abundant genetic resources at the center of origin are yet to be fully exploited for crop improvement, and to date, very few pre-breeding programs exist. Several studies indicate the potential use of the secondary and tertiary gene pools to broaden the narrow genetic base that has been created by the inbreeding nature of the crop. The recent

D. A. Odeny (✉)
The International Crops Research Institute for the Semi-Arid Tropics,
P.O Box 39063 – 00623, Nairobi, Kenya
e-mail: d.odeny@cgiar.org

A. Niazi
Department of Animal Breeding and Genetics, SLU-Global Bioinformatics Centre, Swedish University of Agricultural Sciences, SE75007 Uppsala, Sweden
e-mail: Adnan.Niazi@slu.se

K. Tesfaye
Ethiopian Biotechnology Institute, Ministry of Science and Technology, Ethiopia,
P.O. Box 32853 (Private Box), Addis Ababa, Ethiopia
e-mail: kassahuntesfaye@yahoo.com; kassahun.tesfaye@aau.edu.et

D. Lule
Crop Breeder and Genetics, Oromia Agricultural Research Institute, Nifas Silk Sub City, Addis Ababa, Ethiopia
e-mail: hawinok@gmail.com; dagnachew_lule@iqqo.org

S. Wanyonyi
University of Eldoret, P.O. Box 1125-30100, Eldoret, Kenya
e-mail: sollowanyonyi@yahoo.com

J. S. Kunguni
KALRO-FCRI Kisii, P.O. Box 523, Kisii 40200, Kenya
e-mail: josephinesarah4@gmail.com

C. Kole (ed.), *Genomic Designing of Climate-Smart Cereal Crops*,
https://doi.org/10.1007/978-3-319-93381-8_7

availability of draft whole genome sequences and a robust genetic linkage map, now make it possible to implement large-scale genomics-assisted breeding in finger millet. Comparative mapping with closely related and well-studied crops such as rice, provide an opportunity to understand the complex tetraploid genome more efficiently. With the increasing health awareness, and the growing middle class in Africa and Asia, there is likely to be a higher demand for the nutritious finger millet. Breeders will need to generate relevant populations and increase the yield of finger millet to its full potential in order to meet the demand. The current investment in the generation of genomic resources will need to be matched with investment in phenotyping and germplasm characterization to enable more efficient breeding in finger millet. We discuss the genetic and genomic resources available for finger millet and how they can be exploited to enhance its adaptability to climate change.

Keywords *Eleusine coracana* · Climate change · *Striga* · Blast disease · Tetraploid

7.1 Introduction

Small millets, the earliest domesticated crop species of the world, are a heterogeneous group of cereals that includes finger millet (*Eleusine* spp.), foxtail millet (*Setaria italica*), pearl millet (*Pennisetum glaucum*), little millet (*Panicum miliare*), proso millet (*Panicum miliaceum*), kodo millet (*Paspalum scrobiculatum*), barnyard millet (*Echinochloa colosna*), fonio (*Digitaria exilis*), and teff (*Eragrostis tef*) (Seetharam et al. 1986). They are often grown under harsh environments by small-scale farmers. They are nutritionally rich, genetically diverse and are recognized as crops for new green revolution (Goron and Raizada 2015). Millets are increasingly playing an important role in marginal environments, which are especially vulnerable to climate variability and longer-term climate change scenarios such as increased temperatures and unpredictable rainfall (Padulosi et al. 2009). Finger millet is widely cultivated in the tropical and subtropical regions of Africa and India and ranked third in importance among millets after pearl millet (*Pennisetum glaucum*) and foxtail millet (*Setaria italica*) (Reddy et al. 2009). It is also the most important millet in eastern Africa.

The exact origin of finger millet has been a subject of debate with different authors reporting different locations within eastern Africa. Vavilov (1951) considered finger millet an indigenous crop to Ethiopia and the high plateaus of Abyssinia while Purseglove (1972) reported Uganda as the center of origin. Kenneth and LeRoy (1977) reported the eastern Sudan zone and the highlands stretching from Ethiopia to Uganda as a possible area of domestication of this crop. Biosystematics, ethno-botanical, genetic and linguistic evidence confirmed that the East African highlands, particularly Uganda and Ethiopia, are a possible center of origin for finger millet (Hilu and De Wet 1976). Archaeological record provides evidence for the oldest domesticated finger millet at a prehistoric site in

Axum, Ethiopia, dating back some 5000 years (Hilu et al. 1979). Moreover, vast genetic diversity exists within Ethiopia and Uganda providing further proof that this region is the primary center of origin. From east Africa, the crop was introduced to India over 3000 years ago (Hilu et al. 1979; National Research Council 1996), where it is now an important staple food.

Finger millet belongs to the genus *Eleusine* Gaertn, family Poaceae, subfamily Chloridoideae. The genus comprises eight species including *E. coracana, E. kigeziensis*, *E. indica*, *E. intermedia*, *E. floccifolia, E. tristachya, E. jaegeri* and *E. multiflora. E. coracana* is comprised of two subspecies; *E. coracana* subsp. *coracana,* which is the cultivated finger millet; and *E. coracana* subsp. *africana*, its wild progenitor (Dida et al. 2007). The genus is characterized by three different basic chromosome numbers of $x = 8$, 9 or 10. Dida and Devos (2006) suggested the categorization of the genus *Eleusine* into an *AA* genome group comprising *E. indica* and *E. tristachya* and a *BB* genome group comprising three species namely *E. floccifolia*, *E. intermedia* and *E. multiflora* with tetraploids (*E. coracana* and *E. kigeziensis*) having a combination of AABB. Recent reports (Dwivedi et al. 2012; Vetriventhan et al. 2016) suggest categorization into four genome groups; AA (*E. tristachya* and *E. indica*), BB (*E. floccifolia*), CC (*E. multiflora*) and DD (*E. jaegeri*), with *E. intermedia* being a mixture (AB) and the tetraploids having AABB (*E. coracana*) and AADD (*E. kigeziensis*).

E. intermedia, E. indica, E. floccifolia, and *E. tristachya* are diploids with $2n = 2x = 18$. *E. multiflora* is a diploid with $2n = 2x = 16$ while *E. jaegeri* is a diploid with $2x = 2n = 20$ (Devarumath et al. 2005). *Eleusine coracana* and *E. kigeziensis* are tetraploids with $2n = 4x = 36$. *E. indica* (AA) is believed to be the AA genome donor for *E. coracana* (AABB) while the BB genome donor remains unknown. *Eleusine coracana* subsp. *coracana* is the only cultivated crop of the genus *Eleusine* and has four cultivated races, namely, *elongata, plana, compacta,* and *vulgaris* (Upadhyaya et al. 2010). *Eleusine coracana* subsp. *africana* has two wild races; *africana* and *spontanea* (Upadhyaya et al. 2010). *Eleusine indica* (goosegrass) is categorized as one of the most problematic weeds in the world (Holm et al. 1977; Devarumath et al. 2005).

The cultivated finger millet (*Eleusine coracana* supsp. *coracana*) is a food staple of millions smallholder farmers (Babu et al. 2007; Kumar et al. 2016) and occupies about 12% of the global small millet area, across arid to semi-arid tropics of Asia and Africa (Mirza et al. 2015). It represents one of the critical plant genetic resources for the agriculture and food security of farmers inhabiting arid, infertile and marginal lands (Barbeau and Hilu 1993). India is the largest producer of finger millet in Asia, while in Africa the largest producing region extends from the Rift Valley, Nyanza and Western provinces of Kenya into Uganda. In India, it is cultivated mainly in the terai regions of Himalayas and in the southern peninsula (Bhatt et al. 2011). Other producers of finger millet include Tanzania, Ethiopia, Rwanda, Zaire, Zambia, Zimbabwe, Eritrea and Somalia in Africa, while China, Myanmar (Ekwamu 1991), Nepal, and to some extent, Bhutan and Sri Lanka are the other producers in Asia.

7.2 Prioritizing Climate Smart Traits

Among the merits of finger millet is its ability to adapt to adverse agro-ecological conditions such as soil acidity, moisture stress, minimal inputs and marginal land where other crops cannot perform well (Upadhyaya et al. 2007). Finger millet plants have the advantage of assimilating carbon dioxide through the C4 photosynthetic pathway resulting in more efficient photosynthesis, and therefore more efficient use of water and nutrients. The crop is mainly grown by subsistence farmers in semi-arid areas where it is known to save the lives of poor farmers from starvation during extreme drought (Kotschi 2006). However, despite finger millet's incredible ability to withstand drought, FAO (2005) reports that the anticipated climate change may pose a negative impact on food production and food security, especially in drought-prone regions, where finger millet is mainly grown. Besides, pests and diseases are most likely to be influenced by the changing temperatures (Stireman et al. 2005).

Important abiotic stresses include salinity and heavy metals pollution, declining availability of good quality water, land degradation, and extreme drought. Such abiotic stresses may result into nutritional imbalances in the plant causing reduction in water uptake and increase in toxicity, thereby decreasing the value of the grain. A report by Kukreti (2016) revealed that crop failures in India due to salinity were estimated at approximately 8.6 million per ha. Salts and heavy metals affect the metabolic, physiological and biochemical activities, subsequently inhibiting growth. A reduction in finger millet yield of about 40% has been estimated in both arid and semi-arid areas due to water scarcity.

There are seven major diseases of finger millet namely; blast, seedling blight, wilt or foot rot, Cercospora leaf spot, downy mildew or green ear disease, smut and damping-off (Anilkumar et al. 2003). Finger millet leaf blast (*Magnaporthe grisea*) (T. T. Hebert) M. E. Barr (anamorph *Pyricularia grisea*) was first reported in India, from Tanjore delta of Tamil Nadu by McRae (1920). It affects the leaf, neck and fingers of grown finger millet and causes recurring yield losses. Seedling blight or leaf blight, which was first spotted in India, is caused by *Drechslera nodulosum* (Butler 1918), and is next only to blast in terms of severity and distribution. The other finger millet diseases include *Sclerotium rolfsii* (Coleman 1920), resulting to up to 50% yield loss; Cercospora leaf spot, which causes ear malformation (Pradhanang 1994); smut disease (*Ustilago eleusine*) (Kulkarni 1922) and dumping off disease caused by *Pythium aphanidermatum* (Mehta and Chakravarty 1937).

The most important biotic constraints in finger millet are blast disease and the parasitic weed called *Striga hermonthica.* The blast fungus infects finger millet at all growth stages, causing major losses through neck and panicle infections (Babu et al. 2013). *Striga hermonthica* is specific to African production systems and is considered the greatest biological constraint to food production in sub-Saharan Africa, a more serious problem than insects, birds and plant diseases.

7.2.1 Salinity Tolerance

Finger millet is considered to have a high degree of salt tolerance in comparison to other cereals (Bray et al. 2000; Shailaja and Thirumeni 2007; Rahman et al. 2014). Rahman et al. (2014) undertook a comparison of rice and finger millet accessions in response to salinity and reported superior levels of salinity tolerance in finger millet. Nevertheless, extremely high levels of salinity have been reported to affect the crop's phenology, plant height, shoot biomass and grain yield (Onkware 1986; Krishnamurthy et al. 2014). Natural variation for salinity tolerance among finger millet accessions has been detected through screening of different genotypes under different salinity conditions. At ICRISAT, Krishnamurthy et al. (2014) screened a mini-core collection of 80 finger millet accessions for tolerance to salinity and observed genotypic variation for grain yield, phenology, and shoot biomass. Shailaja and Thirumeni (2007) observed variation among 19 genotypes for their reaction to salt stress.

Rahman et al. (2014) used RNA sequencing of two finger millet genotypes with contrasting response to salinity stress and identified some of the obvious salinity candidates including four aquaporin proteins, sodium/calcium exchanger protein, transporters, signal transducers, and several stress-related transcription factors. In particular, they identified a NAC (no apical meristem) protein (*EcNAC67*) that exhibited differential salinity responsive expression pattern. In a follow-up study, Rahman et al. (2016) undertook functional validation of *EcNAC67* in a rice cultivar ASD16 using *Agrobacterium*-mediated genetic transformation. They confirmed enhanced tolerance against drought and salinity stress. The transformed plants possessed higher relative water content and less reduction in grain yield in comparison to controls. Another *EcNAC1* protein was isolated from finger millet and overexpressed in rice (Ramegowda et al. 2012). The transgenic rice harboring *EcNAC1* proteins were reported to show enhanced tolerance to several abiotic stresses including salinity. More focused studies coupled with functional validation of similar genes have been hindered by the low levels of genetic transformation success in finger millet.

7.2.2 Drought Tolerance

Although finger millet is known to have tolerance to drought, the yield is significantly compromised under extreme drought conditions. Drought stress has been demonstrated to not only cause wilting and leaf rolling in finger millet, but also results in the reduction of leaf solute potential and chlorophyll content with the induction of many drought stress responsive genes (Parvathi et al. 2013). Natural variation for drought stress has been observed in finger millet (Neshamba 2010; Krishnamurthy et al. 2016) although the mechanism of various responses is not entirely known. Puranik et al. (2011) suggested that tolerance to drought might be attributed to an efficient antioxidant potential and increased signal perception in foxtail millet. In response, Bhatt et al. (2011) screened five finger millet varieties against drought and in parallel studied their antioxidant potential. They observed a positive correlation between drought tolerance and the capacity of finger millet variety PR202's antioxidant system to scavenge reactive oxygen species, resulting in a reduced incidence of oxidative damage (Bhatt et al. 2011). When finger millet was further evaluated for its association with arbuscular mycorrhiza (*Rhizophagus intraradices*) and endophyte (*Piriformospora indica*) under drought stress, an enhanced tolerance to drought was observed through a stronger antioxidant defence system, high chlorophyll content and an enriched osmoregulatory network (Tyagi et al. 2017).

The expression of a finger millet dehydration (dehydrin) gene, *EcDehydrin7*, in transgenic tobacco conferred tolerance to drought stress (Singh et al. 2015). Dehydrins are part of a large group of Late Embryogenesis Abundant (LEA) proteins (Rorat 2006), which have been speculated to protect cells against damage caused by cellular dehydration (Ingram and Bartels 1996; Graether and Boddington 2014). Using transcriptomics in finger millet, LEA proteins were recently (Hittalmani et al. 2017) found to be among some of the upregulated genes under low moisture stress in comparison to well-watered conditions. A genome-wide analysis of rice dehydrin family revealed that these genes play an important role in combating dehydration stress (Verma et al. 2017). A more focused study with the aim of understanding the mechanism of drought tolerance and genes involved will go a long way in helping enhance drought tolerance in finger millet and subsequently mitigating the effects of climate change.

7.2.3 Striga hermonthica

The genus *Striga* has over 30 species that occur naturally in most parts of the world (Scholes and Press 2008). The most important *Striga* species for finger millet production is *S. hermonthica*. *Striga* is a parasitic weed to many cereals including maize, sorghum and rice. There are limited studies on the biology of finger millet-*S. hermonthica* interactions and the mechanism of response to *Striga* remains unknown. A major control measure that has been developed to manage *Striga* is the push–pull technology (Midega et al. 2010). The push–pull technology involves the intercropping of cereals with a trap crop (pull), usually Napier grass (*Pennisetum purpureum*), and a forage legume, usually desmodium (*Desmodium* spp.), as a push crop (Khan et al. 2011). Due to the lack of use for desmodium by farmers, the adoption of the push-pull technology has been quite low. Traditionally, farmers have managed *Striga* in cereal fields through intercropping (Aasha et al. 2017) and crop rotations (Oswald and Ransom 2001) with edible legumes such as common bean (*Phaseolus vulgaris* L.), pigeonpea (*Cajanus cajan* (L) Millsp.) and mung bean (*Vigna radiata* (L.) R. Wilczek).

Striga will only germinate upon stimulation by a strigolactone induced by the host plant. Strigolactones are plant hormones that play an important role in plant development. In the case of *Striga*, strigolactones trigger seed germination through receptors called KARRAKIN-INSENSITIVE 2 (KAI2)/HYPERSENSITIVE TO LIGHT (HTL) (Waters et al. 2012; Toh et al. 2015). In sorghum, low germination stimulant (LGS) activity has been exploited for crop improvement with positive results (Gobena et al. 2016). Certain legumes, which are non-hosts to *Striga*, have also been known to induce the germination of *Striga* leading to the death of the germinated *Striga* plants (Odhiambo et al. 2011) as a result of lack of attachment of the *Striga* to the non-host plant. This phenomenon leads to the "suicidal death" of *Striga* and has been used, to some extent, in the control of *Striga*, and especially in the reduction of the seed banks (Fernández-Aparicio 2012).

Despite the limited studies on *Striga*-finger millet interactions, a lot can be learned from other grasses in the management of *Striga* in finger millet fields. Relevant populations will need to be developed to enable the understanding of various mechanisms involved in response to *Striga*. Better characterization of *Striga* should be done in order to establish whether there is gene-for-gene resistance as has been reported in cowpea (Li and Timko 2009). Wild relatives of finger millet will be valuable and should be screened alongside the landraces and cultivated accessions while looking for novel sources of resistance to *Striga*.

7.2.4 Blast Disease

Blast disease has been reported in all finger millet growing regions and is by far the most devastating disease in finger millet. *M. grisea* parasitizes several economically important grasses; destroys rice crops worldwide and now threatens global wheat production (Wang and Valent 2017). Blast disease affects the leaf, neck, and fingers of grown finger millet and can also cause seed discoloration (Panwar et al. 2011). Neck and finger blast are the most destructive forms of the disease (Takan et al. 2012). Studies done in India using isolates from finger millet and rice confirm that the rice- and finger millet-infecting blast populations are distinct (Viji et al. 2000; Takan et al. 2012). Pathogen genetic groups have been reported within finger millet blast populations (Takan et al. 2004; Shanmugapackiam et al. 2015), although there could be regional distinctness. For example, the presence of the *grh* element has been observed in collections from Japan, India and Nepal (Dobinson et al. 1993) and not from Africa. However, there seems to be new introductions of the *grh* element in the east African region (Takan et al. 2004).

In rice, more than 100 *R* genes (Su et al. 2015; Zheng et al. 2016) have been identified and ~500 quantitative trait loci (QTLs) have been mapped for resistance to blast (Ashkani et al. 2016). Most of the *R* genes have been identified in landraces

(Umakanth et al. 2017) or from wild rice species (Das et al. 2012) and most of them belong to the nucleotide-binding site—leusine rich repeat (NBS-LRR) family. Applying the same knowledge to identify resistance genes in finger millet, some functional markers have been developed using comparative genomics (Panwar et al. 2011; Babu et al. 2014a). Association mapping in finger millet has also been attempted but due to limited numbers of markers, the QTLs identified were not highly significant (Babu et al. 2014a; Ramakrishnan et al. 2016). With the availability of the whole genome sequence (Hittalmani et al. 2017; Hatakeyama et al. 2017) and the abundant genetic resources available in gene banks for finger millet, it should be possible to identify major genes and QTLs for blast resistance. Proper characterization of the pathogen will also be required to ensure the mode of resistance in different resistant genotypes is clearly understood.

7.3 Genetic Resources for Climate Smart Traits

East Africa and India are considered the primary and secondary centers of diversity respectively (Bisht and Mukai 2001) for finger millet. Eight species, namely, *E. coracana, E. kigeziensis, E. indica, E. multiflora, E. floccifolia, E. intermedia, E. jaegeri* occur in East Africa (Philips 1972). *E. tristachya* is endemic to South America (Philips 1972; Devarumath et al. 2005; Neves et al. 2005). Most of the *Eleusine* species are quite localized except *E. indica, E. coracana E. tristachya* and *E. multiflora. Eleusine indica,* also known as gooseberry, is considered one of the top ten worst weeds worldwide (Holm et al. 1977). *E. tristachya* has spread from South America to North America, Australia, Africa and Europe (Hilu 1980; Agrawal and Maheshwari 2016). There is also a recent publication reporting the identification of *E. multiflora* in India (Prabhukumar et al. 2017). The primary gene pool of *Eleusine* consists of all the varieties and landraces of the cultivated *E. coracana* subsp. *coracana. Eleusine coracana* subsp. *africana* form the major part of the secondary gene pool as well as any other cross-compatible wild taxa (Agrawal and Maheshwari 2016). All other wild *Eleusine* species belong to the tertiary gene pool.

More than 37,000 wild and cultivated finger millet germplasm has been conserved globally (Vetriventhan et al. 2016) in various gene banks, with the National Bureau of Plant Genetic Resources in India having the highest number of collections (>10,000) followed by ICRISAT (7519). The majority of ICRISAT collections are landraces (7121) with only 205 wild accessions, 143 improved varieties and 50 breeding lines. There are significant collections in Kenya, Uganda, Tanzania, Ethiopia and other neighboring countries (Table 7.1). ICRISAT established a core collection consisting of 622 accessions using 14 quantitative traits (Upadhyaya et al. 2006) and later narrowed this set down to a mini-core collection consisting of 80 accessions (Upadhyaya et al. 2010). Unfortunately, wild accessions are only 5% of all the global collections (CROP TRUST 2012). There is limited use of wild relatives

Table 7.1 A summary of finger millet collections in various countries

Source	No. of collections	Source	No. of collections
ICRISAT	7519	Tanzania	293
USDA (Georgia)	766	Malawi	145
India (genebank)	10,507	Eritrea	120
Nepal	877	Burundi	113
Sri Lanka	393	Ethiopia	71
Bhutan	84	Nigeria	20
Kenya	1902	South Africa	17
Zimbabwe	1158	China	300
Uganda	1155	Russia	110
Zambia	497	Vietnam	52

Source: Summarized from CROP TRUST (2012) ICRISAT record updated from http://genebank.icrisat.org/

in finger millet improvement, except for a currently funded CROP TRUST project in Kenya focusing on the identification of novel genes for resistance to *S. hermonthica* and blast disease from wild relatives. There have been on-going efforts to undertake more expeditions within eastern Africa to collect the vast genetic resources. Bio-Innovate Africa (https://bioinnovate-africa.org/about-us/) recently supported the collection of wild finger millet germplasm in Kenya, Uganda, Tanzania and Ethiopia with great success. In order to identify gaps in collections and the challenges faced for conservation and use of finger millet, the CROP TRUST developed a "Global Strategy for the ex situ Conservation of Finger Millet" in 2012 (file:///Users/dodeny/Downloads/Finger-Millet-Strategy-FINAL-14May2012%20(1).pdf).

Large genetic variation was reported when the mini-core collection at ICRISAT was screened for response to salinity (Krishnamurthy et al. 2014) and drought stress (Krishnamurthy et al. 2016). Genetic diversity analyses of finger millet germplasm from various genebanks revealed a narrower genetic base within the cultivated accessions as compared to wild species (Salimath et al. 1995; Gimode et al. 2016), perhaps due to the self-crossing nature. A distinct clustering between Asian and African germplasm has also been reported in several studies (Dida et al. 2008; Arya et al. 2013; Ramakrishnan et al. 2015). This suggests that higher heterosis is likely to be achieved when crosses are made between Asian and African germplasm. Hybridization based breeding has been going on between Uganda and India since the late 1960s and has led to the release of "*Indaf*" varieties (Dida et al. 2008) with improved yields. In order to efficiently broaden the genetic base and significantly improve the performance of this crop, breeders will need to exploit the different centers of diversities and different genepools in their breeding programs.

7.4 Genetic Mapping in Finger Millet

Genetic mapping in finger millet is in its infancy in comparison with other major cereals like maize, wheat and rice. The first partial finger millet genetic map was constructed by Dida et al. (2007) using restriction fragment length polymorphism (RFLP), amplified fragment length polymorphism (AFLP) and single strand conformation polymorphic (SSCP) expressed sequenced tags. The map covered the nine homoeologous chromosome groups and was constructed using an interspecific F_2 mapping population between *Eleusine coracana* subsp. *coracana* (Okhale 1) and *Eleusine coracana* subsp. *africana* (MD-20). The map contained 327 loci that were mapped to either A or B genomes. One hundred and thirty one (131) markers covered 721.4 cM of 16 linkage groups that were mapped on the A genome. The remaining 196 markers covered 786.8 cM of nine linkage groups of the B genome. More recently, a more robust single nucleotide polymorphism (SNP) linkage map was developed using $F_{2:3}$ families of the same interspecific cross between Okhale 1 and MD-20 (Qi et al. 2018). They mapped 4,453 SNP markers in 18 linkage groups that were designated the same as in Dida et al. (2007) and incorporated a subset of markers that had been mapped in the first linkage map. This linkage map is a significant contribution to finger millet genetics research, especially if it can be used to anchor the whole genome sequence (Hittalmani et al. 2017; Hatakeyama et al. 2017). Future efforts will still need to focus on generating intraspecific maps within the primary gene pool.

The low number of markers has resulted in limited association mapping studies in the past, although some reports are available, especially for association of markers with agro-morphological characters (Table 7.2). Using 113 accessions screened in two sites in India, Sharma et al. (2018) reported significant SNP associations with 14 agro-morphological traits at a p value threshold of $p < 0.01$ and $p < 0.001$. Lule et al. (2018) reported 16 associations between 13 microsatellite markers and six agronomic traits at $p < 0.01$. Babu et al. (2014b) used 46 simple sequence repeats (SSRs) and reported the association of five markers with four agronomic traits (basal tiller number, days to 50% flowering, flag leaf width and plant height) at $p < 0.01$ and $p < 0.001$. Babu et al. (2014a) developed genic SSRs and used 104 markers to detect four QTLs for finger blast and one QTL for neck blast resistance. Two QTLs were found to be associated with tryptophan content and one QTL for protein content in a study aimed at identifying alleles responsible for *opaque2* modifiers (*Opm*) (Babu et al. 2014c). All these mapping studies used relatively low numbers of accessions and extremely low numbers of markers (except in the study by Sharma et al. 2018). Although none of these QTLs has been validated, the recent availability of the whole genome sequence provides more opportunities for using high-throughput markers for association mapping studies.

Comparative mapping of finger millet has been done with rice (Srinivasachary et al. 2007), which revealed high levels of colinearity. A recent phylogenetic study using single copy orthologous genes among closely related grasses confirmed the close relationship between finger millet and rice (Hittalmani et al. 2017). A more

Table 7.2 A summary of QTLs detected in various studies in finger millet using association mapping

Study reference	Traits analysed	Population	QTL identified	P-value cut-off
Sharma et al. (2018)	Days to 50% flowering; Days to maturity; Basal tiller no.; Plant height; Culm thickness; Flag leaf blade length; Flag leaf blade width; Peduncle length; Ear length; Ear width; Length of longest finger; Width of longest finger; Finger number per ear; Grain yield	113 diverse accessions	Several	$P < 0.01$
Lule et al. (2018)	Days to maturity; Finger number; Grain yield per plant; Number of grain per spikelet; Productive tiller number; Thousand grain weight	138 diverse accessions	Several	$P < 0.01$
Babu et al. (2014b)	Basal tiller number; Days to 50% flowering; Flag leaf width; Plant height	190 diverse accessions	Several	$P < 0.01$; $P < 0.001$
Babu et al. (2014a)	Finger blast; neck blast	190 diverse accessions	5 QTLs	$P < 0.01$; $P < 0.001$
Babu et al. (2014c)	Tryptophan content; protein content	190 diverse accessions	2 QTLs	$P < 0.01$; $P < 0.001$

detailed whole genome comparison between finger millet and other grasses will be necessary, not only to confirm the unique structural rearrangements reported by Srinivasachary et al. (2007), but also to understand the genomic relationships between finger millet and other grasses.

7.5 Genomics-Assisted Breeding

Whole genome sequences of three finger millet genotypes PR202 (Hatakeyama et al. 2017), ML365 (Hittalmani et al. 2017), and KNE796 (unpublished data) are available online. There is a consensus in genome size (1.5 Gbp) and GC content of the three genomes although differences prevail in numbers and sizes of the protein coding genes (Table 7.3). The reported gene numbers by Hittalmani et al. (2017) for ML365 and Hatakeyama et al. (2017) for PR202 (Table 7.3) will further need to be re-analyzed in order to come up with a consensus number. A preliminary genome sequence comparison of finger millet with other cereals revealed 95%, 90% and 65% collinear blocks with rice, maize and sorghum, respectively. A large number of genes in finger millet genome form core gene families with other cereal

Table 7.3 Statistics of published finger millet genomes

Source	PR202	ML365
Protein-coding gene number	62,348	85,243
Mean gene length (kbp)	2.50	2.06
Total CDS length (mbp)	68.66	75.38
Mean CDS length (kbp)	1.07	0.88
Number (%) of single-exon genes	16,210	26,449
GC content	43.9	44.9
Repeat content (% of genome)	N/A	49.9

species. In total 9,896 core orthologous groups (COGs) of gene families are shared by rice, maize, sorghum, pearl millet and foxtail millet.

7.6 Climate Smart Genes in the Finger Millet Genome

According to drought stress gene database, 78 genes have been categorized for physiological adaptation under drought stress (Alter et al. 2015). We identified 52 out of the 78 genes in the finger millet genome (Table 7.4), out of which 37 are known to be involved in osmoregulation to balance the ionic and osmotic homeostasis, 12 for detoxification, and three for growth control in plants. For example, *ABCG25* gene is reported to be involved in abscisic acid (ABA) transport and response (Kuromori et al. 2010) as well as enhancing intercellular ABA signaling in plants (Kuromori et al. 2016). Similarly, many other stress-responsive genes found in the finger millet genome such as *LOS5, CPK4, AB1* and *AB2* have been reported

Table 7.4 Genes involved in physiological adaptation in plants under stress conditions

Biological process	Gene	FM ID	Function
Ion and osmotic homeostatis			
AAO3		N.D	Arabidopsis aldehyde oxidase
ABCG25		ECOR024010	ABC-transporter, ABA export from cells
ABCG40		N.D	ABC-transporter, ABA import
ABH1/CBP80		ECOR006018	Subunit of mRNA cap-binding complex
ABI1		ECOR033624	Protein phosphatase 2C 56
ABI2		ECOR049722	Protein phosphatase 2C 77
ABO1/ELP1		ECOR038014	Subunit of Elongator
AQP1/TIP1-1		ECOR020980	PIP1 plasma membrane aquaporin
BG1/BGLU18		ECOR017856	Beta-glucosidase; hydrolyzes glucose-conjugated
ATHB6		N.D	Homeodomain protein, target of ABI1
AtrbohD		N.D	NADPH oxidase catalytical subunit
CBP20		ECOR002984	Cap Binding Protein 20

(continued)

Table 7.4 (continued)

Biological process	Gene	FM ID	Function
CLC-C		ECOR020329	Chloride channel
CPK4		ECOR032246	Calcium-dependent protein kinases
CYP707A1		N.D	ABA catabolism; 8'-hydroxylation of ABA
CYP707A3		ECOR014044	ABA catabolism; 8'-hydroxylation of ABA
DOR		ECOR006224	Drought tolerance repressor, F-box protein
DRB1/HYL1		ECOR009359	Hyponastic leaves 1; RNA binding protein that affects ABA and drought sensitivity
DSM2		N.D	Put. beta-carotene hydroxylase, ABA biosynthesis
GCR1		ECOR045780	Putative G protein-coupled receptor
GORK		N.D	Outward K + channel
GPA1		ECOR028625	Alpha subunit of heterotrimeric GTP-binding protein
GTG1		ECOR036005	PM-ABA receptor, GPCR-type G protein
GTG2		N.D	PM-ABA receptor, GPCR-type G protein
HAB1		N.D	Protein phosphatase 2C 16
KAT2		ECOR033046	K+ channel, inward rectifying
LHCB6		N.D	Light harvesting chlorophyll a/b binding protein
LLA23		N.D	ABA-, stress-, and ripening-induced protein
LOS5/ABA3		ECOR026538	Molybdenum-cofactor sulfurase
MRP4		ECOR012433	Multidrug resistance-associated protein, ABC transporter
NADP-ME1		ECOR009538	NADP-malic enzyme
NCED1		ECOR034474	ABA biosynthesis
NCED3		ECOR042900	ABA biosynthesis key enzyme
OST1/SRK2E		ECOR033474	Kinase-like (open stomata 1), activated by ABA, activates SLAC1
OST2/PMA1		ECOR043481	Plasma membrane proton ATPase
PCKA/PEPCK		ECOR005523	PEP carboxykinase
PED1/KAT2		ECOR038688	3-ketoacyl-CoA thiolase 2
PIP1-1		ECOR032142	Aquaporin
PIP1-4		N.D	Probable aquaporin
PIP2-1		ECOR047129	Aquaporin
PIP2-2		ECOR013882	Aquaporin
PIP2-5		ECOR017811	Aquaporin
PYL9/RCAR1		N.D	Soluble ABA receptor interacts with and regulates PP2Cs ABI1 and ABI2
RBOHF		ECOR039003	NADPH oxidase catalytical subunit
RFP1/SDIR1		ECOR022569	RING-finger protein
RWC3		N.D	Aquaporin
SAD1/LSM5		ECOR001752	Supersensitive to ABA and drought 1

(continued)

Table 7.4 (continued)

Biological process	Gene	FM ID	Function
SLAC1		ECOR033474	Kinase-like (open stomata 1), activated by ABA
SLAH3		ECOR050224	Guard cell S-type anion channel (SLAC1 homolog)
SYP61/OSM1		ECOR000072	Osmotic stress-sensitive; related to mammalian syntaxin
TIP2-2/SITIP2-2		ECOR043324	Probable aquaporin
ACS6		N.D	ACC synthase, first step in ethylene biosynthesis
Growth control			
ANN1		ECOR039452	Annexin1
APX		ECOR040940	Ascorbate peroxidase
EVP1		N.D	Vacuolar pyrophosphatase
HDG11/ROC8		ECOR011713	Enhanced drought tolerance1, HD START TF
RGS1		N.D	Regulator of G-protein signaling
Detoxification			
Osmolyte production			
ADC1		ECOR016738	Arginine decarboxylase; polyamine biosynthesis
CMO		ECOR024646	Choline monooxygenase; glycine betaine biosynthesis
FSPD1/SPDSYN1		ECOR051004	Spermidin synthase
GOLS1		N.D	Galactinol Synthase
GOLS2		ECOR005876	Galactinol Synthase
IMT1		N.D	D-myo-inositol methyltransferase
MYB4		ECOR031142	MYB TF
P5CS		N.D	Pyrroline-5-Carboxylate Synthase; glutamtate → proline
P5CS1/P5CSA		ECOR002007	Pyrroline-5-Carboxylate Synthase; glutamtate → proline
SAMDC		ECOR006068	S-adenosyl methionine decarboxylase, Polyamin synthesis
TPS1		ECOR008673	Trehalose-6-phosphate synthase
Removal of ROS			
APX2		ECOR025343	Ascorbate peroxidase 2, H_2O_2 scavenger
ERD1/CLPD2		ECOR034553	Chloroplast-targeted Clp protease reg SU
GhMT3a		N.D	Metallothionein, ROS scavenger
GPX3		N.D	Glutathione peroxidase3
GSTU17		ECOR035212	Glutathion s-transferase U17
PO2		N.D	Extracellular peroxidase 2
SODCP		ECOR009244	Superoxide dismutase
Protection factor			
HVA1		N.D	Late embryogenesis abundant protein, group 3
TaLEA		N.D	Late embryogenesis abundant
TAS14		N.D	Dehydrin, group2 LEA proteins

to be involved in ABA synthesis and are considered to be important players in abiotic stress tolerance (Meyer et al. 1994; Rodriguez et al. 1998; Xiong et al. 2002).

We also observed several genes involved in growth and development, which have been reported to play a role in stress tolerance. For example, *Annexins,* which act as targets of calcium signals (Mortimer et al. 2008), have been reported to play an important role during growth (Blackbourn et al. 1992; Clark et al. 1992; Carroll et al. 1998). In *Arabidopsis thaliana,* upregulation of *ANN1* (Annexin 1 protein) during abiotic stress conditions was shown to reduce hydrogen peroxide accumulation in guard cells leading to more tolerance to drought as compared to knockout plants (Konopka-Postupolska et al. 2009). A putative homolog of *HDG11*, which encodes for a homeodomain (HD)-START family transcription factor, has also been detected in the finger millet genome. The induction of *HDG11* gene has been shown to increase root growth and reduced stomatal density as well as drought and osmotic stress tolerant characteristics in *Arabidopsis* and other plants (Yu et al. 2008; Zhu et al. 2016; Yu et al. 2016).

There will be need to study these genes individually in finger millet in order to determine their specific roles under various stresses. However, this will need to be done after an improved and anchored assembly and annotation of the genome has been accomplished.

7.7 Future Perspectives

Finger millet remains an important cereal crop in the semi-arid tropics and is likely to gain more importance as more genomic resources become available and more people show interest in healthy eating. The high demand for finger millet in dry areas will require the release of better yielding varieties coupled with unique resistance to biotic and abiotic stresses. The current largely conventional breeding approaches in finger millet will not be sufficient if the full potential of the crop is to be realized. With the availability of a draft whole genome sequence, research will need to focus on characterizing important traits and utilizing genomics-assisted breeding for more efficient release of superior varieties. There is a great opportunity to implement genomic selection in finger millet and breeders will need to work alongside bioinformatics specialists in order to make available the necessary genetic and genomic resources that will enable accurate and more efficient breeding in this unique climate smart crop. Both secondary and tertiary gene pools will be of great value as sources of novel genes and for broadening the genetic base of the crop.

References

Aasha MB, Amani HE, Rashida MAA, Nahid AK (2017) Effects of intercropping pearl millet with some legumes on *Striga Hermonthica* emergence. SSRG Intl J Agri Environ Sci 4(6):65–72

Agrawal R, Maheshwari A (2016) Genetic improvement in the genus *Eleusine*. In: Rajpal VR, Rama Rao S, Raina SN (eds) Gene pool diversity and crop improvement, sustainable development and biodiversity, vol 10. Springer International Publishing, Switzerland, pp 393–414

Alter S, Bader KC, Spannagl M, Wang Y, Bauer E, Schön C-C, Mayer KF (2015) DroughtDB: an expert-curated compilation of plant drought stress genes and their homologs in nine species. Database 2015:bav046. https://doi.org/10.1093/database/bav046

Anilkumar TB, Mantur SG, Madhukeshwara SS (2003) Diseases of finger millet. Project coordination cell. All India Coordinated Small Millets Improvement Project. ICAR, GKVK, Bangalore

Arya L, Verma M, Gupta VK, Seetharam A (2013) Use of genomic and genic SSR markers for assessing genetic diversity and population structure in Indian and African finger millet (*Eleusine coracana* (L.) Gaertn.) germplasm. Plant Syst Evol 299:1395–1401

Ashkani S, Rafii MY, Shabanimofrad M, Ghasemzadeh A, Ravanfar SA, Latif MA (2016) Molecular progress on the mapping and cloning of functional genes for blast disease in rice (Oryza sativa L.): current status and future considerations. Crit Rev Biotechnol 36(2):353–367. https://doi.org/10.3109/07388551.2014.961403

Babu BK, Senthil N, Gomez SM, Biji KR, Rajendraprasad NS, Kumar SS, Babu RC (2007) Assessment of genetic diversity among finger millet (*Eleusine coracana* (L.) Gaertn.) accessions using molecular markers. Genet Resour Crop Evol 54:399–404

Babu TK, Thakur RP, Upadhyaya HD, Reddy PN, Sharma R, Girish AG et al (2013) Resistance to blast (*Magnaporthe grisea*) in a mini-core collection of finger millet germplasm. Eur J Plant Pathol 135:299–311

Babu BK, Dinesh P, Agrawal PK, Sood S, Chandrashekara C et al (2014a) Comparative genomics and association mapping approaches for blast resistant genes in finger millet using SSRs. PLoS ONE 9(6):e99182. https://doi.org/10.1371/journal.pone.0099182

Babu BK, Agrawal PK, Pandey D, Jaiswal JP, Kumar A (2014b) Association mapping of agro-morphological characters among the global collection of finger millet genotypes using genomic SSR markers. Mol Biol Rep 41:5287–5297. https://doi.org/10.1007/s11033-014-3400-6

Babu BK, Agrawal PK, Pandey D, Kumar A (2014c) Comparative genomics and association mapping approaches for opaque2 modifier genes in finger millet accessions using genic, genomic and candidate gene-based simple sequence repeat markers. Mol Breed 34:1261–1279. https://doi.org/10.1007/s11032-014-0115-2

Barbeau WE, Hilu K (1993) Protein, calcium, iron and amino acid content of selected wild and domesticated cultivars of Finger millet. Plant Foods Hum Nutr 43:97–104

Bhatt D, Negi M, Sharma P, Saxena SC, Dobriyal AK, Arora S (2011) Responses to drought induced oxidative stress in five finger millet varieties differing in their geographical distribution. Physiol Mol Biol Plants 17:347–353

Bisht MS, Mukai Y (2001) Genomic in situ hybridization identifies genome donor of finger millet (*Eleusine coracana*). Theor Appl Genet 102:825–832. https://doi.org/10.1007/s001220000497

Blackbourn HD, Barker PI, Huskisson NS, Battey NH (1992) Properties and partial protein-sequence of plant annexins. Plant Physiol 99:864–871

Bray EA, Bailey-Serres J, Weretilnyk E (2000) Responses to abiotic stresses. In: Gruissem W, Buchannan B, Jones R (eds) Biochemistry and molecular biology of plants. American Society of Plant Biologists, Rockville, MD, pp 158–1249

Carroll AD, Moyen C, Van Kesteren P, Tooke F, Battey NH, Brownlee C (1998) Ca2+, annexins, and GTP modulate exocytosis from maize root cap protoplasts. Plant Cell 10:1267–1276

Clark GB, Dauwalder M, Roux SJ (1992) Purification and immunolocalization of annexin-like protein in pea seedlings. Planta 187:1–9

Butler EJ (1918) Fungi and diseases in plants. Thacker Spink and Co, Calcutta, India

Coleman LC (1920) The cultivation of ragi in Mysore. Bull Dep Agri Gen Ser 11

CROP TRUST (2012) Global strategy for the *ex situ* conservation of finger millet and its wild relatives. https://www.croptrust.org/resources/#ex-situ-conservation-strategies. Accessed 18th Dec 2018

Das A, Soubam D, Singh PK, Thakur S, Singh NK, Sharma TR (2012) A novel blast resistance gene, *Pi54rh* cloned from wild species of rice, *Oryza rhizomatis* confers broad spectrum resistance to *Magnaporthe oryzae*. Funct Integr Genom 12:215. https://doi.org/10.1007/s10142-012-0284-1

Devarumath RM, Hiremath SC, Rao SR, Kumar A, Sheelavanthmath SS (2005) Genome interrelationship in the genus *Eleusine* (Poaceae) as revealed through heteroploid crosses. Caryologia 58:300–307

Dida MM, Devos KM (2006) Finger millet. In: Kole C (ed) Genome mapping and molecular breeding in plants, vol 1. Cereals and Millets. Springer, Berlin, Heidelberg, pp 333–343

Dida MM, Srinivasachary Ramakrishna S, Bennetzen JL, Gale MD, Devos KM (2007) The genetic map of finger millet, Eleusine coracana. Theor Appl Genet 114:321–332

Dida M, Wanyera N, Dunn M, Bennetzen J, Devos KM (2008) Population structure and diversity in finger millet germplasms. Trop Plant Biol 1:131–141

Dobinson KF, Harris RE, Hamer JE (1993) Grasshopper, a long terminal repeat (LTR) retroelement in the phytopathogenic fungus *Magnaporthe grisea*. Mol Plant Microbe-Interact 6:114–126

Dwivedi S, Upadhyaya H, Senthilvel S, Hash C, Fukunaga K, Diao X et al (2012) Millets: genetic and genomic resources. In: Janick J (ed) Plant breeding reviews, vol 35. Wiley, Hoboken, NJ, pp 247–375

Ekwamu A (1991) Influence of head blast infection on seed germination and yield components of finger millet (*Eleusine coracana* L. Gaertn). Trop Pest Manage 37:122–123

FAO (2005) The state of food insecurity in the world: eradicating world hunger—key to achieving the millennium development goals. FAO, Rome, Italy

Fernández-Aparicio R (2012) Innovations in parasitic weeds management in legume crops. A review. Agron Sustain Dev 433–449 (Springer/EDP Sciences/INRA)

Gimode D, Odeny DA, de Villiers EP, Wanyonyi S, Dida MM, Mneney EE et al (2016) Identification of SNP and SSR markers in finger millet using next generation sequencing technologies. PLoS ONE 11(7):e0159437. https://doi.org/10.1371/journal.pone.0159437

Graether SP, Boddington KF (2014) Disorder and function: a review of the dehydrin protein family. Front Plant Sci 5:576. https://doi.org/10.3389/fpls.2014.00576

Gobena D, Shimels M, Richa PJ, Ruyter-Spirab C, Bouwmeester H et al (2016) Mutation in sorghum LOW GERMINATION STIMULANT1 alters strigolactones and causes *Striga* resistance. Proc Nat Acad Sci USA 114:4471–44716

Goron TL, Raizada MN (2015) Genetic diversity and genomic resources available for the small millet crops to accelerate a new green revolution. Front Plant Sci 6:157. https://doi.org/10.3389/fpls.2015.00157. PMID: 25852710

Hatakeyama M, Aluri S, Balachadran MT, Sivarajan SR, Patrignani A, Grüter S et al (2017) Multiple hybrid de novo genome assembly of finger millet, an orphan allotetraploid crop. DNA Res 25:39–47

Hittalmani S, Mahesh HB, Shirke MD, Biradar H, Uday G, Aruna YR et al (2017) Genome and transcriptome sequence of finger millet (*Eleusine coracana* (L.) Gaertn.) provides insights into drought tolerance and nutraceutical properties. BMC Genomics 18:465

Hilu KW, De Wet JM (1976) Domestication of *E. coracana*. Econ Bot 30:199–208

Hilu KW, De Wet JMJ, Harlan JR (1979) Archaeobotanical studies of *Eleusine coracana* ssp. *coracana* (finger millet). Amer J Bot 66(3):330–333

Hilu KW (1980) *Eleusine tristachya* (Lam.) Lam. Madroño 27:177–178

Holm LG, Plucknett DL, Pancho JV, Herberger JP (1977) The world's worst weeds: distribution and biology. University Press of Hawaii, Hawaii

Ingram J, Bartels D (1996) The molecular basis of dehydration tolerance in plants. Annu Rev Plant Physiol Plant Mol Biol 47:377–403

Kenneth OR, LeRoy VP (1977) The *Eleusines*. A review of the world literature. International Crops Research Institute for the Semi-Arid Tropics (ICRISAT), Hyderabad, India

Khan Z, Midega C, Pittchar J, Pickett J, Bruce T (2011) Push—pull technology: a conservation agriculture approach for integrated management of insect pests, weeds and soil health in Africa. Intl J Agri Sustain 9:162–170. https://doi.org/10.3763/ijas.2010.0558

Konopka-Postupolska D, Clark G, Goch G et al (2009) The role of annexin 1 in drought stress in *Arabidopsis*. Plant Physiol 150(3):1394–1410. https://doi.org/10.1104/pp.109.135228

Kotschi J (2006) Coping with climate change, and the role of agrobiodiversity. Conference on International Agricultural Research for Development. Tropentag, University of Bonn, October 11–13, 2006

Krishnamurthy L, Upadhyaya HD, Purushothaman R, Gowda CLL, Kashiwagi J, Dwivedi SL et al (2014) The extent of variation in salinity tolerance of the minicore collection of finger millet (*Eleusine coracana* L. Gaertn.) germplasm. Plant Sci 227:51–59. https://doi.org/10.1016/j.plantsci.2014.07.001

Krishnamurthy L, Upadhyaya HD, Kashiwagi J, Purushothaman R, Dwivedi S, Vadez V (2016) Variation in drought-tolerance components and their interrelationships in the minicore collection of finger millet germplasm. Crop Sci 56:1914–1926

Kukreti A (2016) Improvement in seed performance, management of finger millet *(Eleusine coracana* (L.) Gaertn.) through bio-prospecting fluorescent *Pseudomonas* spp. with respect to their plant growth prom. Doctoral dissertation, College of Forestry, Ranichauri campus, VCSG Uttarakhand University of Horticulture and Forestry

Kulkarni GS (1922) The "Murda" disease of chilli (Capsicum). Agri J India 22(1):51–54

Kumar A, Metwal M, Kaur S, Gupta AK, Puranik S et al (2016) Nutraceutical value of finger millet [*Eleusine coracana* (L.) Gaertn.], and their improvement using Omics approaches. Front Plant Sci 7:934. https://doi.org/10.3389/fpls.2016.00934

Kuromori T, Miyaji T, Yabuuchi H et al (2010) ABC transporter AtABCG25 is involved in abscisic acid transport and responses. Proc Natl Acad Sci U S A 107(5):2361–2366. https://doi.org/10.1073/pnas.0912516107

Kuromori T, Fujita M, Urano K, Tanabata T, Sugimoto E, Shinozaki K (2016) Overexpression of AtABCG25 enhances the abscisic acid signal in guard cells and improves plant water use efficiency. Plant Sci 251:75–81. https://doi.org/10.1016/j.plantsci.2016.02.019

Li J, Timko MP (2009) Gene-for-gene resistance in Striga-cowpea associations. Science 325:1094

Lule D, de Villiers SM, Fetene M, Odeny DA, Rathore A, Das RR, Tesfaye K (2018) Genetic diversity and association mapping of Ethiopian and exotic finger millet accessions. Crop Pasture Sci 69(9):879–891. https://doi.org/10.1071/CP18175

McRae W (1920) Detailed administration report of the government mycologist for the year 1919–20. Madras Agric Dept, India

Mehta P, Chakravarty S (1937) A new disease of *Eleusine caracana*. Mysore Agri J 28:783–790

Meyer K, Leube M, Grill E (1994) A protein phosphatase 2C involved in ABA signal transduction in *Arabidopsis thaliana*. Science 264:1452–1455

Midega CAO, Khan ZR, Amudavi DM, Pittchar J, Pickett JA (2010) Integrated management of *Striga hermonthica* and cereal stemborers in finger millet (*Eleusine coracana* (L.) Gaertn.) through intercropping with *Desmodium intortum*. Intl J Pest Manage 56:145–151

Mirza N, Sharma N, Srivastava S, Kumar A (2015) Variation in popping quality related to physical, biochemical & nutritional properties of finger millet genotypes. Proc Natl Acad Sci India Sect Biol Sci 85:507–515

Mortimer JC, Laohavisit A, Macpherson N, Webb A, Brownlee C, Battey NH, Davies JM (2008) Annexins: multifunctional components of growth and adaptation. J Exp Bot 59:533–544

National Research Council (1996) Lost crops of Africa, Volume I, Grains. National Academy Press, Washington, DC

Neshamba SM (2010) Variability for drought tolerance in finger millet [*Eleusine coracana* (L.) Gaertn] accessions from Zambia. MSc thesis, University of Zambia, Lusaka, Zambia

Neves SS, Swire-Clark G, Hilu KW, Baird WV (2005) Phylogeny of *Eleusine* (*Poaceae*: *Chloridoideae*) based on nuclear ITS and plastid trnT-trnF sequences. Mol Phylogenet Evol 35:395–419

Odhiambo J, Vanlauwe B, Tabu I, Kanampiu F, Khan Z (2011) *In Vitro* selection of soybean accessions for induction of germination of *Striga hermonthica* (Del.) Benth seeds and their effect on *Striga hermonthica* attachment on associated Maize. In: Bationo A, Waswa B, Okeyo J, Maina F, Kihara J (eds) Innovations as key to the green revolution in Africa. Springer, Dordrecht, Netherlands, pp 365–372

Onkware AO (1986) The response of finger millet (*Eleusine coracana* (L.) Gaertn) to salinity. MSc Thesis, University of Nairobi, Kenya

Oswald A, Ransom JK (2001) Striga control and improved farm productivity using crop rotation. Crop Protec 20:113–120

Padulosi S, Mal B, Ravi SB, Gowda J, Gowda KTK, Shanthakumar G et al (2009) Food security and climate change: role of plant genetic resources of minor millets. Indian J Plant Genet Resour 22:1–16

Panwar P, Jha AK, Pandey PK, Gupta AK, Kumar A (2011) Functional markers based molecular characterization and cloning of resistance gene analogs encoding NBS-LRR disease resistance proteins in finger millet (*Eleusine coracana*). Mol Biol Rep 38:3427–3436

Parvathi M, Nataraja KN, Yashoda B, Ramegowda H, Mamrutha H, Rama N (2013) Expression analysis of stress responsive pathway genes linked to drought hardiness in an adapted crop, finger millet (*Eleusine coracana*). J Plant Biochem Biotechnol 22:193–201. https://doi.org/10.1007/s13562-012-0135-0

Philips SM (1972) A survey of *Eleusine* Gaertn. (Gramineae) in Africa. Kew Bull 27:251–270

Prabhukumar KM, Sunil CN, Kumar VVN, Bhavadas N, Chhabra T, Balachandran I (2017) *Eleusine multiflora* (Poaceae: Chloridoideae), a new record for India. Nelumbo 59:25–28

Pradhanang PM (1994) Quantification of the relationship between cercospora leaf spot disease (*Cercospora eleusine*) and yield loss of finger millet. Tech Paper-Lumle Agricultural Centre No. 94/3, pp 12

Puranik S, Jha S, Srivastava PS, Sreenivasulu N, Prasad M (2011) Comparative transcriptome analysis of contrasting foxtail millet cultivars in response to short-term salinity stress. J Plant Physiol 168(3):280–287

Purseglove JW (1972) Tropical crops: monocotyledons. Longman Group Limited, London, UK

Qi P, Gimode D, Saha D, Schröder S, Chakraborty D, Wang X, Dida MM, Malmberg RL, Devos KM (2018) UGbS-Flex, a novel bioinformatics pipeline for imputation-free SNP discovery in polyploids without a reference genome: finger millet as a case study. BMC Plant Biol 18:117. https://doi.org/10.1186/s12870-018-1316-3

Rahman H, Jagadeeshselvam N, Valarmathi R et al (2014) Transcriptome analysis of salinity responsiveness in contrasting genotypes of finger millet (*Eleusine coracana* L.) through RNA sequencing. Plant Mol Biol 85:485–503

Rahman H, Ramanathan V, Nallathambi J, Duraialagaraja S, Muthurajan R (2016) Over-expression of a NAC 67 transcription factor from finger millet (*Eleusine coracana* L.) confers tolerance against salinity and drought stress in rice. BMC Biotechnology 16(Suppl 1):35. https://doi.org/10.1186/s12896-016-0261-1

Ramakrishnan M, Ceasar SA, Duraipandiyan V et al (2015) Using molecular markers to assess the genetic diversity and population structure of finger millet (*Eleusine coracana* (L.) Gaertn.) from various geographical regions. Genet Resour Crop Evol. https://doi.org/10.1007/s10722-015-0255-1

Ramakrishnan M, Ceasar SA, Duraipandiyan V, Vinod KK, Kalpana K, Al-Dhabi NA et al (2016) Tracing QTLs for leaf blast resistance and agronomic performance of finger millet (*Eleusine coracana* (L.) Gaertn.) genotypes through association mapping and *in silico* comparative genomics analyses. PLoS One 11(7):e0159264. https://doi.org/10.1371/journal.pone.0159264

Ramegowda V, Senthil-Kumar M, Nataraja KN, Reddy MK, Mysore KS et al (2012) Expression of a finger millet transcription factor, EcNAC1, in Tobacco confers abiotic stress-tolerance. PLoS One 7(7):e40397. https://doi.org/10.1371/journal.pone.0040397

Reddy G, Upadhaya, HD, Gowda, CLL, Sube S (2009) Characterization of East African finger millet germplasm for qualitative and quantitative characters at ICRISAT. J SAT Agri Res 7–9. An open access journal published by ICRISAT

Rodriguez P, Benning G, Grill E (1998) ABI2, a second protein phosphatase 2C involved in abscisic acid signal transduction in *Arabidopsis*. FEBS Lett 421:185–190

Rorat T (2006) Plant dehydrins—tissue location, structure and function. Cell Mol Biol Lett 11:536–556

Salimath SS, Hiremath SC, Murthy HN (1995) Genome differentiation patterns in diploid species of *Eleusine* (Poaceae). Hereditas 122:189–195

Sharma D, Tiwari A, Sood S, Jamra G, Singh NK, Meher PK et al (2018) Genome wide association mapping of agro-morphological traits among a diverse collection of finger millet (*Eleusine coracana* L.) genotypes using SNP markers. PLoS One 13(8):e0199444

Scholes JD, Press MC (2008) Striga infestation of cereal crops–an unsolved problem in resource-limited agriculture. Curr Opin Plant Biol 11:180–186

Seetharam A, Riley KW, Harinarayana G (1986) Small millets in global agriculture. In: Proceedings of the first international small millets workshop, Oct 29–Nov 2, Bangalore, India

Shailaja H, Thirumeni S (2007) Evaluation of salt-tolerance in finger millet (*Eleusine coracana*) genotypes at seedling stage. Indian J Agri Sci 77:672–674

Shanmugapackiam S, Ragupathi N, Raguchander T (2015) Characterization of finger millet blast pathogen population of Magnaporthe grisea by DNA finger printing analysis. Int J Tropical Agric 33:1849–1853

Singh RK, Singh VK, Raghavendrarao S, Phanindra MLV, Raman KV, Solanke AU et al (2015) Expression of finger millet EcDehydrin7 in transgenic tobacco confers tolerance to drought stress. Appl Biochem Biotechnol 177:207–216. https://doi.org/10.1007/s1201

Srinivasachary DMM, Gale MD, Devos KM (2007) Comparative analyses reveal high levels of conserved colinearity between the finger millet and rice genomes. Theor Appl Genet 115:489–499

Stireman JO, Dyer LA, Janzen DH, Singer MS, Lill JT, Marquis RJ, Barone JA (2005) Climatic unpredictability and parasitism of caterpillars: implications of global warming. Proc Natl Acad Sci USA 102(48):17384–17387

Su J, Wang W, Han J, Chen S, Wang C, Zeng L, Feng A, Yang J, Zhou B, Zhu X (2015) Functional divergence of duplicated genes results in a novel blast resistance gene Pi50 at the Pi2/9 locus. Theor Appl Genet 128(11):2213–2225

Takan JP, Akello B, Esele P, Manyasa EO, Obilana AB, Audi PO (2004) Finger millet blast pathogen diversity and management in east Africa: a summary of project activities and outputs. Intl Sorghum Millets Newsl 45:66–69

Takan JP, Chipili J, Muthumeenakshi S, Talbot NJ, Manyasa EO, Bandyopadhyay R et al (2012) *Magnaporthe oryzae* populations adapted to finger millet and rice exhibit distinctive patterns of genetic diversity, sexuality and host interaction. Mol Biotechnol 50(2):145–158

Toh S, Holbrook-Smith D, Stogios Onopriyenko O, Lumba S, Tsuchiya Y, Savchenko A, McCourt P (2015) Structure-function analysis identifies highly sensitive strigolactone receptors in *Striga*. Plant Sci 350:203–206

Tyagi J, Varma A, Pudake RN (2017) Evaluation of comparative effects of arbuscular mycorrhiza (*Rhizophagus intraradices*) and endophyte (*Piriformospora indica*) association with finger millet (*Eleusine coracana*) under drought stress. Eur J Soil Biol 81:1–10

Umakanth B, Vishalakshi B, Sathish Kumar P, Rama Devi SJS, Bhadana VP, Senguttuvel P, Kumar S, Sharma SK, Sharma PK, Prasad MS, Madhav MS (2017) Diverse rice landraces of North-East India enables the identification of novel genetic resources for *Magnaporthe* resistance. Front Plant Sci 8:1500. https://doi.org/10.3389/fpls.2017.01500

Upadhyaya HD, Gowda CLL, Pundir RPS, Reddy VG, Singh S (2006) Development of core subset of finger millet germplasm using geographical origin and data on 14 quantitative traits. Genet Resour Crop Evol 53:679–685

Upadhyaya HD, Gowda CL, Reddy G (2007) Morphological diversity in finger millet germplasm introduced from Southern and Eastern Africa. SAT eJournal|ejournal.icrisat.org. 3(1)

Upadhyaya HD, Sarma NDRK, Ravishankar CR, Albrecht T, Narasimhudu Y et al (2010) Developing a mini-core collection in finger millet using multi-location data. Crop Sci 50:1924–1931

Vavilov N (1951) The origin, variation, immunity and breeding of cultivated plants, pp 37–38. Ronald Press, New York, USA (Translated from the Russian by K. Starrchester)

Verma G, Dhar YV, Srivastava D, Kidwai M, Chauhan PS, Bag SK et al (2017) Genome-wide analysis of rice dehydrin gene family: its evolutionary conservedness and expression pattern in response to PEG induced dehydration stress. PLoS ONE 12(5):e0176399. https://doi.org/10.1371/journal.pone.0176399

Vetriventhan M, Upadhyaya HD, Dwivedi SL, Pattanashetti SK, Kumar Singh S (2016) Finger and foxtail millets. In: Singh M, Upadhyaya HD (eds) Genetic and genomic resources for grain cereals improvement. Academic Press, Cambridge, MA, USA, pp 291–319

Viji G, Gnanamanickam SS, Levy M (2000) DNA polymorphisms of isolates of *Magnaporthe grisea* from India that are pathogenic to finger millet and rice. Mycol Res 104:161–167

Wang G-L, Valent B (2017) Durable resistance to rice blast. Science 355:906–907

Waters MT, Nelson DC, Scaffidi A, Flematti GR, Sun YK, Dixon KW, Smith SM (2012) Specialisation within the *DWARF14* protein family confers distinct responses to *karrikins* and *strigolactones* in *Arabidopsis*. Development 139:1285–1295

Xiong L, Lee H, Ishitani M, Zhu J.-K (2002) Regulation of osmotic stress-responsive gene expression by the LOS6/ABA1 locus in *Arabidopsis*. J Biol Chem 277:8588–8596

Yu H, Chen X, Hong YY, Wang Y, Xu P, Ke SD et al (2008) Activated expression of an *Arabidopsis* HD-START protein confers drought tolerance with improved root system and reduced stomatal density. Plant Cell 20:1134–1151. https://doi.org/10.1105/tpc.108.058263

Zheng W, Wang Y, Wang L, Ma Z, Zhao J, Wang P, Zhang L, Liu Z, Lu X (2016) Genetic mapping and molecular marker development for *Pi65*(t), a novel broad-spectrum resistance gene to rice blast using next-generation sequencing. Theor Appl Genet 129(5):1035–1044

Zhu Z, Sun B, Xu X, Chen H, Zou L, Chen G et al (2016) Overexpression of AtEDT1/HDG11 in Chinese kale (*Brassica oleracea* var. *alboglabra*) enhances drought and osmotic stress tolerance. Front Plant Sci 7:1285. https://doi.org/10.3389/fpls.2016.01285

